1987 FLUID POWER STANDARDS

VOLUME E
CONDUCTORS AND ASSOCIATED PRODUCTS

Shirley C. Seal, Editor

National Fluid Power Association, Inc.
Milwaukee, Wisconsin

Copyright 1987 by the

NATIONAL FLUID POWER ASSOCIATION, INC.

Printed in the USA

All technical reports, citations, references and related data including standards and practices approved and/or recommended are advisory only. Use thereof by anyone for any purpose is entirely voluntary and in any event without risk of any nature to the National Fluid Power Association, Inc., its officers, directors or authors of such work. There is no agreement by or between anyone to adhere to any NFPA Recommended Standard, policy or practice, and related matters. In formulating and approving technical reports, the Technical Board, its councils and committees and/or the National Fluid Power Association, Inc. will not investigate or consider citations, references or patents which may or may not apply to such subject matter since prospective users of such reports and data alone are responsible for establishing necessary safeguards in connection with utilization of such matters, including technical data, proprietary rights or patentable materials.

Recommended standards and/or policies and procedures are subject to periodic review and may be changed without notice. Recommended standards, after publication, may be revised or withdrawn at any time and current information on all approved recommended standards may be received by calling or writing the National Fluid Power Association, Inc.

An approved NFPA Recommended Standard implies a consensus of those substantially concerned with its scope and provisions and is intended as a guide to aid the manufacturer, the consumer and the general public. The publication of the NFPA Recommended Standard does not in any respect preclude anyone, whether they have participated in the development of or approved the recommended standard or not, from manufacturing, marketing purchasing or using products, processes or procedures not conforming to the recommended standard.

Participation by federal agency representative(s) or person(s) affiliated with industry is not to be interpreted as government or industry endorsement of this standard and/or policy and procedure.

This publication may not be reproduced in whole or in part without the written permission of the National Fluid Power Association, Inc.

```
Sci Ref
TJ
840
.N55
1987
v.E
```

ISBN 0-942220-82-X (Set)
ISBN 0-942220-87-0 (Volume E)

NATIONAL FLUID POWER ASSOCIATION, INC., 1987

Contents

NFPA — Providing Services for the U.S. Fluid Power Industry ... v

Excerpts extracted from American National Standard Fluid power systems and products - Glossary, ANSI/B93.2-1986 ... 1

Procedure for Self-Certification by Fluid Power Manufacturers, NFPA/T1.21.1-1978 (R1983) ... 21

Method for Verifying the Fatigue and Static Pressure Ratings of the Pressure Containing Envelope of a Metal Fluid Power Component, NFPA/T2.6.1-1974 (R1982) ... 29

American National Standard Pneumatic fluid power applications - Metal separable tube fittings - Qualifications test, ANSI/B93.48M-1979 ... 59

A Bibliography of Fluid Power Tube Fittings and Conductors Standards, NFPA/T3.8.11-1977 ... 71

American National Standard Fluid power systems and products - Connectors and associated components - Outside diameters of tubes and inside diameters of hoses, ANSI/B93.59M-1982 ... 83

American National Standard Fluid power systems and products - Connectors and associated components - Nominal pressures, ANSI/B93.60M-1982 ... 89

American National Standard Hydraulic fluid power - Line tubing - Electric resistance welded, mandrel drawn, ANSI/B93.4M-1981 ... 95

American National Standard Hydraulic fluid power - Line tubing- Seamless low carbon steel, ANSI/B93.11M-1981 ... 103

American National Standard Method for Testing Hydraulic Fluid Power Quick Action Couplings, ANSI/B93.42M-1977 (R1983) ... 111

American National Standard Pneumatic fluid power - Quick action couplings - Test conditions and procedures, ANSI/B93.51M-1980 ... 141

Glossary For Fluid Power Quick Action Couplings, NFPA/T3.20.1-1973 (R1981) ... 157

A Bibliography of Fluid Power Quick Action Coupling Standards, NFPA/T3.20.7 R1-1983 ... 163

Quick Action Couplings Pressure Rating Supplement No. 5 to NFPA Recommended Standard for Verifying the Fatigue and Static Pressure Ratings of the Pressure Containing Envelope of a Metal Fluid Power Component, NFPA/T2.6.1M S5 (T3.20.8M)-1975 (R1981) ... 167

American National Standard Hydraulic fluid power - Quick action couplings - Surge flow test (short duration flow), ANSI/B93.68M-1983 ... 177

American National Standard Hydraulic fluid power - Quick action couplings - Surge flow test (long duration flow), ANSI/B93.69M-1983 ... 187

A Bibliography of Fluid Power Hose, Hose Fittings and Hose Assemblies Standards, NFPA/T3.26.1 R1-1977 ... 195

- NOTES -

Foreword

FLUID POWER STANDARDS, Seventh Edition

Fluid Power Standards is comprised of ten separate volumes available as a set or separately. The titles of the volumes are:

Volume A	Communication including graphic symbols and metric units
Volume B	Pressure Rating
Volume C	Pumps, Motors, Power Units and Reservoirs
Volume D	Filtration and Contamination
Volume E	Conductors and Associated Products
Volume F	Control Products
Volume G	Cylinders and Accumulators
Volume H	Fluids, Lubricants and Sealing Devices
Volume I	Testing
Volume J	Bibliographies

ORDERING INFORMATION

The volumes of Fluid Power Standards can be ordered by contacting the National Fluid Power Associations Publications Department, 3333 N. Mayfair Rd., Milwaukee, Wisconsin 53222 USA Telephone (414) 778-3344 Telex 26-898.

WHAT IS CONTAINED IN FLUID POWER STANDARDS

NFPA's ten volume set, Fluid Power Standards contains both NFPA and ANSI (American National Standards Institute) Fluid Power standards.

HOW FLUID POWER STANDARDS ARE DEVELOPED

The National Fluid Power Association coordinates development of Fluid Power standards on industry, national and international levels. The association holds the Secretariat for the Fluid Power Committee work of the American National Standards Institute (ANSI) and the International Organization for Standardization (ISO).

On the Industry Level (NFPA)

NFPA standards originate in Product Sections and Technology Committees, and are submitted to the NFPA Technical Board and Board of Directors for approval. Product Sections are generally composed of individuals involved in the design, manufacture, performance and application of specific Fluid Power Products. Technology Committees are comprised of experts in a single broad area of Fluid Power technology, applying to many products.

More than 350 persons currently contribute to the standards writing work of NFPA's technical committees and are helping to develop approximately 140 projects as proposed NFPA Recommended Standards, Information Reports and Recommended Practices. Qualified engineers from NFPA member companies are eligible for participation in NFPA technical committees.

On the National Level (ANSI)

After a standard has been approved by the NFPA, it may be submitted to Standards Committee B93, Fluid Power Systems and Products, accredited by the American National Standards Institute. NFPA serves as secretariat for this committee, which ballots the standard for approval as an American National Standard. To date, more than 60 NFPA Recommended Standards have advanced to become American National Standards.

On the International Level (ISO)

Under authority granted by the International Organization for Standardization and the American National Standards Institute, NFPA administers the Secretariat of ISO Technical Committee 131 (TC 131) Fluid power systems. More than 35 nations participate in standards writing through TC 131. The U. S. Fluid Power Industry is represented through USA TAG to ISO/TC 131, a committee also administered by the NFPA.

On an average, the NFPA coordinates 44 deliberating sessions per year in various parts of the industrialized world.

Participating in Standards Development

Participation in the NFPA, ANSI/B93 and ISO/TC 131 committees is open to qualified engineers. For details on how to become involved, contact the Director of Technical Services at National Fluid Power Association, 3333 North Mayfair Rd., Milwaukee, Wisconsin 53222 USA Telephone (414) 778-3344 Telex 26-898.

Five Year Review of Fluid Power Standards

An NFPA Recommended Standard is subject to revision at anytime by the appropriate technical committee. They are reviewed every five years and if not revised, either reaffirmed or withdrawn. Comments are invited either for revision of any standard or for additional standards. Comments should be addressed to the Director of Technical Services at NFPA Headquarters. Comments will receive careful consideration at a meeting of the appropriate technical committee. Commentators are welcome to attend the meeting.

Obsolete Fluid Power Standards

The 1987 edition of these Fluid Power Standards volumes makes the 1984 edition obsolete. For practical purposes, it is not wise to use obsolete volumes. However, for teaching purposes, the outdated volumes might be useful. Standards contained in each volume have an NFPA or ANSI reference number that includes the year the standard was first approved and the year it was reaffirmed.

Disclaimer

See complete disclaimer on Copyright page (reverse of Title Page).

NATIONAL FLUID POWER ASSOCIATION

The National Fluid Power Association is the trade association for manufacturers of hydraulic and pneumatic products and systems.

Founded in 1953, NFPA has 190 corporate members across the U.S. NFPA members produce more than 75 percent of all U.S. Fluid Power production.

NFPA was founded to pursue activities that advance the performance and application of Fluid Power products, better materials, designs and standards, organize committees to further the art and science of Fluid Power, support activities to advance knowledge and understanding of Fluid Power, and represent the members to the Federal Government, industry and user organizations.

Association activities in the areas of technical standards and services, marketing research and statistics, public affairs, industry promotion and management are directed by volunteer Boards and supported by full-time professional headquarters staff.

NFPA provides administrative services to these organizations in the Fluid Power Industry:

- Fluid Power Committee B93, the committee for national fluid power standards, accredited by American National Standards Institute,
- ISO/TC 131 Fluid power systems, the committee for international fluid power standards, of the International Organization for Standardization,
- USA Technical Advisory Group (USA TAG) to ISO/TC 131, the national committee for international Fluid Power standards,
- Fluid Power Appeals Boards,
- Fluid Power Coordinating Council,
- Fluid Power Educational Foundation,
- International Fluid Power Exposition
- International Fluid Power Exposition Applications Conference.

For information on membership in NFPA, or participation of any of the Fluid Power technical committees, contact the National Fluid Power Association, 3333 North Mayfair Rd., Milwaukee, Wisconsin 53222 USA (414) 778-3344 Telex 26-898.

- NOTES -

ANSI/B93.2-1986

AN INDUSTRY STANDARD FOR FLUID POWER

Excerpts Extracted from ANSI/B93.2

American National Standard

Fluid power systems and products -

Glossary

---

published by
NATIONAL FLUID POWER ASSOCIATION, INC.
3333 N. Mayfair Road / Milwaukee, WI 53222 / 414-778-3344 / TLX 26898

GROUP 01 — PRIMARY TERMS

01.01.080   Fluid Logic

A branch of fluid power associated with digital signal sensing and information processing, using components with or without moving parts.

01.01.100   Fluid Power

Energy transmitted and controlled through use of a pressurized fluid.

01.01.200   Fluid Power System

A system that transmits and controls power through use of a pressurized fluid within an enclosed circuit.

01.01.300   Fluidics

Engineering science pertaining to the use of fluid dynamic phenomena to sense, control process information, and/or actuate.

01.01.400   Hydraulics

Engineering science pertaining to liquid pressure and flow.

01.01.500   Hydrodynamics

The engineering science which governs the movement of liquids and the forces opposing that movement.

01.01.600   Hydrokinetics

Engineering science pertaining to the energy of liquid flow and pressure.

01.01.700   Hydropneumatics

Pertaining to the combination of hydraulic and pneumatic fluid power.

01.01.800   Hydrostatics

Engineering science pertaining to the energy of liquids at rest.

01.01.801   Hydrostatic Transmission

Combination of one or more hydraulic pumps and motors forming a unit.

01.01.850   Moving Parts Logic

The technology of achieving logic control by means of fluid devices having moving parts.

01.01.900   Pneumatics

Engineering science pertaining to gaseous pressure and flow.

01.01.910   Time

An interval comprising a limited but continuous action.

01.01.911   Time, Actuated

The interval in which the component is under the influence of the actuating forces.

01.01.912   Time, Fall

The interval taken in a device for a quantity to change from a specified high level down to a specified lower level.

01.01.913   Time, Operation

The interval of an event measured from "start signal" to final "at rest".

01.01.914   Time, Released

The duration of time in which the component is not under the influence of the actuating forces.

01.01.915   Time, Relative Duty

(Expressed as a percentage)

$$\frac{t \text{ (actuated)}}{t \text{ (actuated)} + t \text{ (released)}} \times 100$$

01.01.916   Time, Response

The elapsed time between the initiation of an action and the resulting reaction (measured under specified conditions).

01.01.918   Time, Rise

The interval in a device for a quantity to change from a specified low level up to a specified high level.

01.01.919   Time, Start Up

The interval needed to reach a steady state operating condition in the system from "start up".

01.01.930   Torque

Turning effort.

01.01.931   Torque, Derived

Torque corresponding to the derived hydraulic power.

01.01.932   Torque, Effective

Actual torque transmitted by the shaft under specified conditions.

01.01.933   Torque, Geometric

Torque corresponding to the geometric hydraulic power.

GROUP 02 - FLUID POWER LAWS AND RELATED TERMS

02.01.100   Bernoulli's Law

If no work is done on or by a flowing frictionless liquid its energy due to pressure and velocity remains constant at all points along the streamline.

02.01.200   Boyle's Law

The absolute pressure of a fixed mass of gas varies inversely as the volume, provided the temperature remains constant.

02.01.300   Charles' Law

The volume of a fixed mass of gas varies directly with absolute temperature, provided the pressure remains constant.

02.01.400   Continuity Equation

The mass rate of fluid flow into a fixed space is equal to the mass flow rate out. Hence, the mass flow rate of fluid past all cross sections of a conduit is equal.

02.01.500   Darcy's Formula

A formula used to determine the pressure drop due to flow friction through a conduit.

$$h_f = \frac{fLv^2}{2Dg}$$

h_f = Head loss, feet (metre)
f = Friction factor (see 02.01.600)
L = Length of conduit, feet (metre)
v = Mean velocity of flow, ft/sec. (metre/sec.)
D = Internal diameter of conduit, feet (metre)
g = Acceleration due to gravity, 32.2 ft./sec.^{2} (9.81 metre/sec.^{2})

02.01.600   Hagen Poiseuille Law

The friction factor of Darcy's Formula is a ratio of 64 to the Reynolds Numbers when flow is laminar.

$$f = \frac{64}{N_r}$$

f = Friction factor
N_r = Reynolds Number

02.01.700   Pascal's Law

A pressure applied to a confined fluid at rest is transmitted with equal intensity throughout the fluid.

02.01.800   Reynolds Number

A numerical ratio of the dynamic forces of mass flow to the shear stress due to viscosity. Flow usually changes from laminar to turbulent between Reynolds Number 2,000 and 4,000.

$$N_r = \frac{pvD}{u} = \frac{vD}{v}$$

N_r = Reynolds Number
p (Rho) = Fluid density, pounds(mass)/ft.^{3} (Kg/m^3)
v = Mean velocity of flow, ft./sec. (metre/sec.)
D = Internal diameter of conduit, feet (metre)
u (Mu) = Absolute viscosity, pounds/ft.sec. (Kg/cm^5)
v (Nu) = Kinematic viscosity, ft.^{2}/sec. (mm^2/sec.)

02.01.900   Toricelli's Theorem

The liquid velocity at an outlet discharging into the free atmosphere is proportional to the square root of the head.

$$v = \sqrt{2gh}$$

v = Velocity (mean), ft./sec. (m/sec.)
g = Acceleration, 32.2 ft./sec.^{2} (9.81 m/sec.^{2})
h = Pressure head, ft(m)

ANSI/B93.2

GROUP 03 — FLOW TERMS

03.01.200 Cavitation

A localized gaseous condition within a liquid stream which occurs where the pressure is reduced to the vapor pressure.

03.01.210 Coanda Effect

The phenomenon, named after its discoverer, of the attachment of a free flowing turbulent jet to an adjacent, possibly curved, wall.

03.01.409 Flow

Movement of fluid generated by pressure differences.

03.01.410 Flow, Laminar (Streamline)

A flow situation in which fluid moves in parallel lamina or layers.

03.01.420 Flow, Metered

Flow at a controlled rate.

03.01.430 Flow, Steady State

A flow situation wherein conditions such as pressure, temperature, and velocity at any point in time do not change.

03.01.450 Flow, Turbulent

A flow situation in which the fluid particles move in a random fluctuating manner.

03.01.460 Flow, Upsteady

A flow situation wherein conditions such as pressure, temperature and velocity at points in the fluid change.

03.01.600 Shock Wave

A pressure wave front which moves at a sonic velocity.

03.01.800 Surge

A transient rise of pressure or flow.

GROUP 04 — MENSURATION TERMS

04.01.005 Air, Free

Air at ambient temperature, pressure, relative humidity, and density.

04.01.010 Air, Standard

Air at a temperature of $68°F$, a pressure of 14.70 pounds per square inch absolute, and a relative humidity of 36% (0.0750 pounds per cubic foot). In gas industries the temperature of "standard air" is usually given as $60°F$.

04.01.011 Amplification

The ratio between the output signal variations and the control signal variations (for analogue devices only).

04.01.012 Amplification, Flow

Ratio between the output flow and the input (control) flow.

04.01.013 Amplification, Power

The ratio between the output power variation and the corresponding input (control) power variation (for analogue devices only).

04.01.014 Amplification, Pressure

Ratio between the outlet pressure and the inlet (control) pressure.

04.01.020 Aniline Point

The lowest temperature at which a liquid is completely miscible with an equal volume of freshly distilled aniline (ASTM Designation D611-64).

04.01.025 Assurance Level

The minimum percentage of pressure containing envelopes of a verified design that will sustain 10 million applications of its Rated Fatigue Pressure.

04.01.028 Bistable

A binary circuit or device which has two stable states and which in each state requires an appropriate impulse to cause a transition to the other state.

04.01.030 Bulk Modulus

The measure of resistance to compressibility of a fluid. It is the reciprocal of the compressibility.

04.01.035 Capacity, Effective

Actual volume displaced under specified conditions.

04.01.036 Capacity, Geometric

Volume displaced, calculated geometrically without reference to tolerances, clearances or deformation.

04.01.040 Compressibility

The change in volume of a unit volume of a fluid when subjected to a unit change in pressure.

04.01.041 Confidence Level

The degree to which a manufacturer can be assured that the desired assurance level is attained.

04.01.043 Displacement, Volumetric

Volume absorbed or displaced per stroke of a cylinder (See 81.01.012) or per cycle of a pump or motor (43.01.005).

04.01.046 Efficiency

Ratio of output to the corresponding input.

04.01.047 Expectancy, Life

The predicted working period during which a component or system will maintain a specified level of performance under specified conditions. Sometimes expressed in statistical terms as a probability.

04.01.048 Fire Point

The temperature to which a fluid must be heated to ignite and burn for five seconds (minimum) in the presence of air when a small flame is applied under controlled conditions.

04.01.050 Flash Point

The temperature to which a liquid must be heated under specified conditions of the test method to give off sufficient vapor to form a mixture with air that can be ignited momentarily by a flame.

04.01.051 Flow Degradation Ratio

The ratio of stabilized flow rate after a contaminant injection to the initial measured pump flow (Qr).

04.01.052 Flow Factor

Characterizes the conductance of a pneumatic or hydraulic device, flowline or connection.

04.01.060 Flow Rate

The volume, mass or weight of a fluid passing through any conductor per unit of time.

04.01.061 Flow Rate, Relief

The rate at which fluid can flow through the unloading device for each specific increase in controlled pressure above the original setting, measured under specified conditions.

04.01.063 Flow, Supply Port

The flow of fluid through the supply ports of the device or system.

04.01.070 Fluid Friction

Friction due to the viscosity of fluids.

04.01.072 Frequency Response

The changes, under steady-state conditions, in the output variable which are caused by a sinusoidal input variable.

04.01.170 Hammer, Liquid

Pressure and depression waves created by relatively rapid flow changes and transmitted through the system.

04.01.100 Head

The height of a column or body of fluid above a given point expressed in linear units. Head is often used to indicate gage pressure. Pressure is equal to the height times the density of the fluid.

04.01.110 Head, Friction

The head required to overcome the friction at the interior surface of a conductor and between fluid particles in motion. It varies with flow, size, type and condition of conductors and fittings, and the fluid characteristics.

04.01.115 Head, Pressure

The pressure due to the height of a column or body of fluid. (It is usually expressed in inches [mm]).

04.01.120 Head, Static

The height of a column or body of fluid above a given point.

04.01.130 Head, Static Discharge

The static head from the centerline of the pump to the free discharge surface.

04.01.140 Head, Static Suction

The head from the surface of the supply source to the centerline of the pump.

04.01.150 Head, Total Suction

The static head from the surface of the supply source to the free discharge surface.

04.01.160 Head, Velocity

The equivalent head through which the liquid would have to fall to attain a given velocity. Mathematically it is equal to the square of the velocity (in feet) divided by 64.4 feet per second squared.

$$h = \frac{v^2}{2g}$$

- h = Head, feet (metre)
- g = Acceleration due to gravity, 32.2 ft./sec.^{2} (9.81 metre/sec.^{2})
- v = Mean velocity of flow, ft./sec. (metre/sec.)

04.01.200 Hydraulic Power

Power computed from flow rate and pressure differential (drop).

Hydraulic Power = .000583 $q_v p$ (0,002234 $q_v p$) expressed as horsepower where:

- q_v = Flow rate, gpm (L/min.)
- p = Pressure, psi (bar)

Alternate formula = $\frac{q_v p}{1714}$ $\left(\frac{q_v p}{447.6}\right)$

04.01.210 Lift

The height of a column or body of fluid below a given point expressed in linear units. Lift is often used to indicate vacuum or pressure below atmosphere.

04.01.220 Lift, Static Suction

The lift from the centerline of the pump to the surface of the supply source. (See Head, Static Suction).

04.01.222 Linear Function

Describes a condition in which the relationship between two interdependent variables is constant.

04.01.223 Linear Region

The region of a given control characteristic over which the linearity remains within specified limits.

04.01.224 Linearity

The faithfulness with which an output signal of an electronic reproducing system reproduces an input signal.

04.01.230 Monostable

A binary circuit or device, which has one stable state and which requires an appropriate change of the input to cause a transition out of its stable state for a specified period of time. The specified period of time at which the circuit stays out of its stable state is independent of the duration of the appropriate change of the input signals.

04.01.250 Neutralization Number

A measure of the total acidity or basicity of an oil; this includes organic or inorganic acids or bases or a combination thereof (ASTM Designation D974-64).

04.01.260 Newt

A unit of kinematic viscosity in the English system. It is expressed in square inches per second (see 04.01.630 Stokes).

04.01.270 Poise

The standard unit of dynamic viscosity in the c.g.s. (centimeter-gram-second) system. It is the ratio of the shearing stress to the shear rate of fluid and is expressed in milli-pascal sec. (= 1 centipoise).

04.01.280 Pour Point

The lowest temperature at which a liquid will flow under specified conditions (ASTM Designation D97-66).

04.01.282 Power Consumption

The total power consumed by the device or system under specified conditions.

04.01.283 Power Supply, Fluid

Energy source which generates and maintains a flow of fluid under pressure.

04.01.290 Precipitation Number

The number of millilitres of precipitate formed when 10 mL of lubricating oil are mixed with 90 mL of ASTM precipitation naptha and centrifuged under prescribed conditions (ASTM Designation D91-61).

04.01.300 Pressure

Force per unit area, usually expressed in pounds per square inch (bar).

04.01.310 Pressure, Absolute

The pressure above zero absolute, i.e., the sum of atmospheric and gage pressure. In vacuum related work it is usually expressed in millimetres of mercury (mm Hg).

04.01.320 Pressure, Atmospheric

Pressure exerted by the atmosphere at any specific location. (Sea level pressure is approximately 14.7 pounds per square inch absolute, 1 bar = 14.5 psi).

04.01.330 Pressure, Back

The pressure encountered on the return side of a system.

04.01.340 Pressure, Breakloose (Breakout)

The minimum pressure which initiates movement.

04.01.360 Pressure, Burst

The pressure which causes failure of and consequential loss of fluid through the product envelope.

04.01.370 Pressure, Charge

The pressure at which replenishing fluid is forced into a fluid power system.

04.01.375 Pressure, Control Range

The permissible limits between which system pressure may be set.

04.01.380 Pressure, Cracking

The pressure at which a pressure operated valve begins to pass fluid.

04.01.385 Pressure, Cyclic Test

A pressure range applied in cyclic tests that are performed to verify a Rated Fatigue Pressure.

04.01.390 Pressure, Differential (Pressure Drop)

The difference in pressure between any two points of a system or a component.

04.01.400 Pressure, Gage

Pressure differential above or below ambient atmospheric pressure.

04.01.412 Pressure, Induced

Pressure generated by an externally applied force.

04.01.414 Pressure, Inlet

The pressure at the apparatus inlet port.

04.01.415 Pressure, Intensified

In a fluid power cylinder, the outlet pressure required to slow the piston rod extending under regulated pressure introduced at the cap end.

04.01.417 Pressure, Maximum Inlet

The maximum rated gage pressure applied to the inlet.

04.01.419 Pressure, Nominal

A pressure value assigned to a component or system for the purpose of convenient designation.

04.01.420 Pressure, Operating

The pressure at which a system is operated.

04.01.425 Pressure, Outlet

Pressure at the apparatus outlet port.

04.01.427 Pressure, Overrange Rating

The pressure to which a device can be subjected for extended time without change in operating characteristics, shift in set point, or damage to the device.

04.01.430 Pressure, Override

The difference between the cracking pressure of a valve and the pressure reached when the valve is passing its rated flow.

04.01.435 Pressure, Peak

The maximum pressure encountered in the operation of a component.

04.01.440 Pressure, Pilot

The pressure in the pilot circuit.

04.01.450 Pressure, Precharge

The pressure of compressed gas in an accumulator prior to the admission of a liquid.

04.01.460 Pressure, Proof

The non-destructive test pressure, in excess of the maximum rated operating pressure, which causes no permanent deformation, excessive external leakage, or other resulting malfuction.

04.01.470 Pressure, Rated

The qualified operating pressure which is recommended for a component or a system by the manufacturer.

04.01.472 Pressure, Recovery

The ratio of output pressure to the supply pressure.

04.01.473 Pressure, Rated Fatigue

A pressure that a pressure containing envelope is represented to sustain 10 million times without failure.

04.01.474   Pressure, Regulation of

Pertains to the control of pressure in a system.

04.01.476   Pressure, Rated Static

A pressure that a component pressure containing envelope is represented to sustain once, under test conditions without failure, after which the component must be discarded.

04.01.477   Pressure, Residual

The value of the output pressure in the "off" state of the device.

04.01.480   Pressure, Shock

The pressure existing in a wave moving at sonic velocity.

04.01.481   Pressure, Shockwave

A pressure pulse which moves at sonic speed in the liquid.

04.01.490   Pressure, Static

The pressure in a fluid at rest.

04.01.495   Pressure, Static Test

A pressure applied in a static test performed to verify a Rated Static Pressure.

04.01.500   Pressure, Suction

The absolute pressure of the fluid at the inlet of a pump.

04.01.505   Pressure, Supply

The pressure at the apparatus inlet port.

04.01.510   Pressure, Surge

The pressure resulting from surge conditions.

04.01.520   Pressure, System

The pressure which overcomes the total resistances in a system. It includes all losses as well as useful work.

04.01.530   Pressure, Vapor

The pressure, at a given fluid temperature, in which the liquid and gaseous phases are in equilibrium.

04.01.540   Pressure, Working

The pressure at which the apparatus is being operated in a given application.

04.01.541   Pressure, Range, Working

The tolerance (plus or minus) range of the working pressure.

04.01.610   Reyn

The standard unit of absolute viscosity in the English system. It is expressed in pound-seconds per square inch.

04.01.611   Rotation

The direction of rotation is always quoted as viewed looking at the shaft end. In dubious cases, provide a sketch.

04.01.612   Rotation, Anti-Clockwise

Rotation in the opposite sense to the clock.

04.01.613   Rotation, Clockwise

Forward rotation of the hands to the clock.

04.01.614   Rotational Frequency

Number of revolutions per unit of time.

04.01.620   Specific Gravity, Liquid

The ratio of the weight of a given volume of liquid to the weight of an equal volume of water.

04.01.630   Stokes

The standard unit kinematic viscosity in the c.g.s. (centimetre-gram-second) system. It is expressed in square centimetres per second; 1 centistokes equals .01 stokes.

04.01.640   Surface Tension

The surface force of a liquid in contact with a fluid by which it tends to assume a spherical form and to present the least possible surface. It is expressed in pounds per foot or dynes per centimetre.

04.01.641   Temperature, Ambient

The temperature of the environment in which the apparatus is working.

04.01.642   Temperature, Equipment

The temperature of the unit at a specified position and measured at a specified point.

04.01.643   Temperature, Fluid

The temperature of the pressure medium measured at a specified point.

ANSI/B93.2

04.01.644 Temperature, Inlet

Fluid temperature at the inlet port.

04.01.645 Temperature, Outlet

Fluid temperature at the outlet port.

04.01.646 Temperature Range

The permissible temperature range within which the apparatus or the fluid can operate satisfactorily.

04.01.647 Torr

A unit of pressure equal to 1/760 of an atmosphere and very nearly equal to 1mm Hg @ $0°C$.

04.01.648 Unistable

A binary circuit or device, which has one stable state and in which the output changes state for the duration of the appropriate change of the input signal.

04.01.650 Vacuum

Pressure less than ambient atmospheric pressure.

04.01.657 Variability Factor

A multiplier applied to the Rated Fatigue Pressure for calculating the Cyclic Test Pressure to account for the variability in fatigue strength of metals. It is also applied to the Rated Static Pressure for calculating the Static Test Pressure.

04.01.700 Viscosity

A measure of the internal friction or the resistance of a fluid to flow.

04.01.710 Viscosity, Absolute

The ratio of the shearing stress to the shear rate of a fluid. It is usually expressed in centipoise.

04.01.720 Viscosity, Kinematic

The absolute viscosity divided by the density of the fluid. It is usually expressed in centistokes.

04.01.730 Viscosity, SAE Number

The Society of Automotive Engineers' arbitrary numbers for classifying fluids according to their viscosities. The numbers in no way indicate the viscosity index of fluids.

04.01.740 Viscosity, SUS

Saybolt Universal Second (SUS), which is the time in seconds for 60 millilitres of oil to flow through a standard orifice at a given temperature (ASTM Designation D88-56).

04.01.800 Viscosity Index

A measure of the viscosity-temperature characteristics of a fluid as referred to that of two arbitrary reference fluids (ASTM Designation D2270-64).

04.01.801 Viscosity Index Improver

A chemical compound added to a fluid to modify its temperature/viscosity relationship.

04.01.810 Vortex

Spiral motion of a fluid resulting in a radial pressure gradient. The trajectories are curves which encircle a single line (axis).

GROUP 05 — SYMBOLS

Section 01 — General
Section 02 — Symbol Types

SECTION 01 — GENERAL

05.01.100 Symbol, Fluid Power

A representation of the characteristics of a fluid power component by means of lines on a flat surface.

SECTION 02 — SYMBOL TYPES

05.02.100 Symbol, Combination

A symbol which combines graphical, cutaway and pictorial representations.

05.02.200 Symbol, Cutaway

A symbol showing principal internal parts, controls and actuating mechanisms, interconnecting lines, and functions of a component.

05.02.300 Symbol, Graphic (Schematic)

A simplified symbol which indicates essential characteristics applicable to all similar components.

05.02.400 Symbol, Pictorial

A symbol showing the actual shape of a component according to the manufacturer's description.

GROUP 10 — GENERAL FLUIDS

Section 01 — General
Section 02 — General Types
Section 03 — Fluid Stability
Section 04 — Fluid Characteristics

SECTION 01 — GENERAL

10.01.001   Additive

A chemical added to a fluid to impart new properties or to enhance those which already exist.

10.01.100   Fluid

A liquid, gas or combination thereof.

10.01.150   Inhibitor

Any substance which, when present in very small proportions, slows, prevents or modifies chemical reactions such as corrosion or oxidation.

SECTION 02 — GENERAL TYPES

10.02.010   Fluid, Aqueous

A fluid which contains water as a major constituent besides the organic material. The fire resistance properties are derived from the water content.

10.02.100   Fluid, Fatty Oil

A fluid composed of fats derived from animal, marine or vegetable origin. It may contain additives.

10.02.200   Fluid Fire Resistant (Non-Flammable)*
***Deprecated**

A fluid difficult to ignite which shows little tendency to propagate flame.

10.02.300   Fluid, Hydraulic

A fluid suitable for use in a hydraulic system.

10.02.350   Fluid, Newtonian

Fluid having a viscosity that is always independent of the rate of shear.

10.02.400   Fluid, Pneumatic

A fluid suitable for use in a pneumatic system.

SECTION 03 — FLUID STABILITY

10.03.000   Fluid Stability

Resistance of a fluid to permanent changes in properties.

10.03.100   Fluid Stability, Chemical

Resistance of a fluid to chemical change.

10.03.150   Fluid Stability, Emulsion

The stability characteristics of an emulsion under defined storage conditions.

10.03.200   Fluid Stability, Hydrolytic

Resistance of a fluid to permanent changes in properties caused by chemical reaction with water.

10.03.300   Fluid Stability, Oxidation

Resistance of a fluid to permanent changes caused by chemical reaction with oxygen.

10.03.350   Fluid Stability, Shear

The ability of a fluid to maintain its viscosity under operating conditions.

10.03.400   Fluid Stability, Thermal

Resistance of a fluid to permanent changes caused soley by heat.

10.03.450   Incompatible Fluids

Fluid which when mixed in a system will have a deleterious effect on that system, its components, or its operation.

SECTION 04 — FLUID CHARACTERISTICS

10.04.010   Fluid, Aeration, Foam

A more or less stable extended air-liquid interface arising when bubbles persist at the surface of a fluid.

10.04.011   Fluid, Air Release

The ability of a fluid to release air bubbles dispersed therein.

10.04.012   Fluid, Anti-Corrosive

A fluid containing metal corrosion inhibitors.

10.04.013   Fluid, Ash Content

Percentage in weight of the residue after calcination of the fluid under defined conditions.

10.04.016 Fluid, Auto Ignition Temperature

The temperature at which a fluid will ignite when dripped by a pipette into a heated flask. A.I.T. is calculated at $5°C$ below the temperature at which ignition takes place.

10.04.100 Fluid Density

Quotient of the mass of a fluid by its volume at a specified temperature ($15°C$).

10.04.110 Fluid, Evaporation Deposits

Percentage of residue obtained after evaporation of the product in free air.

10.04.200 Fluid, Miscibility

Capacity of fluids to be mixed in any ratio without separation into phases.

10.04.270 Fluid, Rust Protection

Capacity of a fluid to prevent the formation of rust under specified conditions.

10.04.280 Fluid, Vapor Pressure

Pressure exerted at any temperature by a vapor from a fluid existing in equilibrium with its liquid phase.

10.04.290 Fluid, Water Content

Quantity of water contained in a mineral oil fluid other than as a contaminant.

GROUP 11 — EMULSIONS

Section 01 — General
Section 02 — Emulsion Types

SECTION 01 — GENERAL

11.01.050 Demulsibility, Water

Capacity of an emulsion of fluid and water to separate into two phases.

11.01.100 Emulsion

A homogeneous dispersion of two immiscible liquids.

11.01.400 Saponification, Value of

A measure of the free and combined acids in oils reacting with potassium hydroxide per gram of fluid.

SECTION 02 — EMULSION TYPES

11.02.525 Emulsion, Oil in Water

A dispersion of oil in a continuous phase of water.

11.02.550 Emulsion, Water in Oil

A dispersion of water in a continuous phase of oil.

GROUP 12 — PETROLEUM FLUIDS

Section 01 — General

SECTION 01 — GENERAL

12.01.100 Petroleum Fluid

A fluid composed of petroleum oil which may contain additives and/or inhibitors.

GROUP 13 — SYNTHETIC FLUIDS

Section 01 — General
Section 02 — Types of Synthetic Fluids

SECTION 01 — GENERAL

13.01.100   Synthetic Fluid

Fluid other than mineral oil which has been artificially compounded for use in a fluid power system.

SECTION 02 — TYPES OF SYNTHETIC FLUIDS

13.02.090   Synthetic Fluid, Chlorinated Hydrocarbon

An aromatic or paraffinic hydrocarbon fluid in which certain hydrogen atoms are replaced by chlorine. The fire resistance is derived from the chlorine present.

13.02.095   Synthetic Fluid, Di-Basic Ester

A fluid manufactured by the reaction of a di-basic acid with a monohydric alcohol.

13.02.100   Synthetic Fluid, Halogenated

A fluid composed of halogenated organic materials. It may contain additives.

13.02.200   Synthetic Fluid, Organic Ester

A fluid composed of esters which are compounds of carbon, hydrogen, and oxygen only (it may contain additives).

13.02.300   Synthetic Fluid, Phosphate Ester

A fluid composed of phosphate esters. It may contain additives.

13.02.400   Synthetic Fluid, Phosphate Ester Base

A fluid which contains a phosphate ester as one of the major components.

13.02.500   Synthetic Fluid, Polyglycol

A non-aqueous fluid composed of polyglycol derivatives. It may contain additives.

13.02.600   Synthetic Fluid, Silicate Ester

A fluid composed of organic silicates. It may contain additives.

13.02.700   Synthetic Fluid, Silicone

A fluid composed of silicones. It may contain additives.

GROUP 14 — WATER-GLYCOLS

Section 01 — General

SECTION 01 — GENERAL

14.01.100   Water Glycol Fluid

A fluid whose major constituents are water and one or more glycols or polyglycols.

GROUP 15 — AIR

Section 01 — General
Section 02 — Types of Air

SECTION 01 — GENERAL

15.01.100   Air

A gas mixture consisting primarily of nitrogen, oxygen, argon, carbon dioxide, hydrogen, neon and helium.

15.01.102   Air, Contamination of

Contaminants in the air supplies to a system or device.

SECTION 02 — TYPES OF AIR

15.02.100   Air, Compressed (Pressure)

Air at any pressure greater than atmospheric pressure.

15.02.200   Air, Dried

Air with moisture content lower than the maximum allowable for a given application.

15.02.800   Air, Saturated

Air at 100 percent relative humidity, with a dew point equal to temperature.

Air, Free (See Mensuration Group)

Air, Standard (See Mensuration Group)

GROUP 16 — WATER

Section 01 — General
Section 02 — Types

SECTION 01 — GENERAL

16.01.100   Water

A fluid compound consisting of primarily two hydrogen atoms and one oxygen atom.

GROUP 21 — RESERVOIRS

Section 01 — General
Section 02 — Reservoir Types
Section 03 — Reservoir Capacities

SECTION 01 — GENERAL

21.01.010   Capacitor

A device capable of storing a signal at a specific point in a pneumatic control or fluidic control circuit.

21.01.100   Reservoir

A container for storage of liquid in a fluid power system.

SECTION 02 — RESERVOIR TYPES

21.02.100   Reservoir, Atmospheric

A container for the storage of a fluid medium at atmospheric pressure.

21.02.200   Reservoir, Hydraulic

A reservoir for storing and conditioning a liquid in a hydraulic system.

21.02.300   Reservoir, Non-Integral

An independent or removable reservoir.

21.02.400   Reservoir, Pressure Sealed

A sealed reservoir for storage of fluids under pressure.

21.02.500   Reservoir, Sealed

A reservoir for storage of fluids isolated from atmospheric conditions.

21.02.600   Reservoir, Top Mounted

A reservoir with provisions for mounting the pump and components on top.

SECTION 03 — RESERVOIR CAPACITIES

21.03.100   Breathing Capacity

A measure of flow rate through an air breather.

21.03.400   Fluid Capacity

The liquid volume coincident with the "high" mark of the level indicator.

21.03.800   Reserve Capcity

The volume of air above the "high" mark of the level indicator.

GROUP 32 — HEAT EXCHANGERS

Section 01 — General
Section 02 — Types of Heat Exchangers

SECTION 01 — GENERAL

32.01.050 Cooler

A heat exchanger which removes heat from a fluid.

32.01.070 Fan, Cooling

A device which mechanically creates a flow of air over a hot surface, usually used with a radiator in order to increase the rate of heat exchange.

32.01.100 Heat Exchanger

A device which transfers heat through a conducting wall from one fluid to another.

32.01.101 Heater

A device which transfers heat through a conducting wall from one fluid to another.

32.01.700 Temperature Controller

A device which maintains the fluid temperature within prescribed limits.

SECTION 02 — TYPES OF HEAT EXCHANGERS

32.02.220 Aftercooler

A device which cools a gas after it has been compressed.

32.02.240 Intercooler

A device which cools a gas between the compressive steps of a multiple stage compressor.

32.02.260 Precooler

A device which cools a gas before it is compressed.

32.02.270 Radiator

Device, usually of honeycomb or multi-tubular construction which transfers heat from a liquid to air, thereby acting as a liquid/air heat exchanger.

GROUP 50 — GENERAL CONDUCTOR TERMS

Section 01 — General
Section 02 — General Components
Section 03 — Functional Lines
Section 04 — Port Types
Section 05 — Port Functions
Section 06 — Port Threads
Section 07 — Physical Characteristics

SECTION 01 — GENERAL

50.01.200 Channel

A fluid passage, the length of which is large with respect to its cross-sectional area.

50.01.201 Channel, Control

Channel through which the control or input signal enters the device.

50.01.202 Channel, Output

The channel through which the output signal leaves the device.

50.01.600 Passage

A machined or cored fluid-conducting path which lies within or passes through a component.

50.01.610 Flow Passage, Controlled

A flow passage whose ability to pass fluid can be changed by the influence of a signal.

50.01.625 Flow Path

A series of conductors and passages which convey fluid.

50.01.700 Port

A terminus of a passage in a component to which conductors can be connected.

50.01.701 Port, Area of

Minimum area of fluid passage through a port.

SECTION 02 — GENERAL COMPONENTS

50.02.100   Conductor

A component whose primary function is to contain and direct fluid.

50.02.200   Conduit

Any confining element employed to transfer fluid.

50.02.300   Line

A tube, pipe, or hose for conducting fluid.

50.02.400   Nipple

A short length of pipe or tube.

SECTION 03 — FUNCTIONAL LINES

50.03.050   Line, Bleed

Line through which air is purged from pipes (conductors) containing liquid.

50.03.100   Line, Drain

A line returning leakage fluid independently to the reservoir or vented manifold.

50.03.200   Line, Exhaust

A line returning power or control fluid back to the reservoir or atmosphere.

50.03.300   Line, Joining

Lines which connect in a circuit.

50.03.350   Line, Make-Up

A pipeline (conductor) to supply working fluid to a circuit to make up losses as required.

50.03.400   Lines, Passing

Lines which cross but do not connect in a circuit.

50.03.500   Line, Pilot

A line which conducts control fluid.

50.03.510   Line, Pump Inlet

A pipe (conductor) connected to the inlet port of a pump and carrying the supply of working fluid to the pump.

50.03.560   Line, Return

A pipe (conductor) to return the working fluid to the reservoir.

50.03.600   Line, Suction

A supply line at sub-atmospheric pressure to a pump, compressor, or other component.

50.03.700   Line, Working

A line which conducts fluid power.

SECTION 04 — PORT TYPES

50.04.200   AND 10050

A United States Air Force-Navy Aeronautical Design Standard in which a straight thread port is used to attach tube fittings to various components. It employs an "O" ring seal compressed in a special cavity.

50.04.400   Pipe

A port which conforms to pipe thread standards.

50.04.600   Plain "O" Ring

A flat-faced port which uses bolts for attaching the conductor coupling and which includes an O-ring in a recessed groove against the flat face of the port.

50.04.800   SAE J 514

A straight thread port used to attach tube and hose fittings. It employs an "O" ring compressed in a wedge-shaped cavity. A standard of the Society of Automotive Engineers J514 and ANSI/B116.1.

50.04.850   Take-Off Point

Auxiliary connection on units or pipes (conductors) for fluid supply or measurement.

SECTION 05 — PORT FUNCTIONS

50.05.030   Port, Bias

The port at which a biasing signal is applied.

50.05.050   Port, Bleed

A port which provides a passage for the purging of gas from a system or components.

50.05.100   Port, Control

A port which provides passage for a control signal.

50.05.150   Port, Cylinder

A port which provides a passage to or from an actuator.

50.05.180   Port, Differential Pressure

A port(s) which provides a passage to the upstream and downstream sides of a component.

50.05.200   Port, Discharge

A port which provides a passage for fluid power to the system.

50.05.250   Port, Drain

A port for removal of fluid from a component, open to atmosphere, or connected to an unrestricted line.

50.05.251   Port, Airline Drain

Port which enables liquid to be drained from pneumatic circuits.

50.05.300   Port, Exhaust

A port which provides a passage to the atmosphere.

50.05.350   Port, Fill

A port which provides a passage for filling purposes.

50.05.360   Port, Flanged

Port arranged to accept flanged connections.

50.05.400   Port, Inlet

A port which provides a passage for the influent.

50.05.416   Port, Manifold

Connection made through a mounting face.

50.05.430   Port, Outlet (Output)

A port which provides a passage for the effluent.

50.05.450   Port, Pressure

A port which provides a passage from the source of fluid.

50.05.600   Port, Suction

A port which provides a passage for atmospheric charging of a pump or compressor.

50.05.620   Port, Supply

The port at which power is provided to an active device.

50.05.650   Port, Tank (Reservoir) (Return)

A port which provides a passage to the fluid source.

50.05.700   Port, Vent

A port which provides a passage to a reference pressure, usually the ambient pressure.

SECTION 06 — PORT THREADS

50.06.000   Port, Threaded

Port arranged to accept screw thread connections.

50.06.400   Pipe Thread

Screw threads for joining pipe.

50.06.420   Pipe Thread, Dryseal

Tapered pipe threads in which sealing is a function of root and crest interference.

50.06.440   Pipe Thread, Tapered

Pipe threads in which the pitch diameter follows a helical cone to provide interference in tightening.

SECTION 07 — PHYSICAL CHARACTERISTICS

50.07.200   Back Connected

Where connections are made to normally unexposed surfaces of components.

50.07.400   Front Connected

Where connections are made to normally exposed surfaces of components.

50.07.600   Port-to-Port Dimension

The distance between two ports measured from face to face or between center lines.

GROUP 51 — FITTINGS

Section 01 — General
Section 02 — Types of Fittings

SECTION 01 — GENERAL

51.01.100   Fitting

A connector or closure for fluid power lines and passages.

SECTION 02 — TYPES OF FITTINGS

51.02.050   Fitting, Bushing

A short externally threaded connector with a smaller size internal thread.

51.02.100   Fitting, Cap

A cover for fluid passage.

51.02.150   Fitting, Closure

A cap or a plug.

51.02.200   Fitting, Compression

A fitting which seals and grips by manual adjustable deformation.

51.02.250   Fitting, Connector

A fitting for joining a conductor to a component port or to one or more other conductors.

51.02.300   Fitting, Coupling

A straight connector for fluid lines.

51.02.350   Fitting, Cross

A fitting with four ports arranged in pairs, each pair on one axis, and the axes at right angles.

51.02.400   Fitting, Elbow

A fitting that makes an angle between mating lines. The angle is always 90 degrees unless another angle is specified.

51.02.440   Fitting, Female Thread

Connection with internal thread.

51.02.450   Fitting, Flange

A fitting which utilizes a radially extending collar for sealing and connection.

51.02.500   Fitting, Flared

A fitting which seals and grips by a pre-formed flare at the end of the tube.

51.02.550   Fitting, Flared AN

A United States Air Force-Navy 37° flared tube fitting Design Standard.

51.02.600   Fitting, Flareless

A fitting which seals and grips by means other than a flare.

51.02.620   Fitting, Male Thread

Connection with external thread.

51.02.650   Fitting, Reusable Hose

A hose fitting that can be removed from a hose and reused.

51.02.700   Fitting, Welded

A fitting attached by welding.

51.02.750   Fitting, Plug

A closure which fits into a fluid passage.

51.02.760   Fitting, Plug, Dryseal Pipe

A plug made with a thread which conforms to Dryseal Pipe Thread Standards.

51.02.770   Fitting, Plug, Short Pipe Thread

A plug which conforms in all respects to standard pipe threads except that the full thread has been shortened one full thread from the small end.

51.02.780   Fitting, Plug, Standard Pipe Thread

A plug with American (National) tapered pipe threads.

51.02.790   Fitting, Plug, Straight Thread

A plug with straight thread conforming to Unified Thread Standards.

51.02.795   Fitting, Pneumatic

Leakproof devices to connect pipelines (conductors) to one another, or the equipment.

51.02.800   Fitting, Reducer

A fitting having a smaller line size at one end than the other.

51.02.840   Fitting, Tailpiece

A fitting inserted into a flexible tube and secured.

51.02.850   Fitting, Tee

A fitting with three ports, a pair on one axis with one side outlet at right angles to this axis.

51.02.860   Fitting, Threaded Union

A straight connector or adaptor with external or internal threads.

51.02.900 Fitting, Union

A fitting which permits lines to be joined or separated without requiring the lines to be rotated.

51.02.950 Fitting, Wye (Y)

A fitting with three ports, a pair on one axis with one side outlet at any angle other than right angles to this axis. The side outlet is usually $45°$, unless another angle is specified.

GROUP 52 — HOSE

Section 01 — General
Section 02 — Types of Hose

SECTION 01 — GENERAL

52.01.100 Hose

A flexible line or conductor whose nomimal size is its inside diameter.

SECTION 02 — TYPES OF HOSE

52.02.800 Hose, Wire Braided

Hose consisting of a flexible material reinforced with woven wire braid. (Other types of hose construction are available.)

GROUP 53 — PIPE

Section 01 — General
Section 02 — Types of Pipe

SECTION 01 — GENERAL

53.01.100 Pipe

A conductor whose outside diameter is standardized for threading. Pipe is available in Standard, Extra Standard, Double Extra Strong or Schedule wall thickness.

53.01.500 Pipe (Conductor) Clamp

Device to hold and support pipe lines (conductors).

GROUP 54 — TUBE

Section 01 — General
Section 02 — Types of Tubing

SECTION 01 — GENERAL

54.01.100 Tube

A conductor whose size is its outside diameter. Tube is available in varied wall thickness and materials.

GROUP 55 — QUICK DISCONNECTS

Section 01 — General
Section 02 — Types of Quick Disconnects

SECTION 01 — GENERAL

55.01.010 Air Inclusion

The ambient atmosphere forced or trapped into the system during connection of the quick disconnect coupling halves.

55.01.020 Break-Away

Automatic separation of a mounted quick disconnect coupling when a force is applied axially to the unmounted coupling half.

55.01.030 Connect Under Pressure

Ability to connect coupling halves with internal line pressure applied to either both sides or one side.

55.01.080 Spillage

The fluid removed from the system during disconnection of a coupling assembly.

55.01.100 Quick Disconnect Coupling

A component which can quickly join or separate a fluid line without the use of tools or special devices.

SECTION 02 — TYPES OF QUICK DISCONNECTS

55.02.200 Quick Disconnect, Break-away

A quick disconnect which provides automatic separation of the coupling halves, when a predetermined axial force is applied.

55.02.220 Quick Disconnect, Claw Type

A connection which is joined by the rotation of one part with respect to the other.

55.02.400 Quick Disconnect, One Valve

A quick disconnect with a shut-off valve in one half only.

55.02.600 Quick Disconnect, Un-valved*
 Deprecated*

A quick disconnect with no shut-off valves.

55.02.800 Quick Disconnect, Valved

A quick disconnect with a shut-off valve in each half.

GROUP 56 — SWIVELS, ROTATING JOINTS & JOINTS

Section 01 — General
Section 02 — Types of Joints

SECTION 01 — GENERAL

56.01.100 Joint

A line positioning connector.

SECTION 02 — TYPES OF JOINTS

56.02.300 Joint, Rotary

A joint connecting lines which have relative operational rotation.

56.02.598 Joint, Spherical

Pipe junction which allows relative movement in any direction about a point.

56.02.600 Joint, Swivel

A joint which permits variable operational positioning of lines.

56.02.602 Joint, Telescopic

A junction consisting of two tubes sliding longitudinally one within the other, to convey the working medium to the equipment.

ANSI/B93.2

GROUP 57 — MANIFOLDS

Section 01 — General
Section 02 — Types of Manifolds

SECTION 01 — GENERAL

57.01.100 Manifold

A conductor which provides multiple connection ports.

57.01.101 Manifold Block, Valve

A base which forms the sub-plate for two or more subplate mounted valves, incorporating the various ports for connection of the external pipelines. It can also embody flow paths for interconnecting the various valves mounted thereon.

SECTION 02 — TYPES OF MANIFOLDS

57.02.800 Manifold, Vented

A manifold which is open to the atmosphere and returns fluid to the reservoir.

GROUP 58 — SUBPLATES

Section 01 — General
Section 02 — Types of Subplates

58.01.100 Subplate (Sub-Base)

Mounting to which a simple valve is fitted and which includes external ports for fluid connections.

58.01.101 Subplate (Sub-Base), Ganged Valve

Similar sub-bases of which two or more can be clamped together by tie bolts or other means. It can be arranged for the mating faces of the sub-bases to have matching ports, thus providing for a common supply and/or exhaust system. The sub-bases incorporate the various ports for connection of the external pipelines.

58.01.102 Subplate (Sub-Base), Multiple Valve

Mounting with appropriate ports matching those of two or more similar valves which are fitted to it and which include external ports for pipe connections.

NFPA Recommended Standard
T1.21.1-1978 (R1983)

AN INDUSTRY STANDARD FOR FLUID POWER

Procedure for Self-Certification

By Fluid Power Manufacturers

Approved as an NFPA Recommended Standard
25 October 1978

---

published by
NATIONAL FLUID POWER ASSOCIATION, INC.

3333 N. Mayfair Road / Milwaukee, WI 53222 / 414-778-3344 / TLX 26898

NFPA/T1.21.1

FOREWORD

This Foreword is not part of NFPA Recommended Standard Procedures for Self-Certification by Fluid Power Manufacturers, NFPA/T1.21.1-1978.

In the early 1970's the Technical Board recognized the need to establish guidelines for self-certification. Work was undertaken, but in 1974 the Technical Board halted its self-certification work on recommendation of legal counsel and pending release of a FTC opinion regarding certification programs. The Technical Board, after consulting legal counsel, reactivated the ad-hoc committee on self-certification at its 7 April 1976 meeting.

The project was initiated and Draft No. 1 was written at the first meeting of the Self-Certification Ad-Hoc group meeting of 14 September 1976. The document was reviewed and Draft No. 2 resulted from the 12 January 1977 meeting. The Technical Board, at its 2 Feburary 1977 meeting approved the Title, Scope and Purpose. Draft No. 2 was reviewed at the 9 August 1977 meeting and the resulting changes were incorporated into a final working draft.

Headquarters prepared the document for General Review on 12 August 1977. The Project Group met on 26 October 1977 to discuss and answer the negative comments received through General Review. Chairman Kay responded to the commentors in writing during December 1977. One comment concerning self-certification to corporate standards remained unresolved until the 18 January 1978 meeting of the NFPA Technical Board. At that meeting, the Board agreed to include corporate standards on the basis that their use would enhance the overall usefulness of the self-certification procedure. The Board also granted approval to ballot. NFPA Headquarters prepared the Ballot Draft on 14 February 1978.

Three negative ballots were received on this document. At it's 21 September 1978 meeting, the NFPA Technical Board discussed the negative comments and concurred that the negative votes could not be resolved. The Technical Board voted to recommend to the Board of Directors that this document be approved as an NFPA Recommended Standard.

The NFPA Board of Directors granted final approval to NFPA/T1.21.1-1978 on 25 October 1978.

NFPA/T1.21.1

PROJECT GROUP MEMBERS WHO DEVELOPED THIS STANDARD

Kay, Robert	Project Chairman	Sperry Vickers
Ratkay, Edward	Technical Auditor	Commercial Shearing, Inc
Luecke, John R.	Director of National Technical Services	National Fluid Power Association*

Chenoweth, R.	Abex/Denison Div.
Johnson, J.	Milwaukee School of Engineering
Sallberg, D.	HUSCO/Pegasus
Stockwell, S.	J I Case Company

REFERENCES

1. American National Standard Glossary of Terms for Fluid Power, ANSI/B93.2-1971, and Supplement ANSI/B93.2A-1978. (ISO/DP 5598)

2. American National Standard, General Requirements for a Quality Program, ANSI/Z1.8-1971.

* Company affiliation has changed since work with the Project Group.

NFPA/T1.21.1

PROCEDURE FOR SELF-CERTIFICATION

BY FLUID POWER MANUFACTURERS

INTRODUCTION

In fluid power systems, power is transmitted and controlled thru a fluid under pressure within an enclosed circuit. Self-certification provides a standard procedure for fluid power manufacturers who wish to certify that their products or services meet specific industry standards and/or specifications.

1. SCOPE

    To include:

    1.1 A standard self-certification procedure for use by fluid power manufacturers.

    1.2 Products and/or services supplied by fluid power manufacturers and certified to technical society, trade association, national, international, government, and publicly available corporate standards or specifications.

2. PURPOSE

    To provide a standardized procedure to be followed when a fluid power manufacturer certifies the product or service to technical society, trade association, national, international, or government and publicly available corporate standards or specification(s).

3. TERMS AND DEFINITIONS

    For definition of terms not defined below, see Reference No. 1.

    3.1 Certification. The procedure by which a product or service becomes certified.

    3.2 Certified. Attested by the manufacturer under the procedures of this recommended standard as satisfying the requirements of the referenced standard(s) and/or specification(s).

3.3 Inspection. The process of examining, measuring, testing, gaging, or otherwise verifying that the product or service conforms with applicable provisions of referenced standard(s) and/or specification(s).

3.4 Self-Certification. Certification to the user by the manufacturer, on his own authority, that a product or service is in compliance with referenced standard(s) and/or specification(s).

3.5 Quality Control. The overall system of activities of the manufacturer to assure that manufactured products or services do in fact comply with requirements of the referenced standard(s) and/or specification(s).

3.6 Specification. A statement of requirements.

3.7 Standard. A document (or an object for physical comparison) for use in defining product characteristics, products, or processes. It is prepared by a consensus of a properly constituted group of those substantially affected and having the qualifications to prepare the standard.

4. STANDARD(S) AND/OR SPECIFICATION(S) REFERENCED

4.1 Select referenced standard(s) and/or specification(s) from the following:

4.1.1 International, national, and foreign national organizations.

4.1.2 Technical societies, trade associations, agencies, or other organizations of national scope and recognition.

4.1.3 Government

4.1.4 Corporations whose standard(s) and/or specification(s) are publicly available.

NOTE: To the fullest extent possible, NFPA will assist the reader in finding sources for referenced standard(s) and/or specification(s).

NFPA/T1.21.1

4.2 Use the referenced standard(s) and/or specification(s) in its entirety, unless full disclosure of limitations is made in the statement of self-certification.

5. QUALITY CONTROL

5.1 Maintain a quality control system sufficient to ensure that product or service provides and consistently meets the criteria contained in the referenced standard(s) and/or specification(s).

5.2 Maintain the quality control system in accordance with American National Standard "Specification of General Requirements for a Quality Program," ANSI/Z1.8-1971, (Reference No. 2) and any other quality control standards agreed to by the manufacturer and buyer.

NOTE: ANSI/Z1.8-1971 includes requirements for:

(A) Quality Management
(B) Design Information
(C) Procurment
(D) Material Control
(E) Manufacture
(F) Acceptance
(G) Measuring Instruments
(H) Quality Information

6. VERIFICATION TESTS AND DOCUMENTATION

6.1 Carry out tests and/or inspection in accordance with the referenced standard(s) and/or specification(s).

6.2 Use managerial controls to assure that <u>samples</u> used for testing and/or inspection are representative of the product or service supplied.

6.3 Maintain a record of results from samples used to verify compliance.

6.4 Make results available to the buyer, upon legitimate and responsible requests.

NFPA/T1.21.1

7. STATEMENT OF SELF-CERTIFICATION

    7.1   Include the following in a formal statement of self-certification.

        7.1.1   Manufacturer's designation for the product or service certified.

        7.1.2   Statement of conformance with applicable referenced standard(s) and/or specification(s).

        7.1.3   Identification of the manufacturer as the certifier.

    7.2   Examples:

    Word the statement of self-certification as follows:

        7.2.1   <u>Valve XYZ-0123</u> conforms
                      (Manufacturer's designation for product or service)

            to <u>ANSI/B93.7-1968 (R1973)</u> as certified
                  (Referenced standard(s) and/or specification(s))

            by <u>XYZ Valve Manufacturing, Inc.</u> in accordance
                  (Identification of Manufacturer)
            with NFPA/T1.21.1-1978.

        7.2.2   <u>Pneumatic FRL Pressure Drop Test</u> conducted per
                      (Manufacturer's designation for product or service)

            <u>NFPA/T3.12.6-1975</u> as certified
            (Referenced standard(s) and/or specification(s))

            by <u>FRL's Inc.</u> in accordance
                  (Identification of Manufacturer)
            with NFPA/T1.21.1-1978

    7.3   Certification extends beyond the initial buyer <u>only</u> when such buyer has not made alterations, changes, or additions to the product or service as produced or received, except by express agreement with the manufacturer. Any such alterations, changes, or additions not expressly approved by the manufacturer invalidates the manufacturer's certification.

7.4 Use of this Recommended Standard does not authorize the use of NFPA Register Trademark.

NOTE: Only NFPA members are permitted to use the trademark on product literature, etc.

8. KEY WORDS

The following Key Words, useful in indexes and in information retrieval systems are suggested for this recommended standard:

self-certification, documentation

self-certification, statement

self-certification, verification

fluid power

NFPA Recommended Standard
T2.6.1-1974 (R1982)

AN INDUSTRY STANDARD FOR FLUID POWER

This Standard is now under review for possible revision see TSP

Method for Verifying the Fatigue and Static

Pressure Ratings of the Pressure Containing

Envelope of a Metal Fluid Power Component

Approved as an NFPA Recommended Standard
18 February 1974

published by
NATIONAL FLUID POWER ASSOCIATION, INC.

3333 N. Mayfair Road / Milwaukee, WI 53222 / 414-778-3344 / TLX 26898

NFPA/T2.6.1

FOREWORD

This Foreword is not part of NFPA Recommended Standard Method for Verifying the Fatigue and Static Pressure Ratings of the Pressure Containing Envelope of a Metal Fluid Power Component, NFPA/T2.6.1-1974.

Early in 1968 producers and users of fluid power components expressed a need for a uniform method for verifying a Pressure Rating for individual components. It was felt that the time had come for industry to establish a suitable set of rules in the form of a standard.

A search of the existing codes in thirty-one states indicated that those quoted were inapplicable to our industry since they covered products not related to hydraulics, pneumatics, fluidics and/or other fluid power areas. This search showed further that there were no known federal regulations covering this subject; and that regulations written by private agencies such as ASME, ASTM, etc., were not applicable. (NFPA will continue to reference material specifications judged applicable and prepared by such societies.)

As the result of the search, the committee developed a belief that there is no applicable reference point suitable for adoption or supplementation. The committee would, therefore, have to prepare its own guidelines and cite references as developed.

The Pressure Rating General Technical (Coordinating) Committee was organized in February 1968. It consisted of Members-at-large who are skilled in the theory of Pressure Rating. In addition, each component section of the NFPA Technical Board was asked to name a representative who would be a specialist for that particular component.

Thru a series of meetings and theoretical investigations, the committee was able to produce their first draft in December, 1970. After a series of five drafts and reviews, committee consensus was reached on 1 February 1972. The General Review Draft was prepared on 6 March 1972.

To aid in understanding the proposal, and to facilitate the General Review process, a general conference was held in Chicago on 23 March. More than 100 representatives of 50 member companies took advantage of the conference to learn the document's history, background and theory and to have their questions answered by those preparing the proposal.

All comments on record were resolved by the Committee at their 18 May 1972 meeting and at their 29 June 1972 meeting when those submitting comments were in attendance or represented. A Ballot Draft was prepared on 11 July 1972 taking into account all the technical refinements and editorial clarifications agreed upon in the process of resolving all comments. This preliminary Ballot Draft was reviewed and revised by the Committee officers and the Ballot Draft was prepared on 19 July 1972.

At their regular meeting on 27 September 1972, the NFPA Technical Board reviewed the proposed Ballot Draft and all the comments which had been submitted. The Board made minor editorial modifications to the Introduction, Section 21 and clause 17.2 and released the proposal for ballot by Member Companies. The Ballot Draft was prepared on 2 October 1972. Balloting closed on 6 November 1972.

Upon recommendation by the Technical Board, the Committee met with representatives of the Cylinder Section, who constituted the majority of the negative voters, to resolve their comments. Agreement was reached by providing for RFP Verification by Similarity. In a related action, the Technical Board Steering Committee proposed, and the Board of Directors and Technical Board approved, a modified sense of direction whereby document T2.6.1 should be immediately approved as an industry-wide philosophy and general intent with limited specific applicability. Document T2.6.1 was modified to facilitate its becoming useable within 12 months of its date of approval thru approximately 15 supplementary NFPA Recommended Standards, each setting forth specific details for testing and extending test data to short-run and one-of-a-kind situations thru verification by similarity.

The Committee met on 28 March 1973 and 6 June 1973 and modified this proposal to incorporate the foregoing and to resolve other comments. An accelerated rereview and reballot procedure was adopted in accordance with the Board of Directors' mandate for "immediate action." A rereview draft was completed on 11 June 1973 and was immediately circulated to all Member Companies, all members of the Pressure Rating Project Groups and members of the NFPA Board of Directors and Technical Board for a review which closed on 25 June 1973.

Eighteen pages of comments were received from 8 Member Companies. All comments were resolved thru meetings on 28 June and 26 July 1973; and thru mail action concluded on 31 August 1973. In addition to many editorial improvements and clarifications, a new fourth paragraph was added to the Introduction suggesting a method for estimating the pressure that an individual component can normally be subjected to based upon its relationship with the rated fatigue and static pressures. Figure 1 was also added. Another significant modification was the addition of clause 14.10, thereby permitting the selection of singular assurance and confidence levels in specific component standards.

A 27 September 1973 draft was circulated for a 10-day review by the committee and others involved in the resolution of comments. As a result of this 10-day review, Figure 1 was improved; the listing of Project T3.8.12 was merged with T3.15.8; and many editorial improvements were made to clarify proposal T2.6.1. Also, Reference A16 was added and improvements made to the Tutorial Reference.

Ballot Draft No. 2 was prepared on 24 October 1973. The balloting of this draft closed on 16 November 1973, and three of the five negative ballots were resolved on 20 November 1973.

The Technical Board recommended approval of this proposed standard to the Board of Directors on 28 November 1973. The Board of Directors approved T2.6.1 as an NFPA Recommended Standard on 18 February 1974.

MEMBERS OF THE PROJECT GROUP THAT DEVELOPED THIS STANDARD:

Schroeder, Reed
Committee Chairman
Schroeder Brothers

Skaistis, Stan
Committee Vice Chairman (May 1972 to present)
Sperry Vickers

Barnes, Robert
Committee Vice Chairman (February 1968 to May 1972)
AMF Cuno Division

Morgan, James I.
Secretariat
National Fluid Power Association

Members-at-large

Cordes, H.
Pall Corporation

Faust, D.
C. A. Norgren

Forster, W.
WABCO Fluid Power Division

Glidden, J.
Hydreco

Johnson, J.
MSOE

Stephens, T.
Gresen Mfg. Co.

Members-Section Representatives

Barthe, H.
Schroeder Brothers
Filter & Separator

Berninger, J.
Parker Hannifin Corp.
Cylinder

Bowbin, J.
Miller Fluid Power
Air Dryer

Dodson, R.
Clayton Mark
Conductor

Jacoby, H.
Webster Electric
Hydraulic Valve

Lamb, T.
Parker Hannifin Corp.
Quick Disconnect

Ratkay, E.
Commercial Shearing
Pump & Motor

Schwarz, A.
ITE Imperial Corp.
Fittings

Sessoms, W.
Scovill Fluid Power
Pneumatic FRL

Zajdler, A.
Superior Hydraulics
Accumulators

Other Participants

Allen, C.
WABCO Fluid Power Division

Bailey, R.
C. A. Norgren Co.

Brake, C.
Scovill Fluid Power

Czarnecki, G.
DeLaval - IMO Division

*Flock, H.
Garlock/OM

Greenwood, M.
ITE Imperial Corp.

Kosmak, M.
Marvel Engineering Co.

Larson, B.
Eaton Corp.

May, R.
Aeroquip

Olson, J.
Applied Power/Dynex

Pauken, D.
Double A Products

Prevallet, D.
Gresen Mfg. Co.

Ritchie, R.
Aeroquip

Roth, R.
Flodar Corp.

Schultz, H.
Garlock/OM

Sethi, I.
Abex Corp.

Smith, H.
U. S. Army - MERDC

Trainor, T.
Battelle Institute

Woodworth, R.
Moog, Inc.

REFERENCES

1. **American National Standard Glossary of Terms for Fluid Power, ANSI/B93.2-1971, and Supplement thereto.** (ISO/TC 131/SC 1 [USA-2] 3)

2. **SI units and recommendations for the use of their multiples and of certain other units, ISO 1000-1973.**

TUTORIAL REFERENCE
(Attached as Appendix)

1. National Fluid Power Association Tutorial Reference, File T2.6.1, 24 October 1973. (Includes listing of all background references.)

*Company affiliation has changed since work with the Project Group.

NFPA/T2.6.1

METHOD FOR VERIFYING THE FATIGUE AND STATIC PRESSURE RATINGS OF THE PRESSURE CONTAINING ENVELOPE OF A METAL FLUID POWER COMPONENT

INTRODUCTION

In fluid power systems, power is transmitted and controlled thru a fluid (liquid or gas) under pressure within an enclosed circuit. A basic requirement of fluid power components is that they should be capable of adequately containing the pressurized fluid.

This recommended standard establishes a group of common requirements intended to provide an industry-wide philosophy and basic standard, providing a rationale for judging a component's ability as a pressure containing envelope. Although this recommended standard's specific applicability is limited, it does immediately establish a uniform base for subsequent, more specific NFPA Proposed Recommended Standards for individual fluid power components. The proposed documents* listed below will implement Recommended Standard NFPA/T2.6.1-1974.

Accumulators	NFPA/T3.4.7-19xx
Hydraulic Valves	NFPA/T3.5.26-19xx
Cylinders	NFPA/T3.6.29-19xx
Cylinders	NFPA/T3.6.31-19xx
Pumps and Motors	NFPA/T3.9.22-19xx
Hydraulic Filters & Separators	NFPA/T3.10.5.1-19xx
Pneumatic Filters, Regulators and Lubricators	NFPA/T3.12.10-19xx
Tube and Fitting Assemblies	NFPA/T3.15.8-19xx
Hydraulic Reservoirs	NFPA/T3.16.8-19xx
Quick Disconnect Couplings	NFPA/T3.20.8-19xx
Pneumatic Valves	NFPA/T3.21.4-19xx
Compressed Air Dryers	NFPA/T3.27.5-19xx
Fluid Logic Devices	NFPA/T3.28.10-19xx
Pressure Switches	NFPA/T3.29.2-19xx

Ratings verified in accordance with this standard do not replace pressure ratings based on considerations such as performance, bearing capacity, leakage and heat rejection. Instead, these new ratings are intended to supplement those that are provided by present practice and may be numerically different from prior ratings.

* NOTE - The above listing is correct as of the date of publication, but is subject to change.

The pressure that an individual component can normally be subjected to has a relationship with the rated fatigue and static pressures. This relationship may be estimated and used as a basis of total life expectancy for the component in an individual application. Such an estimate must be applied by the user and factors such as shock, heat, misuse, etc., must be judged by the user in its application. The selection of a specific pressure and life expectancy for a component in such an individual application may be based upon the rated fatigue and static pressure as prescribed in Figure 1.

Although for test verification, RFP is defined as 10^7 cycles (see clause 3.1), the generally accepted conservative design practice when using this type of S-N curve requires that the locus for the line separating the verified/unverified is based upon 10^6 cycles.

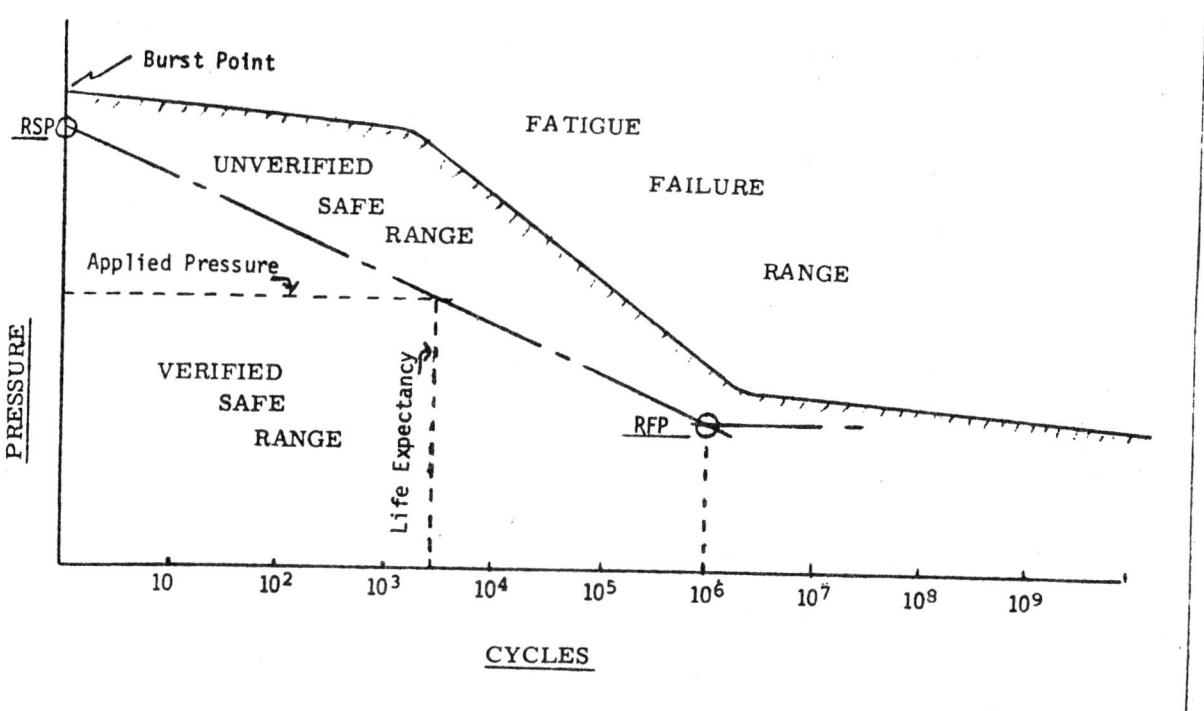

FIGURE 1 - Possible S-N curve method for estimating finite life rating

It should be noted that this document deals solely with verifying the fatigue and static pressure ratings of a component's pressure containing envelope. Separate from this verification procedure, manufacturers have the continuing responsibility to utilize managerial controls necessary to maintain similarity between the test and production pressure containing envelopes.

NFPA/T2.6.1

1. SCOPE

   1.1 To include methods for verifying ratings of the fatigue and static strengths of metal fluid power component pressure containing envelopes by test; or by analysis by geometric similarity (based on tests conducted in accordance with this recommended standard) that are:

   1.1.1 Subjected to pressure induced stresses.

   1.1.2 Used in the -20° to 200°F (-29° to 93°C) fluid temperature range.

   1.1.3 Not subject to loss of strength due to corrosion or other chemical action.

   1.1.4 Made of aluminum, magnesium, steel, iron, copper-based alloys and stainless steels. (Other metals will be included in subsequent revisions as soon as suitable data is available.)

   1.2 Hydraulic hoses and components constructed primarily of non-metals are not included in this recommended standard.

   1.3 This recommended standard does not establish a standard means for determining the magnitude of pressure transients in systems.

   NOTE - These magnitudes are usually determined by the system designer who then selects components with adequate strengths. However, it is generally recognized that there is very little authoritative data available to assist the system designer in making this determination. It is anticipated that the promulagation of this recommended standard will also stimulate the study of the realtionships between peak and operating pressures.

2. PURPOSE

   To provide standard methods for verifying the fatigue and static pressure ratings of a metal fluid power component's pressurized boundaries with regard to sustaining cyclic and steady pressure loads that can be used as part of specific component standards.

NFPA/T2.6.1

3. TERMS AND DEFINITIONS

(For definition of other fluid power terms used, see Reference No. 1.)

3.1 Rated Fatigue Pressure. A pressure that a component pressure containing envelope is represented to sustain 10 million times without failure.

3.2 Rated Static Pressure. A pressure that a component pressure containing envelope is represented to sustain without failure.

3.3 Cyclic Test Pressure. A pressure range applied in cyclic tests that are performed to verify a Rated Fatigue Pressure.

3.4 Static Test Pressure. A pressure applied in static tests performed to verify a Rated Static Pressure.

3.5 Test Duration Factor. A multiplier applied to the Rated Fatigue Pressure for calculating the Cyclic Test Pressure which will permit shortening the qualification test from 10 million to one million cycles.

3.6 Variability Factor. A multiplier applied to the Rated Fatigue Pressure for calculating the Cyclic Test Pressure to account for the variability in fatigue strength of metals. It is also applied to the Rated Static Pressure for calculating the Static Test Pressure.

3.7 Assurance Level. The minimum percentage of pressure containing envelopes of a verified design that will sustain 10 million applications of its Rated Fatigue Pressure.

3.8 Confidence Level. The degree to which a manufacturer can be assured that the desired assurance level is attained.

3.9 Pressure Containing Envelope Failure. A structural fracture, crack or excessive seal leakage caused by deformation resulting from an RFP verification test; or a structural fracture, crack, or excessive seal leakage caused by deformation or permanent deformation which interfered with component functioning resulting from an RSP verification test.

NFPA/T2.6.1

The following terms, common in fluid power terminology, are defined in general usage dictionaries so therefore do not appear in Reference No. 1. Because these terms have more than one accepted meaning, we have set forth below the meaning applied to their use in this document.

3.10 Component. A complete product or assembly (e.g., a valve, pump, fitting).

3.11 Pressure Containing Envelope. That part of the component which contains the pressurized fluid.

3.12 Elements. The integral pieces which make up a part or component (e.g., cylinder end cap, filter element, etc.).

The following terms, not included in this document but found in related reference material, are defined within the meaning that the Committee applied when using the reference.

3.13 Peak Pressure. The maximum pressure actually experienced by a fluid power component in application. It is caused by any number of events, individually or in combination, such as (but not limited to), impact, inertia, shock, heat, etc.

3.14 Nominal Pressure. A pressure value assigned to a component or a system for the purpose of convenient designation. This may be an average of normal values without peaks, or a design figure for stress calculations.

4. UNITS

4.1 The "Customary US" units are used.

4.2 Approximate conversions to the International System of Units (SI) are given per Reference No. 2. These appear after their "Customary US" counterpart in parentheses.

5. LETTER SYMBOLS

RFP	Rated Fatigue Pressure
RSP	Rated Static Pressure
CTP	Cyclic Test Pressure
STP	Static Test Pressure
K_N	Test Duration Factor
K_V	Variability Factor

NFPA/T2.6.1

6. OUTLINE OF PROCEDURES FOR RFP VERIFICATION BY TEST

Select an RFP, then:

6.1 Select a K_N factor per clause 12.2.1.

6.2 Select a K_V factor per clause 12.2.2.

6.3 Calculate the CTP, per clause 12.2.

6.4 Use test equipment per clause 10.1.

6.5 Test the number of pressure containing envelopes upon which the factor K_V was selected in clause 6.2.

6.6 Provide pressure pulses per clause 10.1.

6.7 Subject the pressure containing envelope(s) being tested per section 12 to the number of test cycles upon which the factor K_N was selected in clause 6.1.

6.8 Verify the RFP per section 15.

6.9 Record data per section 17.

7. RFP VERIFICATION BY SIMILARITY

7.1 If it is desired to verify the Rated Fatigue Pressure of components by geometric similarity, base verification upon tests conducted in accordance with proposed recommended standards for individual fluid power components; and use the rules for similarity also included therein.

7.2 Assurance and confidence levels of verification made by geometric similarity are the values derived in the test that provides the basis of the similarity analysis.

8. OUTLINE OF PROCEDURES FOR RSP VERIFICATION BY TEST

Select an RSP, then:

8.1 Select a K_V factor per clause 12.2.2.

8.2 Calculate the STP per clause 13.2.

NFPA/T2.6.1

8.3 Use test equipment per clause 10.2.

8.4 Test the number of pressure containing envelopes upon which the factor K_V was selected in clause 8.1.

8.5 Conduct static tests per section 13.

8.6 Verify the RSP per section 16.

8.7 Record data per section 17.

9. PREPARATIONS FOR TESTING

9.1 Use liquid or gas to pressurize pressure containing envelopes being tested.

9.2 CAUTION: Take special precautions, while conducting any of the pressure tests, in the design of test rigs, and in training the operators to maximize operator safety.

9.3 Bleed the entrapped air from the circuit and from the pressure containing envelope being tested; and bleed gas charges from accumulators before tests are started when liquids are used.

9.4 Apply different ratings to separate portions of the pressure containing envelope as required by operating or component failure mode conditions.

9.5 Allow all drains and low pressure ports that are not part of the pressure containing envelopes to drain freely and keep them at atmospheric pressure.

9.6 Test pressure containing envelopes whose geometries, material properties and manufacturing methods are known and representative of normal production.

9.7 Do not perform manufacturer's proof tests prior to cyclic testing, unless such tests are made on all production pressure containing envelopes.

9.8 Perform cyclic and static tests on separate pressure containing envelopes, if possible.

9.9 Perform the cyclic tests first, when both RFP and RSP verification tests must be made on the same pressure containing envelopes.

9.10 Complete machining and processing of test components to the degree necessary for duplicating operating stress distributions and final strength in the pressure containing envelope.

9.11 It is permissible to make modifications to the test piece to facilitate cyclic or static tests providing that such modifications do not increase the pressure capabilities of the pressure containing envelope being tested.

10. TEST EQUIPMENT

10.1 Cyclic Test

10.1.1 Use a circuit on the test stand which will produce repeatable pressure pulses which cycle between approximately zero and a specified maximum pressure.

NOTE - The specific design of the test stand and circuit is optional.

10.1.2 Use any suitable non-corrosive test fluid.

10.1.3 Mount the pressure measuring instrument directly into the pressure containing envelope being tested, or closely to it, thru a pressurized port that is not being used to supply the test fluid, whenever possible.

10.1.4 Minimize restrictions between the instrument and the pressure containing envelope being tested, when measurements must be made in the pressure supply line.

NOTE - It is recommended that it be verified that the pressure generated in the pressure containing envelope at the actual cycling rate be the intended value.

10.1.5 Provide adequate personnel protection.

10.2 Static Test

10.2.1 Use a circuit on the test stand which provides a stable and controllable hydrostatic fluid pressure.

NOTE - The specific design of the test stand is optional.

NFPA/T2.6.1

    10.2.2   Provide adequate personnel protection.

    10.2.3   Use any suitable non-corrosive test fluid for the test stand.

11. TEST CONDITIONS ACCURACY

Set up and maintain equipment accuracy so that data is accurate within (±) the following:

CTP	3%
STP	2%
Temperature	5°F (3°C)

NOTE - Unless the frequency response of the measurement system or its components is high enough to faithfully reproduce the pressure waveform, the actual Cyclic Test Pressure will be higher than measured, thereby penalizing the component under test.

12. PROCEDURES FOR CYCLIC TEST

12.1 General Provisions

    12.1.1   Place metal shot or loosely fitting metal pieces in the pressure containing envelope being tested to reduce the volume of the pressurized fluid when higher cycling rates are desired.

    12.1.2   Subject all pressure containing envelopes to the same pressure loadings when testing multiple samples.

    12.1.3   In each complete cycle, apply test pressure to all parts of the pressure containing envelope being tested, in combination or sequentially, in a manner that simulates operating modes occurring in service.

    12.1.4   Replace gaskets, seals and other expendable items that fail due to wear during the test so long as pre-loads in stressed elements are not affected such that clause 9.11 is not violated.

    12.1.5   Verify with straingages that the ratio of induced stresses to pressure under static loading is attained at the test cycling rate where pressures must penetrate between close fitting parts, very large components are tested, or where hysteresis in joints can significantly affect stresses.

12.2 Cyclic Test Pressure

The Cyclic Test Pressure (CTP) is a product of two test parameter factors (K_N and K_V) and the Rated Fatigue Pressure (RFP), calculated as follows:

$$CTP = RFP\, (K_N \times K_V)$$

12.2.1 Test Duration Factor (K_N).

12.2.1.1 Select Test Duration Factor from Table 1.

12.2.1.2 Use factors specified for non-ferrous metals in testing all pressure containing envelopes with non-ferrous elements.

TABLE 1 - Test duration factor, K_N

Test Cycles (minimum)	Ferrous	Non-Ferrous
10 million	1.0	1.0
1 million	1.15	1.25

12.2.2 Variability Factor (K_V).

12.2.2.1 Select Variability Factor from Table 2.

12.2.2.2 Use the Variability Factor for the materials contained in the pressure containing envelope being tested which have the highest value.

TABLE 2 - Variability factor, K_V

Assurance Levels*	Number of Test Units		
	1	2	5
Irons			
99.9%	1.90	1.42	1.22
99	1.55	1.30	1.15
90	1.30	1.15	1.07
Aluminums, Magnesiums & Steels			
99.9%	1.55	1.30	1.15
99	1.45	1.25	1.12
90	1.30	1.15	1.06
Copper Based Alloys			
99.9%	1.50	1.20	1.09
99	1.35	1.15	1.07
90	1.20	1.10	1.02
Stainless Steels			
99.9%	1.30	1.17	1.09
99	1.25	1.15	1.07
90	1.17	1.10	1.02

* Assurance levels are based on a 90 percent confidence level. Test twice the specified number of pressure containing envelopes to achieve a 99 percent confidence level.

12.3 Pressure Pulses

12.3.1 Provide a test pressure that decreases to below 0.05 CTP in each cycle.

12.3.2 Provide a maximum test pressure that exceeds the minimum test pressure (per clause 12.3.1) by an amount equal to the CTP.

12.3.3 Provide pressure pulses having any shape with maximum values that repeat cycle to cycle, as required by section 11, and do not have superimposed oscillations whose range exceeds 0.6 CTP.

12.3.4 Apply pressure pulses at any rate, up to 30 per second, that permits stable operation in which the pressure range repeats within ($\pm$) 2 percent.

13. PROCEDURES FOR STATIC TEST

13.1 General Provisions

13.1.1 Apply test pressure to all elements, individually or in combination, in a manner that simulates pressure containing envelope operating modes occurring in actual service.

13.1.2 Apply test pressure slowly and maintain for one minute.

13.2 Static Test Pressure

13.2.1 Calculate the Static Test Pressure as follows:

$$STP = K_V \times RSP$$

14. ADDITIONAL PROVISIONS

NOTE - There are other provisions, such as those listed below, that pertain to the verifying of presssre ratings for individual components. Each of these provisions will be taken into account as more detailed supplementary standards are prepared for individual components.

Give consideration to the following additional provisions:

14.1 Integrating the pressure rating test procedures into performance tests, where possible.

14.2 Specifying sequences or combinations to be used in applying pressures to specific components.

14.3 Permissible methods for restraining functional motions (essentially to attain higher cyclic test rates).

14.4 Positioning of internal parts to produce maximum stresses.

NFPA/T2.6.1

14.5 Parts of components that may be omitted for test purposes.

14.6 Component modifications that will be permitted to facilitate testing.

14.7 Additional criteria for judging failure.

14.8 Additional designated information.

14.9 Criteria and procedures required for verifying Rated Fatigue Pressures by geometric and material similarity to components having pressure ratings that were previously verified by testing in accordance with this recommended standard.

14.10 Specifying singular assurance and confidence levels for specific components or component usage.

15. CRITERIA FOR RFP VERIFICATION

NOTE - More specific details for this requirement will be set forth in the individual component standards.

15.1 Consider structural fracture a failure.

15.2 Consider any crack produced by metal fatigue due to pressure cycling, as verified by magnetic particle or fluorescent penetrant techniques after testing, a failure.

15.3 Consider excessive leakage at seals or sealing surfaces, that result from deformation of the metallic portion of the pressure containing envelope, a failure.

16. CRITERIA FOR RSP VERIFICATION

NOTE - More specific details for this requirement will be set forth in the individual component standards.

16.1 Consider structural fracture a failure.

16.2 Consider any crack produced by internal static pressure, as verified by magnetic particle or fluorescent penetrant techniques after testing, a failure.

NFPA/T2.6.1

16.3 Consider excessive leakage at seals or sealing surfaces, that result from deformation of the metallic portion of the pressure containing envelope, a failure.

16.4 Consider any permanent deformation, which interferes in any way with the proper functioning of the pressure containing envelope, a failure.

17. DATA PRESENTATION

17.1 Include the following minimum information in all component manufacturer's specifications, test reports, drawings, catalogs and sales literature referencing this recommended standard:

17.1.1 Rated Fatigue Pressure, RFP, and the assurance and confidence categories upon which the RFP is based.

17.1.2 Rated Static Pressure, RSP, and the assurance and confidence categories upon which the RSP is based.

17.2 Have available a record of all the following minimum test data for all component manufacturer's test reports referencing this recommended standard.

17.2.1 All physical values pertaining to the test.

17.2.2 All additional provisions or modifications pertaining to the test (see Section 14).

18. SUMMARY OF DESIGNATED INFORMATION

The following designated information is needed when applying this recommended standard to a particular application or use:

18.1 Desired Assurance Level

18.2 Desired Confidence Level

18.3 Desired RFP

18.4 Desired RSP

18.5 Expected ambient and fluid temperature range

NFPA/T2.6.1

19. JUSTIFICATION STATEMENT

This recommended standard verification procedure is based upon the combined expert experiences of those who have participated in its preparation and/or review.

Further justification is set forth in the Tutorial Reference, attached hereto as an Appendix.

Additional justification data will be obtained and compiled as the procedures set forth in this document become more widely used. Recommendations for the further improvement of this recommended standard are always welcome. Forward them directly to NFPA at the following address:

> National Fluid Power Association
> 3333 North Mayfair Road
> Milwaukee, Wisconsin 53222

20. TEST/PRODUCTION SIMILARITY

Utilize managerial controls necessary to maintain substantial similarity between test and production components or elements.

21. IDENTIFICATION STATEMENT

Use the following statement in test reports, catalogs and sales literature when electing to comply with this voluntary standard:

> "Method of verifying rated fatigue and rated static pressures of the pressure containing envelope conforms to NFPA Recommended Standard, NFPA/T2.6.1-1974, category _____*."

*NOTE - Category code numbers are as follows:

		Assurance Level		
		90%	99%	99.9%
Confidence Level	90%	1/90	2/90	3/90
	99%	1/99	2/99	3/99

(For explanation of percent values, see clause 12.2.2.)

EXAMPLE - RFP (1/90) = _____ psi (bar, kPa)
This means that the component's Rated Fatigue Pressure is _____ psi (bar), based upon a 90% Assurance Level and 90% Confidence Level. (The use of the code 1/90 in this example does not suggest that these levels are preferred.)

NFPA/T2.6.1

APPENDIX

NFPA/T2.6.1

TUTORIAL REFERENCE

FOR USE WITH

NFPA RECOMMENDED STANDARD

METHOD FOR VERIFYING THE FATIGUE AND STATIC

PRESSURE RATINGS OF THE PRESSURE CONTAINING ENVELOPE

OF A METAL FLUID POWER COMPONENT

> NOTE. - This appendix is intended for reference purposes only. It is not a part of NFPA Recommended Standard Method for Verifying the Fatigue and Static Pressure Ratings of the Pressure Containing Envelope of a Metal Fluid Power Component, NFPA/T2.6.1-1974. Its contents do not necessarily represent an agreed-upon position by NFPA or its Pressure Rating Coordinating Committee. However, its material may be helpful towards a fuller understanding of the philosophy which led to the development of the basic standard.

Published by

NATIONAL FLUID POWER ASSOCIATION, INC.

MILWAUKEE, WISCONSIN 53222

NFPA/T2.6.1

REFERENCES

A1. "The Statistical Aspects of Fatigue Strength" by S. J. Skaistis, September 14, 1971. (Unpublished)

A2. "Some Quantitative Aspects of Fatigue of Materials" by Harold N. Cummings, WADD Technical Report 60-42, July 1960.

A3. Fatigue Design Handbook by J. A. Graham, Editor, SAE.

A4. "Reliability Prediction - Mechanical Stress/Strength Interference" by Lipson, Sheth and Disney, Technical Report RADC-TR-66-710, March 1967, AD 813574.

A5. "Reliability Prediction - Mechanical Stress/Strength Interference (Nonferrous)" by Lipson, Sheth, Disney and Altun, Technical Report RADC-TR-68-403, February 1969, AD 856021.

A6. "Relating Probabilistic Methods to Reliability Considerations in Product Design" by Gerhard Reethof, ASME Paper 70 DE 70, May 1970.

A7. "Estimation of Fatigue Life with Particular Emphasis on Cumulative Damage" by Milton A. Miner, Chapter 12 of Metal Fatigue, Sines and Waisman Editors, McGraw-Hill Book Co., 1959.

A8. "The Fatigue of Metals and Structures" by Grover, Gordon and Jackson, NAVWEPS 00-25-534, revised June 1960.

A9. "NFPA Pressure Cycling Test Study - Evaluation of Endurance Test" by E. A. Ostreicher, August 17, 1970. (Unpublished)

A10. "Feasibility of Including 10^5 Cycle Tests in the Pressure Rating Standard" by S. J. Skaistis, January 27, 1972. (Unpublished)

A11. "Evaluating Component Fatigue Performance Under Programmed Random, and Programmed Constant Amplitude Loading" by S. R. Swanson, SAE Paper 690050, January 1969.

A12. "The Effect of Test Pulse Oscillations" by S. J. Skaistis, November 20, 1970. (Unpublished)

A13. "Aluminum and Aluminum Alloys", ASM Metals Handbook, Vol. 1, 8th Edition, 1961.

NFPA/T2.6.1

A partial listing of other sources of information not referenced in this appendix:

A14. "Statistical Design and Analysis of Engineering Experiments" by Lipson and Sheth, McGraw-Hill Book Co., 1973.

A15. "Fatigue-Based Pressure Standards Introduced by Fluid Power Group" Product Engineering, June 1973.

A16. "Fatigue Testing of Hydraulic Equipment Components", by R. Stephens, Sperry Vickers.

TUTORIAL REFERENCE FOR USE WITH

NFPA RECOMMENDED STANDARD

METHOD FOR VERIFYING THE FATIGUE AND STATIC

PRESSURE RATINGS OF THE PRESSURE CONTAINING ENVELOPE

OF A METAL FLUID POWER COMPONENT

INTRODUCTION

The following is tutorial and is presented to aid in developing a greater understanding of rating specific fluid power components and in utilizing the ratings verified by recommended standard T2.6.1-1974.

These discussions are summaries of many of the technical studies that were made in preparing recommended standard T2.6.1-1974. Most of them deal with metal fatigue technology because it is believed that most structural failures in the Fluid Power Industry are due to metal fatigue. Static strength is believed to be less important.

1. THE STATISTICAL NATURE OF FATIGUE

    The fatigue strengths of components are subject to considerable variability. Because of this it could be expected that extensive testing would be necessary to verify a Rated Fatigue Pressure. However, it was found that by using statistical techniques it is possible to develop a standard method for verifying ratings with a single sample test and still attain a specified level of confidence (Reference No. A1).

    With this method, the allowance for the statistical nature of fatigue is made by increasing test pressures by a specified Variability Factor, K_V. This factor was determined by using a statistical analysis of fatigue strength data found in the referenced literature (References No. A2, A3, A4, and A5). The values selected were based on attaining a 90, 99, or 99.9 percent Assurance Level with a 90 percent Confidence Level. When desired, a 99 percent Confidence Level can be achieved by successfully testing twice the specified number of units.

NFPA/T2.6.1

Variability Factors are also given for multiple sample tests. These permit the use of lower test pressures. Tests with multiple samples using these factors are statistically equivalent to the single sample tests.

2. ASSURANCE LEVEL

As used in recommended standard T2.6.1-1974, the Assurance Level is the minimum percentage of pressure containing envelopes of a verified design that will sustain 10 million applications of its Rated Fatigue Pressure. In practice, a given design must have a fatigue strength greater than that implied by the Assurance Level to reduce the risk of its failing in its cyclic test. This conservatism is implicit in the use of a small number of samples to verify fatigue strengths. It can be decreased when the number of samples used to verify the rating is increased.

Even without this conservatism, a 90 percent Assurance Level does not imply a 10 percent service failure rate. The reason for this is that only a portion of the service pressure containing envelopes of a given design will be subjected to the equivalent of 10 million applications of the Rated Fatigue Pressure (References No. A3, A4, A6). In the case of a 90 percent Assurance Level, if only 5 percent of the pressure containing envelopes encounter the rated loading, the service failure rate will be less than 1/2 percent.

Service failure rates, even with the lowest Assurance Level, can be acceptably low where only a small percentage of the pressure containing envelopes will be subjected to the rated loading. The higher Assurance Levels are needed where a larger percentage of service pressure containing envelopes are used at the rated loading or where the consequences of service failure require that failure rates be lower.

3. RATED FATIGUE PRESSURE

Recommended standard T2.6.1-1974 in verifying the Rated Fatigue Pressure for a component's pressure containing envelope provides a datum point for judging whether or not a component has adequate strength for use in a specific application.

Fatigue failures are caused by the actual extreme values of stresses induced by changing pressures, regardless of whether or not the stresses remain at their extremes for hours or for a few micro-seconds. Therefore, the peak transition pressures, rather than the normal system or working pressures, determine the strength requirements of pressure containing envelopes in a given system.

Usually pressure transients are a characteristic of complete circuits rather than single components, so it must be left to the system designer to determine their magnitudes and to select components with adequate strengths. There is very little data available to assist the system designer in making this determination. It is anticipated that the promulgation of recommended standard T2.6.1-1974 will stimulate the study of the relationships between peak and working pressures.

In most cases the Rated Fatigue Pressure of components will have to equal the system pressure peaks to provide adequate strength. In cases where some of the materials in the pressure containing envelope, like aluminum, do not have true endurance limits and where more than 10 million cycles can occur, the Rated Fatigue Pressure must be greater than the pressure peaks to allow for the higher cycle life.

Pressure peaks can be allowed to exceed the Rated Fatigue Pressure where it can be proved that less than 1,000,000 of these peaks will occur. When such peaks are all of the same magnitude, the effect can be judged by using the "method for estimating finite life capability", given in the Introduction. This method can also be used as a basis to evaluate the cumulative fatigue damage effect of peaks of various magnitudes. (References A3 and A7)

4. REDUCED TEST CYCLE

Rated Fatigue Pressure is based on sustaining 10 million cycles. It can be verified with tests consisting of one million cycles, if the test pressure is increased by the factor K_N (the ratio of the fatigue strengths at one and at ten million cycles). This factor was determined by using a large volume of data found in the referenced literature (Reference No. A8). Since most of this data was for reversed stressing, the modified Goodman relationship was used to convert this to the equivalent zero-to-max stressing usually occurring in pressurized structures (Reference No. A9).

The values used in recommended standard T2.6.1-1974 are those that equal or exceed 90 percent of the ratios calculated. This was done to eliminate data points that did not appear to be consistent. While this action tends to reduce the true Assurance Levels in a small percentage of cases, the use of the Test Duration Factor is still felt to be conservative.

Further decreases in the required number of test cycles were considered. It was found that this might affect the validity of the tests, therefore, shorter tests have not been utilized in recommended standard T2.6.1-1974 (See Reference No. A10). Such tests may still be used for production monitoring where they are first correlated with the tests prescribed in recommended standard T2.6.1-1974.

NFPA/T2.6.1

5. TEST RATE

Pressure changes travel at the speed of sound which approximate 50,000 IPS in oil and 13,000 IPS in air at atmospheric pressure. Stresses travel through metals at even higher speed.

However, exceptions to this rule can occur. These would include: (1) very large pressure containing envelopes where pressure may be measured far from the point where maximum stresses occur, (2) assemblies where appreciable joint slippage must occur before maximum stressing can be attained, and (3) pressure containing envelopes where large deflections can produce high internal inertial forces. Where it is suspected that such factors are having an effect, strain gages on the test pressure containing envelope should be used to verify whether or not the ratio of the measured stress to the measured pressure remains constant from a static condition to the actual test cycling rate.

The specified maximum cycling rate of 30 cycles per second is not based on stress propagation limits. Experience determined this to be the approximate practical limit of test hydraulic and measuring systems under very favorable conditions. Therefore, in most cases, limits for specific tests will be well below this rate.

Testing economics favor using the highest cycling rate permitted by the test circuit used. For this reason, good practice includes restraining all operational motions and displacing fluid in test pressure containing envelope cavities with metal to reduce the pressurized oil volume. These procedures reduce the fluid flow required for each pressure cycle so that a higher cycling rate can be used with a given test circuit flow capacity.

6. PULSE SHAPE

Fatigue failure is governed by the extreme stresses experienced by a pressure containing envelope and by the fact that the shape of the stress-time curve is usually not significant (Reference No. A11). However, in some cases, the presence of high amplitude oscillations superimposed on the test pulses can have an effect. This possibility has been minimized by using tests of one million cycles or more. Analyses show that in such tests oscillations must have amplitudes greater than 60 percent of the test pulse to contribute to fatigue failure (Reference No. A12).

NFPA/T2.6.1

The standard recognizes that the pressure in a test piece decays very slowly as it approaches zero. To make it possible to achieve practical cycling rates, it does not require that the pressure return to zero in each cycle (12.3.1). The effect of this deviation is minimized by requiring the pressure range to equal the Cyclic Test Pressure (12.3.2). When the minimum and maximum cyclic pressures are measured individually, the pressure range is subject to the total errors in the two measurements. Since the cyclic test pressure range is of prime importance, it is desirable to use instrumentation that measures it directly so that these measurements are only subject to one set of errors.

It is also recognized that the minimum and maximum pressures may not repeat exactly from cycle to cycle so a limit in the variability of the cycle to cycle pressure range is also specified (12.3.4). This tolerance should not be confused with the accuracy specified for measuring pressure range (Section 11). In the presence of variations, accuracy refers to how closely the measured average of the pressure range agrees with the actual value of this average. In general, it is the resultant of instrumentation and reading errors only and does not include the effect of cyclic variation.

7. TEMPERATURE

The tests specified in recommended standard T2.6.1-1974 are to be run at room temperature and the results are stated to be applicable to operating temperatures up to $200°F$ ($93°C$). This upper limit was selected because some aluminum alloys only retain their strength to this temperature (Reference No. A13). For many other materials, this limit is conservative and it did not seem practical to provide the temperature-strength characteristics of all materials. For applications above $200°F$ ($93°C$) special tests at the application temperature are considered necessary.

8. STATIC TESTS

Burst tests have often been used to evaluate the strength of fluid power components because they are simple to perform. The interpretation sometimes given to such tests is that the fatigue strength of a component is one-fourth or one-fifth its burst strength. No substantiation for this relationship was found in the referenced literature. Therefore, cyclic, rather than burst tests, are employed in recommended standard T2.6.1-1974 to verify fatigue strength.

NFPA/T2.6.1

The static test included in recommended standard T2.6.1-1974 is only for the purpose of verifying the Rated Static Pressure. Generally, this pressure will be higher than the Rated Fatigue Pressure. In addition to being a pressure limit for static applications, it may also provide a datum for estimating the effect of a limited number of load cycles as indicated in the "method for estimating finite life capability" given in the Introduction.

In most cases, the duration of the static load is not important as failure will occur almost as fast as pressure is applied. A duration of one minute was specified to be conservative and allow for possible cases where appreciable yielding or buckling must take place before failure becomes apparent. It is still possible that some components will require longer times for failure to be completed. Since it is in the best interests of the producer to detect such cases, longer durations are advised.

Static strength generally has less variability than fatigue strength. However, sufficient data for calculating allowances for this variability were not found. To be conservative, it is assumed that this variability is the same as that for fatigue strength.

When the static test pressure is applied, it will often cause localized yielding at points of higher stress concentration such as in the roots of threads or in sharp corners. So little metal is affected that such yielding cannot be detected by inspection methods, but it results in residual compressive stresses in the highest stressed areas, after the pressure is removed. These stresses subtract from the stresses that will be produced in subsequent loadings so the fatigue strength is improved. Care must be taken, therefore, that units used in cyclic tests are not previously strengthened by static testing unless such prestressing is also performed regularly in the production of these units.

NOTES

ANSI/B93.48M-1979

AN INDUSTRY STANDARD FOR FLUID POWER

American National Standard

Pneumatic fluid power applications -

Metal separable tube fittings -

Qualifications test

Approved as an ANSI Standard
18 July 1979

Descriptors: fluid power; testing, burst pressure; testing, leakage; testing, repeated assembly; testing, tensile; testing, tube fitting; testing, vibration; tube fitting, pneumatic fluid power; tube fitting, separable.

NOTES

ANSI/B93.48

FOREWORD

(This Foreword is not part of American National Standard Pneumatic fluid power applications — Metal separable tube fittings — Qualification test. ANSI/B93.48-1979).

This project was initiated at the 7 December 1970 meeting of the National Fluid Power Association Fittings Section, T3.8. J. Hiszpanski (ITE Imperial) agreed to serve as Project Chairman. The Project Group was formed and agreed to review the Hydraulic Tube Fitting Test standard (NFPA/T3.8.3-1970) with a view toward the development of a parallel and consistent pneumatic tube fitting standard.

A TSP was prepared and submitted to the NFPA Technical Board for approval on 20 January 1971. Draft No. 1 was written on 26 July 1972 and was reviewed by the Project Group at their 6 March 1973 meeting. After several changes in the draft, consensus was reached at that meeting. The Final Working Draft was submitted to Headquarters on 14 June 1973, at which time J. Hiszpanski resigned as Project Group Chairman due to the demands of his new corporate position. He was replaced by Al Schwarz (ITE Imperial).

On 1 May 1974, the NFPA Technical Staff prepared a Pre-General Review Draft of this document, incorporating a number of editorial and stylistic changes required by the NFPA Style Guide. This draft was sent to Chairman Schwarz for his review, with a request that he complete several unfinished sections. On 3 March 1975, Chairman Schwarz returned the draft to NFPA Headquarters with changes and complementary information noted in the body of the document.

The NFPA Technical Staff prepared the General Review Draft on 19 March 1975. General Review comments were resolved on 16 December 1975, and Technical Board approval to ballot was obtained on 4 February 1976. The Ballot Draft was prepared by the NFPA Technical Staff on 24 February 1976. All negative ballots were resolved by 12 July 1976.

On 11 August 1976, the NFPA Technical Board voted unanimously to recommend that this document be approved as an NFPA Recommended Standard. The NFPA Board of Directors granted final approval to NFPA/T3.8.9-1976 on 22 September 1976.

Members of the NFPA Project Group Responsible for the development of this standard included:

Al Schwarz
Project Chairman
Gould/Fittings & Valves

Jan Hiszpanski
Project Chairman
(December 1971-June 1973)
Gould/Fittings & Valves

Larry O'Sickey
Section Chairman
Parker Hannifin Corp.

John Colter
Technical Auditor
Watts Regulator Co.

John Luecke
Director of National Technical Services
National Fluid Power Association

J. Leber
Eaton Corp.

P. Robinson
Parker Hannifin Corp

E. Saloum
Snap-Tite, Inc.

A. Schmidt
The Weatherhead Co.

T. Soltis
The Weatherhead Co.

The membership roster for Standards Committee B93 at the time of ballot reaffirmation of ANSI/B93.48-1979 was comprised of:

MELVIN E. LONG, Chairman
JACK McPHERSON, Vice Chairman
JOHN R. LUECKE, Co-Secretary
WILLIAM TOTH, Co-Secretary

American Society of Agricultural Engineers
Ed Fletcher

American Society for Engineering Education
William R. Smith

American Society for Lubrication Engineers
M. M. Gurgo

American Society of Mechanical Engineers
Robert Hildebrandt
Thomas R. Curran (alternate)
Frank Yeaple (alternate)

Compressed Air & Gas Institute
David E. Bonn
John Addington (alternate)

Construction Industry Manufacturers Association
Glenn Stewart
H. T. Larmore (alternate)

Fluid Controls Institute
H. H. Kaemmer
Eric Bianchi (alternate)

Fluid Power Distributors Association
Thomas Neff

Fluid Power Society
Marsh Allen
R. D. Burgess, Sr.
William H. Dreher
Robert W. Hanpeter
Anton Hehn
Richard Read
Allen Tucker

Fluid Sealing Association
Ronald Prachel
John Scannell (alternate)

Industrial Truck Association
C. D. Gibson

Instrument Society of America
Aaron I. Kutz

Joint Industry Council
Robert Muhl

Material Handling Institute
Jack C. McPherson
W. L. Chichester (alternate)

Motor Vehicle Manufacturers Association
Jim Phillipson

National Fluid Power Association
James L. Fisher, Jr.
Walter Forster
Z. J. Lansky
Melvin Long

Otto Maha
John H. Peppenger
A. O. Roberts

National Machine Tool Builders Association
Edward Loeffler

Power Crane and Shovel Association
(to be named)

Rubber Manufacturers Association
William A. Hertel
William J. Atwell (alternate)

Society of Automotive Engineers
William Hertel
Eugene Falendysz
Henry Schultz
D. B. Shore
W. L. Snyder
David Prevallet
Robert W. White

Society of Manufacturing Engineers
Kevin Miller

U.S. Department of Defense
William Coyne (alternate)

Individual Member
E. C. Fitch, Jr.
Jack Johnson

Favorable ANSI/B93 ballot and ANSI Public Review were completed and on 2 June 1979 ANSI/B93.48-1979 was submitted to ANSI Board of Standards Review. Final approval was granted for ANSI/B93.48-1979 on 18 July 1979.

Pneumatic fluid power applications - Metal separable tube fittings - Qualification test

0 INTRODUCTION

In pneumatic fluid power systems, power is transmitted and controlled thru a gas under pressure within an enclosed circuit. Fittings are connectors or closures for fluid power lines or passages.

Although the use of plastic tubing in pneumatic systems is becoming quite common, it is not included in this recommended standard. The exclusion of plastic tubing from this document should not discourage its use in pneumatic systems.

1 SCOPE AND FIELD OF APPLICATION

1.1 This national standard includes:

- requirements and methods of testing for metal separable tube fittings used with metal tubing in compressed air applications.

- metal separable tube fittings made of carbon steel, stainless steel, brass and aluminum alloy.

NOTE - Fittings or tubings made of plastics are not considered in the scope of this national standard.

1.2 This national standard is intended to establish:

- minimum requirements for metal separable tube fittings used in compressed air applications.

- uniform methods of testing metal separable tube fittings used in compressed air applications.

NOTE - This national standard applies only to compressed air applications and does not include other pressurized gaseous media.

2 REFERENCES

ANSI/B93.2-1971, and Supplements B93.2A (ISO/DP 5598), **American National Standard, Glossary of Terms of Fluid Power.**

NFPA/T2.10.1-1978, **National Fluid Power Association Recommended Standard, Metric Units for Fluid Power Applications.**

NFPA/T3.8.2 R1-1970, **National Fluid Power Association Recommended Standard, Requirements of Separable Fluid Power Tube Fittings.**

MIL-H-5606, **Military Specification, Hydraulic Fluid, Petroleum Base, Aircraft and Ordinance.**

ANSI/B93.4-1969, **American National Standard, Electric Resistance Welded Mandrel Drawn Hydraulic Line Tubing.**

ASTM A 269, **American Society for Testing and Materials Standard, Specifications for Seamless and Welded Austenitic Stainless Steel Tubing for General Service.**

ASTM B 75, **American Society for Testing and Materials Standard, Specifications for Seamless Copper Tubing.**

ASTM B 210, **American Society for Testing amd Materials Standard, Specifications for Aluminum Alloy Drawn Seamless Tubing.**

ISO/R 370-1964, **International Standard, Conversion of Toleranced Dimensions from Inches into Millimetres and Vice Versa.**

ISO 2944-1974, **International Standard, Fluid Power Systems and Components - Nominal Pressures.**

3 TERMS AND DEFINITIONS

For terms and definitions used, see ANSI/B93.2 and ANSI/B93.2A (ISO/DP 5598).

3.1 test specimen: A component in the "as received" condition and not assembled for testing.

3.2 tube fitting test assembly: A tube fitting assembled to tubing and ready for testing.

4 UNITS

4.1 Units of measurement are used in accordance with NFPA T2.10.1. This document agrees with ISO 1000.

4.2 Approximate conversions to metric units are given in accordance with NFPA/T2.10.1 following their Customary US counterparts.

5 LETTER SYMBOLS

The following letter symbols are used in this national standard:

L   leakage test

RA repeated assembly test

B   burst test

T   tensile test

V   vibration test

Lg gauge length of test assembly

E   modulus of elasticity

D   tube outside diameter

S   tube fiber stress

p   test pressure

t   tube wall thickness

6 TEST EQUIPMENT

6.1 Test Specimens

6.1.1 Select test specimens at random from production lots to insure that the physical and chemical characteristics of these specimens are representative of the material and quality workmanship used in standard production practice.

6.1.2 Use a total number of eight carbon steel or stainless steel test specimens and six brass or aluminum alloy test specimens for each specimen size and material.

6.1.3 Number test specimens as samples one thru eight for carbon steel or stainless steel, and one thru six for brass or aluminum alloy.

6.1.4 Inspect test specimens for conformity to clause 8 of NFPA/T3.8.2 R1.

6.1.5 Inspect test specimens for conformity to manufacturer's dimensions and tolerances.

6.1.6 Replace non-conforming test specimens with randomly-selected test specimens that conform to 6.1.4 and 6.1.5.

6.2 Test Assembly Tubing

6.2.1 Use tubing of desired material(s) in accordance with Table 1.

6.2.2 Use tubing wall thickness in accordance with Table 1.

— TABLE 1 — **Tubing for test assemblies** —

Fitting Material	Tubing Description	Tubing Specification
Carbon Steel	Electric Resistance Welded Hydraulic Line Tubing	ANSI/B93.4
Stainless Steel	TP304 Welded Austenitic Stainless Steel Tubing (Annealed)	ASTM A 269
Brass	Seamless Copper Tubing	ASTM B 75
Aluminum Alloy	6061-T6 Aluminum Alloy Drawn Seamless Tubing	ASTM B 210

6.3 Test Equipment for Vibration Testing

6.3.1 Reversed binding test method.

6.3.1.1 Use fatigue testing machine of cantilever beam, reversed bending type. (See Figure 1.)

6.3.1.2 Obtain reversed, planar bending by applying a concentrated load on the free swiveling end of the test assembly.

6.3.1.3 Use a vibration frequency of 29 cps (29 Hz.)

6.3.2 Rotary bending test method

6.3.2.1 Use a fatigue testing machine of cantilever beam rotary bending type.

6.3.2.2 Use one of the following two arrangements:

a) Test assembly held stationary.

1) Mount one end in a fixed support.

2) Mount the other end in a rotary eccentric drive causing the test assembly end to deflect in multi-planar, $360°$ directions. (See Figure 2).

b) Test assembly rotated.

1) Mount test specimen next to the main bearings.

2) Mount the other end in a selfaligning bearing and hold deflected. (See Figure 3).

6.3.2.3 Use a direction of rotation which tends to loosen the nut on the test specimen.

6.3.2.4 Use a rotation frequency of 1750 RPM (1750 r/min).

7 TEST CONDITIONS ACCURACY

— TABLE 2 — **Test conditions accuracy** —

Test Conditions	Customary US Unit	SI Unit	Maintain Within + of True
Pressure[1]	psig	bar, gauge	5.0 psi (0.3)
Pressure[2]	psi	bar	5%
Frequency	cps	Hz	1.6 cps (1.6 Hz)
Rotational Speed	RPM	r/min	100 RPM (100 r/min)
Force	lb (f)	N	5%
Amplitude	in	mm	5%

1) Use with compressed air in Leakage and Repeated Assembly tests.
2) Use with water or hydraulic fluid in Tensile and Vibration tests.

8 TEST PROCEDURES

8.1 Test Sequence

8.1.1 Use mandatory test sequence set forth in Table 3.

ANSI/B93.48

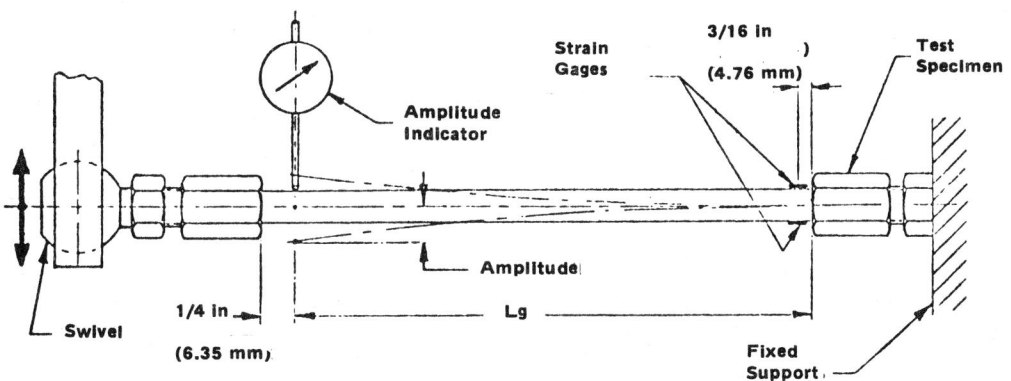

Figure 1 — Reversed bending test method; Test assembly arrangement.

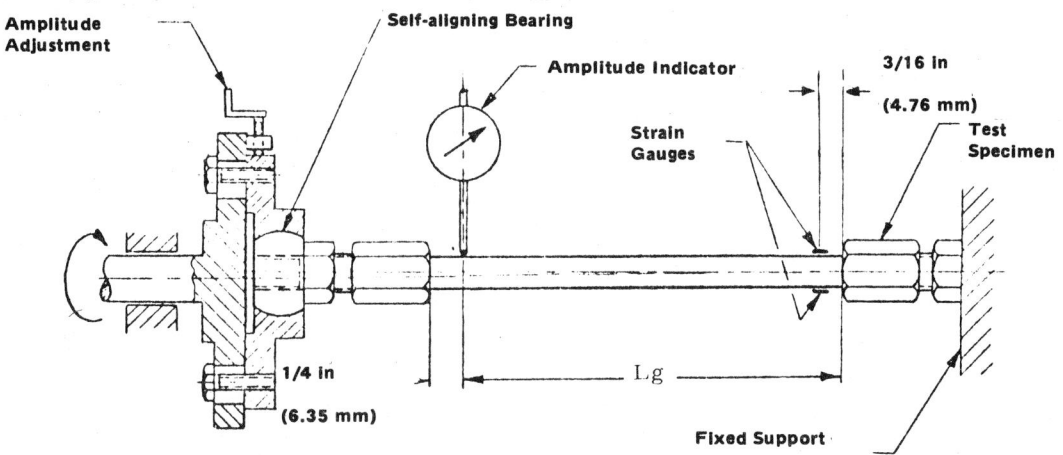

Figure 2 — Rotary bending test method; test specimen stationary.

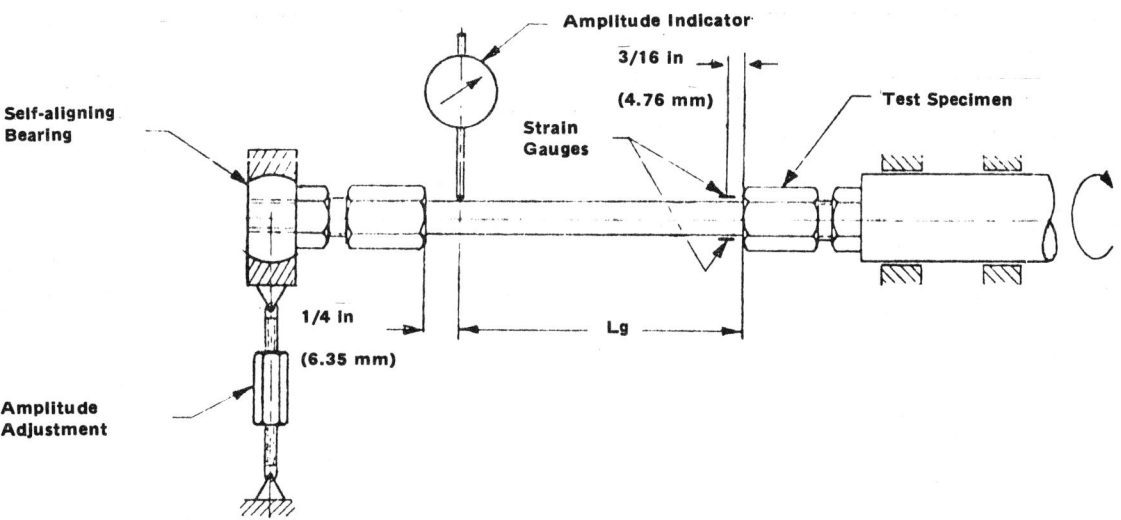

Figure 3 — Rotary bending test method; test specimen rotated.

TABLE 3 — Number of test specimens and sequence of tests.

Test Specimen Number	Sequence of Tests		
	1	2	3
1	L	RA	T
2	L	RA	T
3	L	RA	B
4	L	RA	B
5	L	T	B
6	L	T	B
7*	L	V	—
8*	L	V	—

* Test only those assemblies made exclusively of carbon steel and/or stainless steel. Brass/copper and aluminum alloy tubing or fittings are not subjected to the vibration test due to inherent reduces fatigue resistance properties of the tube material.

8.2 General Test Procedures

8.2.1 Use Table 3 to determine the number of samples to be subjected to a series of tests.

8.2.2 Observe manufacturer's instructions when assembling, reassembling and disassembling test specimens.

8.2.3 Perform all tests at atmospheric pressure and ambient room temperature unless otherwise specified in individual test procedures.

8.2.4 Use clean, dry air or nitrogen unless otherwise specified in individual test procedures.

8.3 Leakage Test

8.3.1 Assemble test specimens to tubing in accordance with 6.2 and 8.2.2.

8.3.2 Submerge test assemblies in water.

8.3.3 Use test medium in accordance with 8.2.4.

8.3.4 Pressurize test assembly at 20 psig (1.38 bar) for two minutes.

8.3.5 Pressurize test assembly at test pressure specified in Table 4 for five minutes.

8.3.6 Observe test specimens for visual signs of leakage.

8.3.7 Accept test specimens having no visual signs of leakage.

8.3.8 Record data in accordance with clause 10.

8.4 Repeated Assembly Test

8.4.1 Assemble and disassemble test assemblies 10 times.

8.4.2 Check for leakage in accordance with 8.3 after the third, sixth, and tenth reassembly.

8.4.3 Observe test specimens for visual signs of leakage.

8.4.4 Accept test specimens having no visual signs of leakage.

8.4.5 Record data in accordance with clause 10.

8.5 Burst Pressure Test

8.5.1 Construct a test assembly consisting of a piece of tubing having a minimum length of eight times the tubing outside diameter and with a test specimen mounted on each tubing end.

8.5.2 Use water or a hydraulic fluid in accordance with Military Specification MIL-H-5606.

8.5.3 Connect one end to the pressure source and cap the other end.

8.5.4 Pressurize the test assembly at a rate not exceeding 20,000 psi/min (1379 bar/min) until the tubing bursts.

8.5.5 Observe test specimens for visual signs of leakage or damage before the tubing bursts.

8.5.6 Accept test specimens having no visual signs of leakage or damage before the tubing bursts.

8.5.7 Record data in accordance with clause 10.

8.6 Tensile Test

8.6.1 Construct a test assembly consisting of a piece of tubing with a test specimen mounted at each tubing end.

8.6.2 Use water or hydraulic fluid in accordance with MIL-H-5606.

8.6.3 Connect one end to the pressure source and cap the other end.

8.6.4 Pressurize test assembly in accordance with Table 4.

8.6.5 Apply an axial, tensile test load in accordance with Table 4.

NOTE — Since these tests are performed with the test assembly pressurized to a specific level, the results may or may not provide an indication of in-service superiority between different types of fittings. Users should be cautioned against over-generalization from laboratory results to field expectations.

8.6.6 Observe test specimens for visual signs of leakage, damage or relative movement between the test specimens and the tube.

8.6.7 Accept test specimens showing no signs of visual leakage, damage or relative movement between the test specimen and the tube.

8.6.8 Record data in accordance with clause 10.

8.7 Vibration Test

8.7.1 Use one of the following vibration test methods:

a) Reversed bending in accordance with 6.3.1.

b) Rotary bending in accordance with 6.3.2.

8.7.2 Construct a test assembly consisting of a test specimen assembled at one end of a prescribed length of tubing (Lg in accordance with Table 5) and with an adapted closure which is connected to the driving device of the testing machine.

8.7.3 Use straight connector type test specimens.

8.7.4 Use port side connectors which are NPTF type or straight thread with O ring as specified in Table 5.

8.7.5 Use water or hydraulic fluid in accordance with Military Specification MIL-H-5606.

8.7.6 Pressurize test assembly to test pressure in accordance with Table 4.

8.7.7 Close the fluid circuit.

8.7.8 Set the initial amplitude in accordance with Table 5.

NOTE — If strain gauges are used, set the amplitude necessary to induce tube fiber tensile stresses shown as "initial" in Table 5.

8.7.9 Insure symmetrical amplitude within 0.002 in (0.05mm) in regard to the original undeflected tube axis.

8.7.10 Run the vibration test for 1 million cycles or until failure (in accordance with 8.7.14) occurs.

8.7.11 After 1 million cycles vibration period at initial amplitude, set "increased" amplitude value in accordance with Table 5.

NOTE - If strain gauges are used, set the amplitude necessary to induce tube fiber tensile strengths shown as "increased" in Table 5.

8.7.12 Continue test at "increased" amplitude to 10 million cycles (additional 9 million) or until failure (in accordance with 8.7.14) occurs.

8.7.13 Observe for visual signs of leakage or damage to the test specimen or to the tube.

8.7.14 Accept test specimens which have attained a minimum of 1 million cycles at "initial" amplitude or stress without leakage or damage to the test specimen or to the tube, and which have attained a minimum of 10 million cycles at "increased" amplitude or stress without leakage or damage to the test specimen.

NOTE - Tube failure occuring at the increased amplitude testing is permissible if external to the test specimen or at outward tube support. Such failure is deemed to be equivalent to successful completion of 10 million cycles at "increased" amplitude or stress.

8.7.15 Record data in accordance with clause 10.

9 DATA ACCURACY

— TABLE 6 — Data accuracy

Quantity	Customary US Unit	ISO 1000 Unit	Maintain Within (+) of Actual
Burst Pressure	psi	bar	2%

NOTE — All other testing is based on "pass/fail" criteria, utilizing defined test conditions, and therefore does not produce numerical data.

10 DATA PRESENTATION

10.1 Have available a record of all the following minimum test data in all test reports referencing this recommended standard:

a) fitting identification.

b) manufacturer's name.

c) catalog number.

d) tubing description.

1) material.

2) diameter.

3) wall thickness.

e) All physical values pertaining to the tests.

f) All additional provisions or modifications pertaining to the tests.

11 SUMMARY OF DESIGNATED INFORMATION

11.1 The following designated information is needed when applying this national standard to a particular application:

a) Size, style , and material of fitting.

b) Size, wall thickness, and material of tubing.

c) Burst pressure capability.

ANSI/B93.48

— TABLE 5 – Vibration test data —

Tube O.D.		Gauge Length of the Test Assembly Lg		Port Side Connection		Carbon Steel Tubing						Stainless Steel Tubing					
				NPTA Pipe Thread	Straight thread with O-ring seal per SAE J514	Amplitude				Tube Fiber Stress		Amplitude				Tube Fiber Stress	
						Initial		Increased		Initial	Increased	Initial		Increased		Initial	Increased
in	mm	in	mm			in	mm	in	mm	18,000 lb/in² (12,4 daN/mm²)	38,000 lb/in² (26,2 daN/mm²)	in	mm	in	mm	25,000 lb/in² (17,2 daN/mm²)	50,000 lb/in² (34,5 daN/mm²)
1/8	3,18	4	102	1/8	5/16-24	0,046	1,17	0,103	2,62			0,067	1,70	0,138	3,50		
3/16	4,76	4	102	1/8	3/8-24	0,030	0,76	0,068	1,73			0,043	1,09	0,090	2,29		
1/4	6,35	4	102	1/8	7/16-20	0,022	0,56	0,050	1,27			0,032	0,81	0,067	1,70		
5/16	7,94	4	102	1/4	1/2-20	0,017	0,43	0,040	1,02			0,025	0,63	0,053	1,35		
3/8	9,52	4	102	3/8	9/16-18	0,013	0,33	0,032	0,81			0,020	0,51	0,048	1,22		
1/2	12,70	5	127	1/2	3/4-16	0,016	0,41	0,038	0,96			0,024	0,61	0,051	1,30		
5/8	15,88	5	127	1/2	7/8-14	0,013	0,33	0,031	0,79			0,019	0,48	0,046	1,17		
3/4	19,05	5	127	3/4	1-1/16-12	0,010	0,25	0,025	0,63			0,015	0,38	0,034	0,86		
7/8	22,22	6	152	3/4	1-3/16-12	0,012	0,30	0,030	0,76			0,018	0,46	0,041	1,04		
1	25,40	6	152	1	1-5/16-12	0,011	0,28	0,028	0,71			0,016	0,41	0,036	0,09		
1-1/4	31,75	7	178	1-1/4	1-5/8-12	0,014	0,37	0,036	0,91			0,022	0,56	0,049	1,25		
1-1/2	38,10	8	203	1-1/2	1-7/8-12	0,009	0,23	0,023	0,58			0,014	0,37	0,032	1,81		

1) The above amplitude values correspond to the tube fiber stresses as indicated in the table.
2) The amplitude values have been calculated using the formula:

$$y = \frac{2 L_g^2}{3 E D} \left[S - \frac{pD}{4T} \right]$$

where y = amplitude
Lg = gauge length of the test assembly
E = modulus of elasticity
D = tube outside diameter
S = tube fiber stress
p = test pressure
t = tube wall thickness

ANSI/B93.48

JUSTIFICATION STATEMENT

This document formalizes practices and equipment requirements that have gained general acceptance and are currently being used by the majority of the manufacturers who participated in the development of this national standard.

TEST/PRODUCTION SIMILARITY

Utilize managerial controls necessary to maintain substantial similarity between test and production components or elements.

IDENTIFICATION STATEMENT

Use the following statement in catalogs and sales literature when electing to comply with this voluntary national standard:

Qualification data for brass, aluminum, stainless steel, and steel tubing obtained and presented in accordance with ANSI/B93.48-1979, "Pneumatic fluid power applications — metal separable tube fittings — Qualification test."

NOTE — Include only those tubing materials actually tested.

NOTES

NFPA Bibliography
T3.8.11-1977

AN INDUSTRY STANDARD FOR FLUID POWER

A Bibliography of

Fluid Power Tube Fittings and Conductors

---

published by
NATIONAL FLUID POWER ASSOCIATION, INC.
3333 N. Mayfair Road / Milwaukee, WI 53222 / 414-778-3344 / TLX 26898

SOURCE OF DOCUMENTS

ANSI   American National Standards Institute
1430 Broadway
New York, NY 10018

ASTM   American Society for Testing and Materials
1916 Race Street
Philadelphia, PA 19103

ISO   International Standards Organization
1, Rue de Varembe
1211 Geneva 20 SWITZERLAND

NFPA   National Fluid Power Association, Inc.
3333 North Mayfair Road
Milwaukee, WI 53222

NPFC   Commanding Officer, Naval Publications and Forms Center
5801 Tabor Avenue
Philadelphia, PA 19120
Attn: NPFC 105

SAE   Society of Automotive Engineers
400 Commonwealth Drive
Warrendale, PA 15096

Note: This bibliography provides a list of standards related to fluid power tube fittings and conductors which have been issued by NFPA and other standards-writing organizations. Listing in this bibliography is only for reference and does not imply Association endorsement.

NFPA/T3.8.11

Identification Number	Title	Source	Date of Issue
ANSI/B1.1	Unified Inch Screw Threads (UN and UNR Thread Form)	ANSI	1974
ANSI/B1.2	Gages & Gaging for Unified Screw Threads	ANSI	1974
ANSI/B2.1	Pipe Threads (Except Dryseal)	ANSI	1968
ANSI/B1.20.3	Dryseal Pipe Threads	ANSI	1976
ANSI/B31.1	Power Piping (with Addenda)	ANSI	1973
ANSI/B93.4	Electric Resistance Welded Mandrel Drawn Hydraulic Line Tubing	NFPA (ANSI)	1969
ANSI/B93.5	Practice for the Use of Fire Resistant Fluids for Fluid Power Systems	NFPA (ANSI)	1973
ANSI/B93.11	Seamless Low Carbon Steel Hydraulic Line Tubing	NFPA (ANSI)	1969
ANSI/B115.1	Pipe Unions for Flammable and Combustible Fluids, and Fire-Protection Service, Safety Standard for	ANSI	1969
ANSI/B116.1	Hydraulic Tube Fittings	ANSI	1974
ANSI/B117.1	Hydraulic Flanged Tube, Pipe, and Hose Connections, 4-Bolt Split Flange Type	ANSI	1974
ASTM A105	Forgings, Carbon Steel, for Piping Components	ASTM	1971
ASTM A181	Forged or Rolled Steel Pipe Flanges, Forged Fittings and Valves and Parts for General Service	ASTM	1968

NFPA/T3.8.11

Identification Number	Title	Source	Date of Issue
ASTM A182	Forged or Rolled Alloy-Steel Pipe Flanges, Forged Fittings, and Valves and Parts for High-Temperature Service	ASTM	1972
ASTM A334	Seamless and Welded Carbon and Alloy-Steel Tubes for Low-Temperature Service	ASTM	1973
ASTM A350	Forged or Rolled Carbon and Alloy-Steel Flanges, Forged Fittings, and Valves and Parts for Low-Temperature Service	ASTM	1965
ASTM A404	Forged or Rolled Alloy-Steel Pipe Flanges, Forged Fittings and Valves and Parts Specially Heat Treated for High-Temperature Service	ASTM	1968
ASTM A423	Seamless and Electric Welded Low-Alloy Steel Tubes	ASTM	1973
ASTM A513	Electric Resistance Welded Carbon and Alloy Steel Mechanical Tubing	ASTM	1975
ASTM A519	Seamless Carbon and Alloy Steel Mechanical Tubing	ASTM	1977
ASTM A539	Electric-Resistance-Welded Coiled Steel Tubing for Gas and Fuel Oil Lines	ASTM	1973
ASTM E213	Ultrasonic Inspection of Metal Pipe and Tubing for Longitudinal Discontinuities	ASTM	1968
ASTM E273	Ultrasonic Inspection of Longitudinal and Spiral Welds of Welded Pipe and Tubing	ASTM	1968

NFPA/T3.8.11

Identification Number	Title	Source	Date of Issue
ISO/R7	Pipe Threads for Gas List Tubes and Screwed Fittings Where Pressure - Tight Joints Are Made on the Threads (1/8 inch to 6 inches)	ISO	1955
ISO/64	Steel Tubes, - Outside Diameters	ISO	1973
ISO/68	General Purpose Screw Threads - Basic Profile	ISO	1961
ISO/R228	Pipe Threads Where Pressure - Tight Joints Are Not Made on the Threads (1/8 Inch to 6 Inches)	ISO	1973
ISO/965/I	ISO General Purpose Metric Screw Threads, Tolerances, Principles and Basic Data	ISO	1970
ISO/965/II	General Purpose Metric Screw Threads - Tolerances - Limits of Sizes for Commercial Bolt and Nut Threads - Medium Quality	ISO	1973
ISO/R1943	Coupling Threads for Hydraulic or Pneumatic Piping (Pipe Threads)	ISO	1970
ISO/R1944	Pipe Couplings for Hydraulic Piping (Pipe Threads)	ISO	1970
ISO/R2037	Pipes and Fittings - Stainless Steel Tubes for the Food Industry	ISO	1971
MIL-F-1224C	Fittings, Tube, Brass or Bronze, 45 Degree Flared	NPFC	1976

NFPA/T3.8.11

Identification Number	Title	Source	Date of Issue
MIL-F-5509	Fitting, Flared Tube, Fluid Connection	NPFC	1973
MIL-F-15049A	Fittings, Tube, Brass or Copper 45 Degree Flared	NPFC	1969
MIL-F-18280A	Fitting, Flareless Tube, Fluid Connection	NPFC	1971
MIL-F-18866D	Fitting, Hydraulic Tube, Flared, 37 Degree and Flareless; Steel	NPFC	1977
MIL-F-45911A	Fitting, Automotive, Tube, Air Brake Type	NPFC	1973
MIL-J-5513C	Joint, Hydraulic Swivel	NPFC	1976
MIL-T-5695D	Tubing, Steel, Corrosion Resistant (304), Cold Drawn	NPFC	1973
MIL-T-6737B	Tubing, Steel, Corrosion and Heat Resisting (18-8 Stabilized), Welded	NPFC	1969
MIL-T-6845C	Tubing, Steel, Corrosion Resistant (304) Aerospace Vehicle Hydraulic System, 1/8 Hard Condition	NPFC	1970
MIL-T-8504A	Tubing, Steel, Corrosion Resistant (304) Aerospace Vehicle Hydraulic Systems, Annealed, Seamless and Welded	NPFC	1971
MIL-T-8506A	Tubing, Steel Corrosion Resistant (304) Annealed Seamless and Welded	NPFC	1966
MIL-T-8606C	Tubing, Steel, Corrosion Resistant (18-8 Stabilized)	NPFC	1974
MIL-T-8808A	Tubing, Steel, Corrosion Resistant 18-8 Stabilized Aircraft Hydraulic Quality	NPFC	1969

NFPA/T3.8.11

Identification Number	Title	Source	Date of Issue
MIL-T-8887C	Tubing, Steel, Corrosion Resistant, Heat Resistant, Stabilized, Welded, Thin Wall	NPFC	1961
MIL-T-20157	Tube and Pipe, Carbon Steel, Seamless	NPFC	1964
MS-21344	Fitting, Installation of Flared Tube, Straight Threaded Connectors, Design Standard for	NPFC	1971
MS-24385C	Fitting End, Flared Tube Connection and Gasket Seal Precision Type, Standard Dimensions	NPFC	1974
MS-24386D	Fitting End, Bulkhead Flared Tube Connection and Gasket Seal, Precision Type, Standard Dimensions	NPFC	1976
MS-27850	Fitting, Installation of Straight Threaded Fluid Connection	NPFC	1967
MS-27856	Fitting Assembly, Straight Thread, Fluid Connection, Low Pressure	NPFC	1967
MS-33514F	Fitting End, Standard Dimensions for Flareless Tube Connection and Gasket Seal	NPFC	1969
MS-33515J	Fitting End, Standard Dimensions for Bulkhead Flareless Tube Connections	NPFC	1973
MS-33566B	Fitting, Installation of Flareless Tube, Straight Threaded Connector	NPFC	1975

NFPA/T3.8.11

Identification Number	Title	Source	Date of Issue
MS-33583B	Tubing End, Double Flare, Standard Dimensions for	NPFC	1969
MS-33584B	Tubing End, Standard Dimensions for Flared	NPFC	1960
MS-33611	Tube Bend Radii	NPFC	1958
MS-33656F	Fitting End, Standard Dimensions for Flared Tube Connection and Gasket Seal	NPFC	1974
MS-33657F	Fitting End, Standard Dimensions for Bulkhead Flared Tube Connections	NPFC	1975
MS-33658A	Fitting End, Hose Connection, Standard Dimension for	NPFC	1959
MS-33677A	Fitting End, Taper Pipe Thread, Standard Dimensions for	NPFC	1961
MS-39321A	Fitting, Flared, Tube, 37 Degree, Steel	NPFC	1970
NFPA/T2.6.1	Recommended Standard Method for Establishing and Verifying the Fatigue and Static Pressure Ratings of the Pressure Containing Envelope of a Metal Fluid Power Component	NFPA	1974
NFPA/T3.8.2 R1	Recommended Standard Requirements of Separable Fluid Power Tube Fittings	NFPA	1970
NFPA/T3.8.3 R2	Recommended Standard Qualification Test for Steel Separable Tube Fittings for Hydraulic Fluid Power Applications	NFPA	1977

NFPA/T3.8.11

Identification Number	Title	Source	Date of Issue
NFPA/T3.8.9	Recommended Standard Qualification Test for Separable Tube Fittings for Pneumatic Fluid Power Applications	NFPA	1976
SAE AIR 737	Aerospace Hydraulic and Pneumatic Specifications and Standards	SAE	1966
SAE ARP 24B	Determination of Hydraulic Pressure Drop	SAE	1968
SAE ARP 600	Torque Determination, Method of, for Tube or Hose Fitting Connections, Flared, Flareless or Miscellaneous Screw Thread Style	SAE	1968
SAE ARP 683	Installation Procedures and Torques for Fluid Connections	SAE	1964
SAE AS 756	Fitting End Assembly, Universal	SAE	1963
SAE AS 758	Fittings, Installation in Straight Threaded Boss	SAE	1963
SAE AS 759	Fittings, Universal or Internally Threaded, Flared and Flareless	SAE	1963
SAE J 246b	Spherical Sleeve (Compression) Tube Fittings	SAE	1974
SAE J 356a	Welded Flash Controlled Low Carbon Steel Tubing Normalized for Bending Double Flaring and Beading	SAE	1972
SAE J 512h	Automotive Tube Fittings	SAE	1974

NFPA/T3.8.11

Identification Number	Title	Source	Date of Issue
SAE J 513c	Refrigeration Tube Fittings	SAE	1974
SAE J 514g	Hydraulic Tube Fittings	SAE	1974
SAE J 515a	Hydraulic "O" Ring	SAE	1962
SAE J 518c	Hydraulic Flanged Tube, Pipe, and Hose Connections, 4-Bolt Split Flange Type	SAE	1972
SAE J 524b	Seamless Low Carbon Steel Tubing Annealed for Bending and Flaring	SAE	1972
SAE J 525b	Welded and Cold Drawn Low Carbon Steel Tubing Annealed for Bending and Flaring	SAE	1972
SAE J 526b	Welded Low Carbon Steel Tubing	SAE	1972
SAE J 527b	Brazed Double Wall Low Carbon Steel Tubing	SAE	1972
SAE J 528b	Seamless Copper Tube	SAE	1968
SAE J 530a	Automotive Pipe Fittings	SAE	1973
SAE J 531a	Automotive Pipe, Filler and Drain Plugs	SAE	1973
SAE J 532a	Automotive Straight Thread Filler and Drain Plugs	SAE	1973
SAE J 533b	Flares for Tubing	SAE	1972
SAE J 846e	Coding System for Identification of Tube, Pipe, and Hose Fittings	SAE	1974
SAE J 926	Hydraulic Pipe Fittings	SAE	1965

NFPA/T3.8.11

Identification Number	Title	Source	Date of Issue
SAE J 1065	Pressure Ratings for Hydraulic Tubing and Fittings	SAE	1974
SAE J 1149	Metallic Air Brake Systems Tubing and Pipe	SAE	1977
WW-P-351-A	Pipe: Red Brass (Copper Alloy No. 230) Seamless, Standard Pipe Size, Regular and Extra Strong	NPFC	1975
WW-P-460B	Pipe Fittings: Brass or Bronze (Threaded) 125 and 150 Pound	NPFC	1977
WW-P-521-F	Pipe Fittings, Flange Fittings and Flanges Steel and Iron (Threaded and Butt Welded) 150 Pound	NPFC	1968
WW-T-775a	Tube, Copper Seamless (for Refrigeration and General Use)	NPFC	1967

NOTES

ANSI/B93.59M-1982

AN INDUSTRY STANDARD FOR FLUID POWER

American National Standard

Fluid power systems and products -

Connectors and associated components -

Outside diameters of tubes and inside diameters of hoses

(NFPA/T3.8.22M-1981)
(Technically identical to ISO 4397)
(Metric Only)

Approved as an ANSI Standard
9 February 1982

Descriptors: hydraulic equipment, pneumatic equipment, tubes (pipes), hoses, pipe joints, dimensions, diameters.

---

published by
NATIONAL FLUID POWER ASSOCIATION, INC.
3333 N. Mayfair Road / Milwaukee, WI 53222 / 414-778-3344 / TLX 26898

ANSI/B93.59

FOREWORD

This Foreword is not part of American National Standard - Fluid power systems and products - Connectors and associated components - Outside diameters of tubes and inside diameters of hoses, ANSI/B93.59M-1982 (technically identical to ISO 4397).

NFPA, thru its dual administration of the NFPA and ISO Technical Committees for Fluid Power, encouraged the participation of NFPA members in the development of ISO standards. While this work was underway at ISO, on ISO 4397 (Fluid power systems and components - Connectors and associated components - Outside diameters of tubes and inside diameters of hoses), NFPA participants in cooperation with NFPA Headquarters, developed a similar NFPA proposal. Thru the close liaison resulting from many of the same participants working at both the NFPA and ISO levels, identical documents were developed and are being promulgated.

As a result of these efforts, a Ballot Draft of Fluid power systems and products - Connectors and associated components - Outside diameters of tubes and inside diameters of hoses, NFPA/T3.8.22M was prepared on 31 January 1980 by Headquarters Staff. Favorable Ballot closed 14 March 1980. The document was submitted to the Technical Board and granted approval on 11 February 1981.

Subsequently, T3.8.22M was submitted and granted final approval by the Board of Directors on 4 March 1981.

TAG SC 4 and Project Group members who developed this standard:

Larry O'Sickey
Section Chairman
Parker Hannifin

Adam Schmidt
Section Secretary
Dana Corp./Weatherhead

Pete Mardosa
Technical Auditor
Dayton T. Brown

James C. White[**]
Director of Technical Services
National Fluid Power Association

W. Atwell
Uniroyal

R. Bailey
C. A. Norgren Co.

H. Burns
Stratoflex

M. Chermak
Imperial Clevite Inc.

W. Currie
Parker Hannifin

E. English
Uniroyal

J. Glidden
Hydreco

C. Grigsby
Schrader Bellows

W. Hertel
Parker Hannifin

G. Herzan
Parker Hannifin

J. Hinske
Parker Hannifin

N. Johnston
International Harvester

B. Keister
Eaton/Samuel Moore

K. Koch
Imperial Clevite Inc.

Z. Lansky
Parker Hannifin Corp.

R. Lobmeyer
John Deere

O. Maha [*]
Parker Hannifin

E. Maroney
W. H. Nichols

R. May
Aeroquip Corp.

E. McCarthy
Rubber Mfgs. Association

J. Mueller [**]
Weatherhead Co.

H. Newman
Caterpillar Tractor

R. Orr
Fairchild Industries

J. Pippenger [*]
W. H. Nichols

R. Rider
Imperial Clevite Inc.

R. Rogers
Aeroquip Corp.

E. Saloum
Snap-Tite Inc.

A. Schwarz
Imperial Clevite Inc.

E. Wolf
Amalga Corp.

---
[*] Retired
[**] Company affiliation has changed.

It was intended that NFPA Recommended Standard, T3.8.22M, be submitted to ANSI for promulgation as an American National Standard. This would accomplish total harmonization between identical ISO and ANSI standards.

International Standard ISO 4397 was developed by Technical Committee ISO/TC 131, "Fluid Power Systems" and by TAG SC 4, "Connectors & Similar Products." It was circulated to the Member Bodies in November 1976 and published as a standard in 1978.

It has been approved by the Member Bodies of the following countries:

Australia	Korea, Rep. of
Austria	Mexico
Belgium	Netherlands
Brazil	Poland
Chile	Romania
Czechoslovakia	South Africa, Rep. of
Finland	Spain
France	Sweden
Germany	Turkey
Hungary	USA
India	USSR
Italy	Yugoslavia
Japan	

The member bodies of the following countries expressed disapproval of the document on technical grounds:

Switzerland   United Kingdom

On 31 July 1981, ANSI/B93.59M was submitted to ANSI Committee B93 for ballot. Balloting closed on 5 October 1981 with unanimous approval.

ANSI/B93.59M was forwarded to ANSI Board of Standards Review on 13 January 1982 and granted approval on 9 February 1982.

The membership roster for the Standards Committee B93 at the time of ballot:

Jack McPherson
Chairman

Daniel B. Shore
Vice Chairman

James C. White
Co-Secretary

American Society of Agricultural Engineers
Ed Fletcher

American Society for Engineering Education
Wm. R. Smith

American Society of Lubrication Engineers
To be named

Compressed Air & Gas Institute
David E. Bonn
John Addington (alternate)

Construction Industry Manufacturers
Glen Stewart

Fluid Controls Institute
Jude Pauli
Eric Bianchi (alternate)

Fluid Power Distributors Association
Thomas Neff

Fluid Power Society
Edward C. Briggs
Ronald Brettnacher
Robert L. Firth
Carroll Grigsby
Robert W. Hanpeter
Marty King
Richard Read
Ronald Smith
Alan Tiedman

Fluid Sealing Association
John Scannell
Alex Pilecki (alternate)

Instrument Society of America
Aaron I. Kutz

Joint Industry Council
Robert Muhl

Material Handling Institute
Jack McPherson
Willard Chichester (alternate)

National Fluid Power Association
Richard N. Bailey
John Bowbin
James L. Fisher, Jr.
Walter Forster
Z. J. Lansky
John H. Mueller
A. O. Roberts

National Machine Tool Builders Association
John B. Deam

Rubber Manufacturers Association
William A. Hertel
John R. Loder
E. J. McCarthy (alternate)

SAE
William A. Hertel
Eugene Falendysz
Henry Schultz
Daniel B. Shore
W. L. Snyder
Robert W. White

Society of Manufacturing Engineers
To be named

U. S. Dept. of Defense
William P. Coyne

Company Member
L. L. Schmaltz
Don McGeachy
John Welker (alternate)

Individual Member
Dr. E. C. Fitch, Jr.
Robert Hildebrandt
Otto Maha
John J. Pippenger
Frank Yeaple
Tom Wanke

ANSI/B93.59

Fluid power systems and products -
Connectors and associated components -
Outside diameters of tubes and inside diameters of hoses

0 INTRODUCTION

In fluid power systems, power is transmitted and controlled through a fluid (liquid or gas) under pressure within an enclosed circuit. Components are interconnected through their ports and associated fluid conductor fitting ends. Tubes are rigid conductors; hoses are flexible conductors.

1 SCOPE AND FIELD OF APPLICATION

This National Standard establishes the following select series for use within hydraulic and pneumatic fluid power products and associated components.

a) a select series of outside diameters for hydraulic and pneumatic rigid tubes, irrespective of material composition;

b) a select series of inside diameters for hydraulic and pneumatic hoses made of rubber or plastic.

2 REFERENCES

ANSI/B93.2-1971 and Supplement ANSI/B93.2A-1978 American National Standard, Glossary of Terms for Fluid Power.

3 TERMS AND DEFINITIONS

For defintions of other terms used, see ANSI/B93.2 and Supplement ANSI/B93.2A.

3.1 tube: A pipeline of metal or plastic, used for connecting fixed assemblies, the size of which is defined by its nominal outside diameter and is available in various wall thicknesses.

3.2 hose: A flexible pipeline, usually of wire-reinforced rubber or plastic, the size of which is defined by its nominal inside diameter and is available in various wall thicknesses.

4 DIMENSIONS

Select outside diameters of tubes and inside diameters of hoses from the dimenisons in tables 1 and 2.

TABLE 1 - Series of outside diameters of tubes	TABLE 2 - Series of inside diameters of hoses
Dimensions in millimetres	Dimensions in millimetres
4	3,2
5	5
6	6,3
8	8
10	10
12	12,5
16	16
20	19*
25	20
32	25
38*	31,5
40	38*
50	40
	50
	51*

* For some flanged connection applications.

* For hydraulic purposes only.

5 IDENTIFICATION STATEMENT

Use the following statement in test reports, catalogs and sales literature when electing to comply with this National Standard:

"Outside diameters of tubes and inside diameters of hoses selected in accordance with ANSI/B93.59M-1982, **Fluid power systems and products - Connectors and associated components - Outside diameters of tubes and inside diameters of hoses.**"

ANSI/B93.59

APPENDIX
to ANSI/B93.59M-1982

2   REFERENCES

ISO 5598[1], Fluid power systems and components - Vocabulary.

NFPA/T2.10.1M-1978, National Fluid Power Association Recommended Standard, Metric Units for Fluid Power Applications.

ISO 1000-1981, SI units and recommendations for the use of their multiples and of certain other units.

---

[1] Presently at the stage of draft.

NOTE:

This document uses the ISO method of writing metric units (a comma as the decimal marker). The standard US custom is to use the period or dot as the decimal marker.

TABLE 1 — Series of outside diameters of tubes

mm	in[1]
4	.16
5	.20
6	.24
8	.31
10	.39
12	.47
16	.63
20	.79
25	.98
32	1.26
38*	1.50
40	1.57
50	1.97

*For some flanged connection applications.

TABLE 2 - Series of inside diameters of hoses

mm	in[1]
3,2	.13
5	.20
6,3	.25
8	.31
10	.39
12,5	.49
16	.63
19*	.75
20	.79
25	.98
31,5	1.24
38*	1.50
40	1.57
50	1.97
51*	2.00

*For hydraulic purposes only.

1) NOTE: Inch dimensions for reference purposes and are approximate.

NOTES

ANSI/B93.60M-1982

AN INDUSTRY STANDARD FOR FLUID POWER

American National Standard

Fluid power systems and products -

Connectors and associated components -

Nominal pressures

(NFPA/T3.8.23M-1981)
(Technically identical to ISO 4399)
(Metric Only)

Approved as an ANSI Standard
26 February 1982

Descriptors: fluid power, hydraulic fluid power, pneumatic fluid power, pipe fittings, ratings, pressure.

---

published by

NATIONAL FLUID POWER ASSOCIATION, INC.

3333 N. Mayfair Road / Milwaukee, WI 53222 / 414-778-3344 / TLX 26898

ANSI/B93.60

FOREWORD

This Foreword is not part of American National Standard Fluid power systems and products - Connectors and associated components - Nominal pressures, ANSI/B93.60M-1982 (technically identical to ISO 4399).

NFPA, thru its dual administration of the NFPA and ISO Technical Committtes for Fluid Power, encouraged the participation of NFPA members in the development of ISO standards. While this work was underway at ISO, on ISO 4399 (Fluid power systems and components - Connectors and associated components - Nominal pressures), NFPA participants in cooperation with NFPA Headquarters, developed a similar NFPA proposal. Thru the close liaison resulting from many of the same participants working at both the NFPA and ISO levels, identical documents were developed and are being promulgated.

As a result of these efforts, a Ballot Draft of Fluid power systems and products - Connectors and associated components - Nominal pressures, NFPA/T3.8.23M was prepared on 23 January 1980 by Headquarters Staff. Favorable ballot closed 14 March 1980. The document was submitted to the Technical Board and granted approval on 11 February 1981.

Subsequently, T3.8.23M was submitted and granted final approval by the Board of Directors on 4 March 1981.

TAG SC 4 and Project Group members who developed this standard:

Larry O'Sickey
Section Chairman
Parker Hannifin

Adam Schmidt
Section Secretary
Dana Corp./Weatherhead

John Bowbin
Technical Auditor
Miller Fluid Power

James C. White[*]
Director of Technical Services
National Fluid Power Association

W. Atwell
Uniroyal

R. Bailey
C. A. Norgren Co.

J. Berninger (alternate)
Parker Hannifin

H. Burns
Stratoflex

M. Chermak
Imperial Clevite Inc.

W. Currie (alternate)
Parker Hannifin

E. English
Uniroyal

J. Glidden
General Signal - Hydreco

C. Grigsby
Schrader Bellows Div.

W. Hertel (alternate)
Parker Hannifin

G. Herzan (alternate)
Parker Hannifin

J. Hinske (alternate)
Parker Hannifin

N. Johnston
International Harvester Co.

B. Keister
Eaton/Samuel Moore

K. Koch
Imperial Clevite Inc.

Z. Lansky
Parker Hannifin

R. Lobmeyer
John Deere Product Engineering Center

O. Maha
Individual Member

E. Maroney
W. H. Nichols Co.

R. May
Aeroquip Corp.

E. McCarthy
Rubber Manufacturers Association

J. Mueller[*]
Dana Corp./Weatherhead

J. Newman
Caterpillar Tractor Co.

R. Orr
Fairchild Industries Inc.

J. Pippenger
Individual Member

T. Rider
Imperial Clevite Inc.

R. Rogers
Aeroquip Corp.

E. Saloum
Snap-Tite Inc.

A. Schwarz
Imperial Clevite Inc.

E. Wolf
Amalga Corp.

---

*Company affiliation has changed.

It was intended that the NFPA Recommended Standard, T3.8.23M, be submitted to ANSI for promulgation as an American National Standard. This would accomplish total harmonization between identical ISO and ANSI Standards.

International Standard ISO 4399 was drawn up by Technical Committee ISO/TC 131, "Fluid Power Systems" and by TAG SC 4, "Connectors & Similar Products." It was circulated to the Member Bodies in January 1976 and published as a standard in 1977.

It has been approved by the Member Bodies of the following countries:

Belgium	Poland
Bulgaria	Romania
Finland	South Africa, Rep of
Germany	Spain
Hungary	Sweden
Italy	Switzerland
Japan	Turkey
Korea, Rep of	USA
Mexico	USSR
Netherlands	Yugoslavia

The member bodies of the following countries expressed disapproval of the document on technical grounds:

Australia	France
Austria	United Kingdom

On 14 August 1981, ANSI/B93.60M was submitted to ANSI Committee B93 for ballot. Balloting closed 23 October 1981 with unanimous approval.

ANSI/B93.60M was forwarded to ANSI Board of Standards Review on 29 January 1982 and granted approval 26 February 1982.

The membership roster for the Standards Committee B93 at the time of ballot:

Jack McPherson
Chairman

Daniel B. Shore
Vice Chairman

James C. White
Co-Secretary

William Wagner
Co-Secretary

American Society of Agricultural Engineers
Ed Fletcher

American Society for Engineering Education
Wm. R. Smith

American Society of Lubrication Engineers
To be named

Compressed Air & Gas Institute
David E. Bonn
John Addington (alternate)

Construction Industry Manufacturers
Glen Stewart

Fluid Controls Institute
Jude Paull
Eric Bianchi (alternate)

Fluid Power Distributors Association
Thomas Neff

Fluid Power Society
Edward C. Briggs
Ronald Brettnacher
Robert L. Firth
Carroll Grigsby
Robert W. Hanpeter
Marty King
Richard Read
Ronald Smith
Alan Tiedman

Fluid Sealing Association
John Scannell
Alex Pliecki (alternate)

Instrument Society of America
Aaron I. Kutz

Joint Industry Council
Robert Muhl

Material Handling Institute
Jack McPherson
Willard Chichester (alternate)

National Fluid Power Association
Richard N. Bailey
John Bowbin
James L. Fisher, Jr.
Walter Forster
Z. J. Lansky
John H. Mueller
A. O. Roberts

National Machine Tool Builders Association
John B. Deam

Rubber Manufacturers Association
William A. Hertel
John R. Loder
E. J. McCarthy (alternate)

SAE
William A. Hertel
Eugene Falendysz
Henry Schultz
Daniel B. Shore
W. L. Snyder
Robert W. White

Society of Manufacturing Engineers
To be named

U.S. Department of Defense
William P. Coyne

Company Member
L. L. Schmaltz
Don McGeachy
John Welker (alternate)

Individual Member
Dr. E. C. Fitch, Jr.
Robert Hildebrandt
Otto Maha
John J. Pippenger
Frank Yeaple
Tom Wanke

ksg

ANSI/B93.60

**Fluid power systems and products -
Connectors and associated components -
Nominal pressures**

0 INTRODUCTION

In fluid power systems, power is transmitted and controlled through a fluid (liquid or gas) under pressure within an enclosed circuit. Systems and components are generally designed and marketed for a specific fluid pressure.

Components are interconnected through their ports and associated fluid conductor fitting ends.

1 SCOPE AND FIELD OF APPLICATION

This National Standard establishes a selection of nominal pressures for hydraulic and pneumatic fluid power connectors and associated components.

NOTE: There may be a need to provide a selection of nominal pressures for connectors and associated components used in applications where the external pressure on the components is greater than the internal pressure, for example, vacuum service. A document which deals with this subject will be established in due course.

2 REFERENCES

ISO 2944-1974, **Fluid power systems and components - Nominal pressures.**

ISO 5598[1), **Fluid power systems and components - Vocabulary.**

3 DEFINITIONS

3.1 **nominal pressure:** a pressure value assigned to a component or a system for the purpose of convenient designation.

NOTE: This designation is the same as that used in ISO 2944 and is intended solely to complete this NFPA Recommended Standard. A more comprehensive definition for general purposes may be established subsequently.

3.2 For definitions of other terms used, see ISO 5598.

4 UNITS

4.1 The pressure unit used is the bar.

$1 \text{ bar} = 100 \text{ kPa}^* \approx 14.5 \text{ lbf/in}^2$

4.2 Express nominal pressures as "pressure of . . . bar."

4.3 Assume the nominal pressure to be "gage" pressure (i.e. the pressure above atmospheric) when no modifier is given.

4.4 Select any other values required from ISO 2944.

5 NOMINAL PRESSURES

Select from values in the table.

TABLE - Nominal pressures -
Gage pressure in bars

2,5
6,3
10
16
25
40
100
160
200
250
315
400
630

---

1) **Presently at the stage of draft.**

* $1 \text{ Pa} = 1 \text{ N/m}^2$

6 IDENTIFICATION STATEMENT

Use the following statement in test reports, catalogs and sales literature when electing to comply with this National Standard:

"Nominal pressures selected from ANSI/B93.60M-1982, Fluid power systems and products - Connectors and associated components - Nominal pressures."

APPENDIX
to ANSI/B93.60M-1982

2 REFERENCES

ANSI/B93.2-1971 and Supplement ANSI/B93.2A-1978, American National Standard, Glossary of Terms for Fluid Power.

NFPA/T2.10.1M-1978, **National Fluid Power Association Recommended Standard, Metric Units for Fluid Power Applications.**

ISO 1000-1981, **SI Units and recommendations for the use of their multiples and of certain other units.**

NOTE:
This document uses the ISO method of writing metric units (a comma as the decimal marker). The standard US custom is to use the period or dot as the decimal marker.

TABLE - Nominal pressures -*
Gage pressures in bars/psi

Bars	psi
2,5	36.25
6,3	91.35
10	145
16	232
25	362.5
40	580
100	1450
160	2320
(200)	2900
250	3625
(315)	4567.5
400	5800
630	9135

NOTE: **Non-preferred values are in parentheses**

* NOTE: **Exact conversions**

ANSI/B93.4M-1981

AN INDUSTRY STANDARD FOR FLUID POWER

American National Standard

Hydraulic fluid power -

Line tubing -

Electric resistance welded, mandrel drawn

Approved as an ANSI Standard
11 February 1981

---

published by
NATIONAL FLUID POWER ASSOCIATION, INC.
3333 N. Mayfair Road / Milwaukee, WI 53222 / 414-778-3344 / TLX 26898

ANSI/B93.4

FOREWORD

This Foreword is not part of American National Standard - Hydraulic fluid power - Line tubing - Electric resistance welded, mandrel drawn, ANSI/B93.4M-1981.

Low carbon steel pressure tubing has been used extensively for hydraulic lines in fluid power systems. In 1963, NFPA became aware of the need to update existing hydraulic line tubing specifications, and assigned the project to a Project Group of the Conductors Section, T3.15. The Project Group developed a document which was approved by the Board of Directors on 29 January 1965 as NFPA Recommended Standard T3.15.1-1965. The document was then introduced into the American National Standards Institute, which approved it as ANSI Standard B93.4-1966.

In March 1966, the NFPA Conductors Section began the work of revising T3.15.1-1965. The revision was approved as an NFPA Recommended Standard on 12 November 1967. The document was then submitted to ANSI, which approved it as ANSI Standard B93.4-1969 on 5 September 1969.

On 6 March 1974, in accordance with ANSI requirements, NFPA Section T3.15 initiated a five-year review of B93.4-1969. After discussion, the Section recommended that the standard be revised.

On 5 June 1974, Section T3.15 merged with Section T3.8, Fittings, to form an expanded Section T3.8, called the Fittings and Conductors Section. On 16 October 1974, T3.8 reviewed B93.4-1969 and appointed Richard Dodson (Unarco - Leavitt) as Chairman of the Project Group to undertake the revision of the document.

In revising the existing standard, the Project Group sought to bring it into conformity with a similar standard, SAE J525-1972, as well as to specify the number of tests required for hydraulic line tubing.

Chairman Dodson submitted the Final Working Draft of the revision, called T3.15.1 R2 - 19xx, to NFPA Headquarters on 3 December 1975. The NFPA Technical Staff prepared the General Review Draft on 27 June 1977.

General Review comments were resolved on 22 November 1977 by letters from new Project Group Chairman, Larry O'Sickey (Parker Hannifin) and by phone on 11 January 1978. The NFPA Technical Board granted approval to ballot on 18 January 1978, and the NFPA Technical Staff prepared the Ballot Draft on 8 February 1978.

The ballot closed on 24 March 1978 with one negative vote. This negative was discussed at the 19 April 1978 meeting of the Fittings and Conductors Section and was resolved by several editorial clarifications and recognition that the section on mechanical properties will require additional study and, ultimately, may result in further amplification in subsequent editions of the document.

The NFPA Technical Board recommended final approval at it 4 May 1978 meeting and the NFPA Board of Directors approved T3.15.1M R2 as an NFPA Recommended Standard at its 7 June 1978 meeting.

Project Group Members who developed this standard:

O'Sickey, Larry
Section Chairman and Project Chairman
Parker Hannifin

Dodson, Richard †
Project Chairman (1974 - 1977)
Unarco - Leavitt

Campbell, Robert
Section Vice Chairman
L & L Manufacturing

Schmidt, Adam
Section Secretary
The Weatherhead Co.

---

† Deceased
* Retired
** Company affiliation changed.

Saloum, Edward
Technical Auditor
Snap - Tite, Inc.

Luecke, John R.**
Director of National Technical Services
National Fluid Power Association

On 28 March 1979, the revision to B93.4M (NFPA/T3.15.1M R2-1978), was submitted to ANSI Committee B93 for ballot. Negative comments were received and subsequent changes were made to the document. On 20 January 1980, the changes were submitted to ANSI Committte B93. All of the additions and corrections were approved except for one negative comment received through Standards Action which was concerned with the use of metric units. ANSI Committee B93 met on 9 December 1980 to clarify the metric unit usage and the negative comment was withdrawn. ANSI B93.4M was approved by ANSI's Board of Standards Review on 11 February 1981.

The membership roster for the Standards Committee B93 at the time of ballot for revision was:

Melvin E. Long
Chairman

Jack McPherson
Vice Chairman

William Toth
Co-Secretary

Dixie Prevost
Co-Secretary

American Society of Agricultural Engineers
Ed Fletcher

American Society for Engineering Education
William R. Smith

American Society for Lubrication Engineers
M. M. Gurgo

American Society of Mechanical Engineers
Robert Hildebrandt
Thomas R. Curran (alternate)
Frank Yeaple (alternate)

Compressed Air & Gas Institute
David E. Bonn
John Addington (alternate)

Construction Industry Manufacturers Association
Glenn Stewart

Fluid Controls Institute
H. H. Kaemmer
Eric Bianchi (alternate)

Fluid Power Distributors Association
Thomas Neff

Fluid Power Society
Marsh Allen
Edward C. Briggs
Ray Fiedler
Robert W. Hanpeter
Anton Hehn
Richard Read
Allen Tucker

Fluid Sealing Association
Ronald Prachel
John Scannell (alternate)

Industrial Truck Association
C. D. Gibson

Instrument Society of America
Aaron I. Kutz

Joint Industry Council
Robert Muhl

Material Handling Institute
Jack C. McPherson
Willard Chichester (alternate)

Motor Vehicle Manufacturing Association
Jim Phillipson

National Fluid Power Association
James L. Fisher, Jr
Walter Forster
Z. J. Lansky
Melvin Long
Otto Maha *
John H. Pippenger *
A.O. Roberts

National Machine Tool Builders Association
J. Deam

Power Crane and Shovel Association
(to be named)

Rubber Manufacturers Association
William A. Hertel
John R. Loder
E.J. McCarthy (alternate)

Society of Automotive Engineers
Eugene Falendysz
William Hertel
Henry Schultz
D. B. Shore
W. L. Snyder
David Prevallet
Robert W. White

Society of Manufacturing Engineers
Kevin Miller

US Department of Defense
Henry Schaefer
William Coyne (alternate)

Individual Member
E. C. Fitch, Jr.
Jack Johnson

mm

ANSI/B93.4

Hydraulic fluid power - Line tubing - Electric resistance welded, mandrel drawn

0 INTRODUCTION

In hydraulic fluid power systems power is transmitted and controlled thru a liquid under pressure within an enclosed circuit. Hydraulic line tubing conducts liquid within a hydraulic system.

1 SCOPE AND FIELD OF APPLICATION

1.1 This American National Standard includes the following for electric resistance welded, mandrel drawn low carbon steel hydraulic line tubing of a quality suitable for bending and flaring:

— Chemical composition.

— Mechanical properties.

— Dimensional tolerances.

— Quality.

— Testing.

1.2 This American National Standard is intended:

— to establish minimum requirements for electric resistance welded, mandrel drawn hydraulic line tubing.

— to increase reliability of electric resistance welded, mandrel drawn hydraulic line tubing.

2 REFERENCES

ANSI/B93.2M-1971 and Supplement ANSI/B93.2A-1978, **American National Standard, Glossary of Terms for Fluid Power.**

NFPA/T2.10.1M-1978, **National Fluid Power Association Recommended Standard, Metric Units for Fluid Power Applications.**

ASTM/A370-1974, **American Society for Testing and Materials Standard, Methods and Definitions for Mechanical Testing of Steel Products, Supplement II.**

SAE J409c-1972, **Permissible Variations from Specified Ladle Chemical Ranges, and Limits for Steels ..**

SAE J533b, **Flares for Tubing.**

3 TERMS AND DEFINITONS

For definitions of terms used, see ANSI/B93.2M and Supplement ANSI/B93.2A.

4 UNITS OF MEASUREMENT

4.1 Units of measurement are used in accordance with NFPA/T2.10.1M.

4.2 Approximate conversions to Customary US units are given in parentheses following their metric counterparts and are made in accordance with ISO 1000.

5 MATERIAL

5.1 Manufacture

Produce tubing from a single strip of steel made by the open hearth, basic oxygen or electric furnace process.

5.2 Chemical Composition (Ladle Analysis)

5.2.1 Manufacture the steel to conform to the chemical composition requirements prescribed in table 1.

TABLE 1 - Chemical requirements

Element	Percentage by Weight
Carbon	0.18 max
Manganese	0.30 to 0.60
Phosphorus	0.040 max
Sulfer	0.050 max

5.2.2 Make a ladle analysis of each heat to determine the percentages of the elements specified.

5.2.3 Report the chemical composition thus determined to the purchaser (or his representative) when requested.

5.2.4 Do not use a check analysis unless misapplication is apparent. Check analysis tolerance in accordance with SAE J409c, table 3.

NOTE: Rimmed or capped steels used for tubing are not uniform.

5.3 Mechanical Properties

Manufacture tubes to conform to mechanical properties requirements prescribed in table 2.

TABLE 2 - Mechanical properties

Property	Criterion
Tensile Strength (1)	310 MPa min (45,000 psi)
Yield Strength (2)	172 MPa min (25,000 psi)
Elongation in 50.8 mm (2.000 in)	35% min (3)
Rockwell Harness	B65 max (4)

NOTES:

1. See ASTM/A370, Section S6 - Tension Test.
2. See ASTM/A370, Section S9 - Elongation Test.
3. For tubes with O.D. of 9.5 mm (0.375 in) or less, or wall thickness of 0.9 mm (0.035 in) or less, minimum elongation of 25% is permitted.
4. The hardness requirement does not apply to tubes with less than 1.65 mm (0.065 in) wall thickness. Such tubes shall meet all other mechanical properties and all mechanical tests of this recommended standard.

6 DIMENSIONAL TOLERANCES

6.1 Apply tolerances shown in table 3 when tubing is specified by outside diameter (O.D.) and inside diameter (I.D.)

Table 3 - O.D. and I.D. Tolerances

Nominal Outside Diameter	O.D. Tolerance	I.D. Tolerance
Up to 9.5 mm (0.375 in), inclusive	± 0.05 mm (0.002 in)	± 0.12 mm (0.005 in)
Over 9.5 mm (0.375 in) to 15.9 mm (0.625 in), inclusive	± 0.06 mm (0.0025 in)	± 0.06 mm (0.0025 in)
Over 15.9 mm (0.625 in) to 50.8 mm (2.000 in), inclusive	± 0.08 mm (0.003 in)	± 0.08 mm (0.003 in)
Over 50.8 mm (2.000 in), to 63.5 mm (2.500 in), inclusive	± 0.10 mm (0.004 in)	± 0.10 mm (0.004 in)
Over 63.5 mm (2.500 in) to 76.2 mm (3.000 in), inclusive	± 0.12 mm (0.005 in)	± 0.12 mm (0.005 in)
Over 76.2 mm (3.00 in) to 101.6 mm (4.000 in), inclusive	± 0.15 mm (0.006 in)	± 0.15 mm (0.006 in)

NOTE: Permit an additional ovality tolerance of ± 1/2% of the mean outside diameter or inside diameter when the wall thickness is less than 3% of the O.D.

6.2 Apply the tolerances in table 3 for the specified diameters when tubing is specified by the outside diameter (or the inside diameter) and the nominal wall thickness, allowing variation of wall thickness of ± 10% for tubes 9.5 mm (0.375 in) in over, and ± 15% for tubes under 9.5 mm (0.375 in) O.D.

7 QUALITY

7.1 Manufacture

7.1.1 Normalize tubing at a temperature above the upper critical temperature and follow by a cold working operation.

7.1.2 Ensure that the cold working provides a minimum reduction of area of 15%, of which at least 8% is reduction of wall thickness.

7.1.3 Anneal the tubing after cold working in such a manner that the resultant product meets this recommended standard.

7.1.4 Treat tubing, which has been pickled to remove scale, to eliminate pickle brittleness as necessary.

7.2 Workmanship

7.2.1 Straighten finished tubing to a reasonable degree.

7.2.2 Smooth and remove burrs from ends

7.2.3 Prevent dimensional indication of welding flash.

7.2.4 Prevent injurious imperfections to tubing.

NOTE: Surface discontinuities on the outside such as handling marks, staightening marks, light die marks, or shallow pits are not to be considered injurious provided that the discontinuities are within the tolerance specified for diameter and wall thickness, and are not detrimental to the function of the tube. It is not required to remove such surface discontinuities.

7.2.5 Manufacture inside surface free of any discontinuities injurious to the sealing of a flare.

7.3 Flaring Ability

7.3.1 Manufacture tubing capable of withstanding a standard single flare and the flared joint, in accordance with figure 3, table 2 SAE J533b, capable of withstanding a hydrostatic test at a material stress of 138 MPa (20 000 psi) or at a pressure of 34.5 MPa, 345 bar (5 000 psi) whichever is less, based on the applicable minimum wall.

7.3.2 Ensure that there is no evidence of cracking in the flare area.

7.4 Cleanliness

7.4.1 Manufacture the inside of the tubing commercially bright, clean and free from grease, drawing compounds, oxide scale, carbon deposits and any other contamination that cannot be readily removed with an alkaline cleaner normally used in manufacturing plants.

7.4.2 Manufacture the outside of the tubing commercially bright.

7.5 Protective Coatings

7.5.1 Protect the inside and outside fo the tubing by a coating of clean oil to protect these surfaces against corrosion during shipment and normal storage periods.

7.5.2 Remove the corrosion preventative after extended storage periods with an alkaline cleaning solution normally used in manufacturing plants.

8 TESTING

8.1 Flattening Test

8.1.1 Obtain a section of tubing at least 76.2 mm (3.00 in) in length taken from **every 457 metres (1 500 feet) or less of finished tubing.**

8.1.2 Place the weld at a point 90 degrees from the direction of applied force.

8.1.3 Flatten the length of tubing between parallel plates to three times the wall thickness of the tube.

8.1.4 Reject any tubing with cracks or flaws revealed by this test.

NOTE: Superficial ruptures resulting from minor surface discontinuities are not to be considered cause for rejection.

8.2 Reverse Flattening Test

8.2.1 Obtain a section of tubing at least 101.6 mm (4.00 in) in length taken from every 457 metres (1 500 feet) or less of finished tubing.

8.2.2 Split the length of tubing longitudinally 90 degrees on each side of the weld, and open and flatten it.

8.2.3 Ensure that there is no evidence of cracks, metal flaking or lack of weld penetration.

8.2.4 Reject any lengths of tubing having overlaps in the weld resulting from flash removal.

8.3 Expansion Test

8.3.1 Obtain a section of tubing taken from every 457 metres (1500 feet) or less of finished tubing.

8.3.2 Expand the outside diameter of the tubing 25% over a hardened tapered plug having a slope of 25.4 mm (1.00 in) in 254 mm (10.00 in).

8.3.3 Ensure that there is no evidence of cracking or flaws in the flare area.

8.4 Pressure or Electrical Test

8.4.1 Test each tube in one of the following manners:

8.4.1.1 Hydrostatically at a pressure causing a material stress of 138 MPa (20 000 psi) but not exceeding a hydraulic pressure of 34.5 MPa, 345 bar (5 000 psi) whichever is less, based on the applicable minimum wall as determined in the following formula:

Hydrostatic test Pressure:

$$P = \frac{2ST}{D}$$

Where: P = Hydrostatic test pressure, megapascal bar (psi) 34.5 MPa or 345 bar max (5 000 psi)

S = Material stress = 138 MPa (20 000 psi)

T = Minimum wall thickness, mm (in)

D = Nominal outside diameter, mm (in)

8.4.1.2 Electrically by use of a non-destructive electrical or electronic test capable of detecting defects that would prevent the tubing from passing the hydrostatic pressure proof test.

8.4.2 Calculate the average wall, in the case of tubing being ordered to O.D. and I.D. dimensions, by substracting the mean I.D. from the mean O.D. and dividing by two.

8.4.3 Reject a tube if it leaks during the hydrostatic test or does not meet the nondestructive electrical test.

8.5 Hardness Test

8.5.1 Make one Rockwell B Hardness Test on a tubing specimen from each 1 524 metres (5 000 feet) of tubing or fraction thereof, with a minimum of two tests per production lot.

NOTE: It is not required to make a hardness test on tubes less than 1.65 (0.065 in) in wall thickness.

8.5.2 Make the hardness test on the inside of a specimen cut from the tube.

NOTE: If the inside diameter of the tube is less than 6.4 mm (0.250 in), the hardness test may be made on the outside of the tube.

8.6 Tensile Test

8.6.1 Make one tension test on a tubing specimen from each 1 524 metres (5 000 feet) of tubing or fraction thereof, with a minimum of two tests per production run.

8.6.2 Determine the yield strength corresponding to a permanent offset of 0.2% of the gage length of the specimen, or to a total extension of 0.5% of the gage length under load.

8.6.3 Allow a retest if the percentage of elongation of any test specimen is less than that specified and any part of the fracture is more than 19 mm (0.750 in) from the center of the gage length, as indicated by scribe marks on the specimen before testing.

8.7 Method of Mechanical Testing

8.7.1 Make specimens and mechanical tests for hardness and tensile requirements in accordance with ASTM/A370. In case of conflict, the requirements of this recommended standard prevail over those of ASTM/A370.

8.7.2 Test the specimens at room temperature, 15.5 - 32.2° C (60 - 90°F).

8.8 Test Specimens

8.8.1 Smooth the ends and remove flaws and burrs from specimens for mechnical tests.

8.8.2 Discard any test specimen which shows a flaw or defective marking, and substitute with another specimen.

8.9 Retests

If the result of the mechanical tests of any production lot does not conform to the requirements specified, make retests on additional tubes of double the original number from the same production lot, requiring each to conform to the requirements specified.

9 SHIPPING INFORMATION

Mark the following information on a tag securely attached to the bundle or box in which the tubes are shipped:

9.1   Name or brand of manufacturer.

9.2   Identification statement in accordance with section 10.

9.3   The letters ERW - hydraulic line to ANSI/B93.4M - size and thickness.

10   IDENTIFICATION STATEMENT

Use the following statement in catalogs and sales literature when electing to comply with this voluntary standard:

"Specification for electric resistance welded, mandrel drawn hydraulic line tubing conform to American National Standard, ANSI/B93.4M-1981, **Hydraulic fluid power - Line tubing - Electric resistance welded, mandrel drawn.**"

ANSI/B93.4

APPENDIX

to ANSI/B93.4M-1981

2 REFERENCES

ISO 5598[1], Fluid power systems and components - Vocabulary.

ISO 1000, SI units and recommendations for the use of their multiples and of certain other units.

NFPA/T3.8.11, A Bibliography of Fluid Power Tube Fittings and Conductors.

---

[1] At present at the stage of draft.

ANSI/B93.11M-1981

AN INDUSTRY STANDARD FOR FLUID POWER

American National Standard

Hydraulic fluid power -

Line tubing -

Seamless low carbon steel

Approved as an ANSI Standard
11 February 1981

---

published by
NATIONAL FLUID POWER ASSOCIATION, INC.

3333 N. Mayfair Road / Milwaukee, WI 53222 / 414-778-3344 / TLX 26898

ANSI/B93.11

FOREWORD

This Foreword is not part of American National Standard, Hydraulic fluid power - Line tubing - Seamless low carbon steel, ANSI/B93.11M-1981.

Low carbon steel pressure tubing has been used extensively for hydraulic lines in fluid power systems. In 1963, NFPA became aware of the need to update existing hydraulic line tubing specifications, and assigned the project to a Project Group of the Conductors Section, T3.15. The Project Group developed a document which was approved by the Board of Directors on 12 November 1967 as NFPA Recommended Standard T3.15.2-1967. The document was then introduced into American National Standards Institute, which approved it as ANSI Standard B93.11-1969.

On 6 March 1974, in accordance with ANSI requirements, NFPA Section T3.15 initiated a five-year review of B93.11-1969. After discussion, the Section recommended that the standard be revised.

On 5 June 1974, Section T3.15 merged with Section T3.8, Fittings, to form an expanded Section T3.8, called the Fittings and Conductors Section. On 16 October 1974, T3.8 reviewed B93.11-1969 and appointed Richard Dodson (Unarco-Leavitt) as Chairman of the Project Group to undertake the revision of the document.

In revising the existing standard, the Project Group sought to bring it into conformity with a similar standard, SAE J524b-1972, as well as to specify the number of tests required for hydraulic line tubing.

Based on Chairman Dodson's work and on document T3.15.1 R1, Headquarters prepared the Final Working Draft on the revision, called T3.15.2 R1-19xx, on 24 June 1977. The NFPA Technical staff prepared the General Review Draft on 27 June 1977.

General Review comments were resolved on 22 November 1977 by letters from new Project Group Chairman Larry O'Sickey (Parker Hannifin) and by phone on 11 January 1978. The NFPA Technical Board granted approval to ballot on 18 January 1978, and the NFPA Technical Staff prepared the Ballot Draft on 8 February 1978.

The ballot closed on 24 March 1978 with one negative vote. This negative was discussed at the 19 April 1978 meeting of the Fittings and Conductors Section and was resolved by several editorial clarifications and recognition that the section on mechanical properties will require additional study and ultimately, may result in further amplification in subsequent editions of the document.

The NFPA Technical Board recommended final approval at its 4 May 1978 meeting and the NFPA Board of Directors approved T3.15.2 R1 as an NFPA Recommended Standard at its 7 June 1978 meeting.

Project Group members who developed this standard:

Larry O'Sickey
Section Chairman and Project Chairman
Parker Hannifn

Richard Dodson †
Project Chairman (1974-1977)
Unarco-Leavitt

Robert Campbell
Section Vice Chairman
L & L Manufacturing

Adam Schmidt
Section Secretary
The Weatherhead Co.

---

† Deceased
* Retired
** Company affiliation changed

Edward Saloum
Technical Auditor
Snap-Tite, Inc.

John Luecke **
Director of National Technical Services
National Fluid Power Association

On 28 March 1979, the revision to B93.11 was submitted to ANSI Committee B93 for ballot. Negative comments were received and subsequent changes were made to the document. On 20 January 1980, the changes were submitted to ANSI Committee B93. All of the additions and corrections were approved except for one negative comment received through Standards Action which was concerned with the use of metric units. ANSI Committee B93 met on 9 December 1980 to clarify the metric units usage and the negative comment was withdrawn. ANSI/B93.11M-1981 was approved by ANSI Board of Standards Review on 11 February 1981.

The membership roster for the Standards Committee B93 at the time of ballot for revision was:

Melvin E. Long
Chairman

Jack McPherson
Vice Chairman

William Toth
Co-Secretary

Dixie Prevost
Co-Secretary

American Society of Agricultural Engineers
Ed Fletcher

American Society for Engineering Education
William R. Smith

American Society for Lubrication Engineers
M. M. Gurgo

American Society of Mechanical Engineers
Robert Hildebrandt
Thomas R. Curran (alternate)
Frank Yeaple (alternate)

Compressed Air & Gas Institute
David E. Bonn
John Addington (alternate)

Construction Industry Manufacturers Association
Glenn Stuart

Fluid Controls Institute
H. H. Kaemmer
Eric Bianchi (alternate)

Fluid Power Distributors Association
Thomas Neff

Fluid Power Society
Marsh Allen
Edward C. Briggs
Ray Fiedler
Robert W. Hanpeter
Anton Hehn
Richard Read
Allen Tucker

Fluid Sealing Association
Ronald Prachel
John Scannell (alternate)

Industrial Truck Association
C. D. Gibson

Instrument Society of America
Aaron I. Kutz

Joint Industry Council
Robert Muhl

Material Handling Institute
Jack C. McPherson
Willard Chichester (alternate)

Motor Vehicle Manufacturing Association
Jim Phillipson

National Fluid Power Association
James L. Fisher, Jr.
Walter Forster
Z. J. Lansky
Melvin Long
Otto Maha *
John H. Pippenger *
A. O. Roberts

National Machine Tool Builders Association
J. Deam

Power Crane and Shovel Association
(to be named)

Rubber Manufacturers Association
William A. Hertel
John R. Loder
E. J. McCarthy (alternate)

Society of Automotive Engineers
William Hertel
Eugene Falendysz
Henry Schultz
D. B. Shore
W. L. Snyder
David Prevallet
Robert W. White

Society of Manufacturing Engineers
Kevin Miller

US Department of Defense
Henry Schaefer
William Coyne (alternate)

Individual Member
E. C. Fitch, Jr.
Jack Johnson

mm

---

† Deceased
* Retired
** Company affiliation changed

ANSI/B93.11

Hydraulic fluid power - Line tubing - Seamless low carbon steel

0 INTRODUCTION

In hydraulic fluid power systems, power is transmitted and controlled thru a liquid under pressure within an enclosed circuit. Hydraulic line tubing conducts liquid within a hydraulic system.

1 SCOPE AND FIELD OF APPLICATION

1.1 This American National Standard includes the following for seamless low carbon steel hydraulic line tubing of a quality suitable for bending and flaring:

— Chemical composition.

— Mechanical properties.

— Dimensional tolerances.

— Quality.

— Testing

1.2 This American National Standard is intended:

— to establish minimum requirements for seamless low carbon steel hydraulic line tubing.

— to increase reliability of seamless low carbon steel hydraulic line tubing.

2 REFERENCES

ANSI/B93.2M-1971, and Supplement ANSI/B93.2A-1978, Amercan National Standard, Glossary of Terms for Fluid Power.

NFPA/T2.10.1M-1978, National Fluid Power Association Recommended Standard, Metric Units for Fluid Power Applications.

ASTM/A370-1974, American Society for Testing and Materials Standard, Methods and Definitions for Mechanical Testing of Steel Products, Supplement II.

SAE J409c-1972, Permissible Variations from Specified Ladle Chemical Ranges, and Limits for Steels.

SAE J533b, Flares for Tubing.

3 TERMS AND DEFINITIONS

For definitions of terms used, see ANSI/B93.2M and Supplement ANSI/B93.2A.

4 UNITS OF MEASUREMENT

4.1 Units of measurement are used in accordance with NFPA/T2.10.1M.

4.2 Approximate conversions to Custmary US units are shown in parentheses after their metric counterparts and are made in accordance with ISO 1000.

5 MATERIAL

5.1 Manufacture

5.1.1 Produce tubing by the seamless process using steel made by the open hearth, basic oxygen or electric furnace process.

5.2 Chemical Composition (Ladle Analysis)

5.2.1 Manufacture the steel to conform to the chemical composition requirements prescribed in table 1.

TABLE 1 - **Chemical requirements**

Element	Percentage by Weight
Carbon	0.18 max
Manganese	0.30 to 0.60
Phosphorus	0.040 max
Sulfur	0.050 max

5.2.2 Make a ladle analysis of each heat to determine the percentages of the elements specified.

5.2.3 Report the chemical composition thus determined to the purchaser (or his representative) when requested.

5.2.4 Do not use a check analysis unless misapplication is apparent. Check analysis tolerance in accordance with SAE J409-c, table 1.

NOTE: Rimmed or capped steels used for tubing are not uniform.

5.3 Mechanical Properties

5.3.1 Manufacture tubes to conform to mechanical properties requirements prescribed in table 2.

TABLE 2 - Mechanical properties

Property	Criterion
Tensile Strength (1)	310 MPa min (45 000 psi)
Yield Strength (2)	172 MPa min (25 000 psi)
Elongation in 50.8mm (2.000 in)	35% min (3)
Rockwell Hardness	B65 max (4)

NOTES:

1. See ASTM/A370, Section S6 - Tension Test.
2. See ASTM/A370, Section S9 - Elongation Test.
3. For tubes with O.D. of 9.5 mm (0.375 in) or less, or wall thickness of 0.9 mm (0.375 in) or less, minimum elongation of 25% is permitted.
4. The hardness requirement does not apply to tubes with less than 1.65 mm (0.065 in) wall thickness. Such tubes shall meet all other mechanical properties and all mechanical tests of this recommended standard.

6 DIMENSIONAL TOLERANCES

6.1 Apply tolerances shown in table 3 when tubing is specified by outside diameter (O.D.) and inside diameter (I.D.).

TABLE 3 - O.D. and I.D. tolerances

Nominal Outside Diameter	O.D. Tolerance	I.D. Tolerance
Up to 12.7 mm (0.500 in), inclusive	± 0.0762 mm (0.003 in)	-----
Over 12.7 mm (0.500 in) to 38.1 mm (1.50 in), inclusive	± 0.127 mm (0.005 in)	± 0.127 mm (0.005 in)
Over 38.1 mm (1.50 in) to 88.9 mm (3.50 in), inclusive	± 0.254 mm (0.010 in)	± 0.254 mm (0.010 in)

NOTE: Permit an additional ovality tolerance of ± 1/2% of the mean outside diameter or inside diameter when the wall thickness is less than 3% of the O.D.

6.2 Apply the tolerances in table 3 for the specifed diameters when tubing is specified by the outside diameter (or the inside diameter) and the nominal wall thickness, allowing variation of wall thickness of ± 10% for tubes 9.5 mm (0.375 in) and over, and ± 15% for tubes under 9.5 mm (0.375 in) O.D.

7 QUALITY

7.1 Manufacture

7.1.1 Cold work the tubing.

7.1.2 Anneal the tubing after cold working in such a manner that the resultant product meets this recommended standard.

7.2 Workmanship

7.2.1 Staighten finished tubing within 0.254 mm per 304 mm (0.010 inches per foot) of length.

7.2.2 Smooth and remove burrs from ends.

7.2.3 Prevent injurious imperfections to tubing.

NOTE: Surface discontinuities on the outside such as handling marks, staightening marks, light die marks, or shallow pits are not to be considered injurious provided that the discontinuities are within the tolerance specified for diameter and wall thickness, and are not detrimental to the function of the tube. It is not required to remove such surface discontinuities.

7.2.4 Manufacture inside surface free of any discontinuities injurious to the sealing of a flare.

7.3 Flaring Ability

7.3.1 Manufacture tubing capable of withstanding a standard single flare and the flared joint in accordance with figure 3, table 2-SAE J533b, capable of withstanding a hydrostatic test at a material stress of 138 MPa (20 000 psi) or at a pressure of 34.5 MPa, 345 bar (5 000 psi) whichever is less, based on the applicable minimum wall.

7.3.2 Ensure that there is no evidence of cracking.

7.4 Cleanliness

7.4.1 Manufacture the inside of the tubing commercially bright, clean and free from grease, drawing compounds, oxide scale, carbon deposits and any other contamination that cannot be readily removed with an alkaline cleaner normally used in manufacturing plants.

7.4.2 Manufacture the outside of the tubing commercially bright.

7.5 Protective Coatings

7.5.1 Protect the inside and outside of the tubing by a coating of clean oil to protect these surfaces against corrosion during shipment and normal storage periods.

7.5.2 Remove the corrosion preventative after extended storage periods with an alkaline cleaning solution normally used in manufacturing plants.

8 TESTING

8.1 Flattening Test

8.1.1 Obtain a section of tubing at least 76.2 mm (3.00 in) in length taken from every 457 metres (1 500 feet) or less of finished tubing.

8.1.2 Flatten the length of tubing between parallel plates to three times the wall thickness of the tube.

8.1.3 Reject any tubing with cracks or flaws revealed by this test.

NOTE: Superficial ruptures resulting from minor surface discontinuities are not to be considered cause for rejection.

8.2 Expansion Test

8.2.1 Obtain a section of tubing taken from every 457 metres (1 500 feet) or less of finished tubing.

8.2.2 Expand one end of the section over a polished tapered mandrel having an angle of $60°$, until the actual average inside diameter is increased by 30%.

8.2.3 Ensure that there is no evidence of cracking or flaws.

8.3 Pressure or Electrical Test

8.3.1 Test each tube in one of the following manners:

8.3.1.1 Hydrostatically at a pressure causing a material stress of 138 MPa (20 000 psi) but not exceeding a hydraulic pressure of 34.5 MPa, 345 bar (5 000 psi) whichever is less, based on the applicable minimum wall as determined in the following formula:

Hydrostatic test pressure

$$P = \frac{2ST}{D}$$

where: P = Hydrostatic test pressure, megapascal bar (psi) 34.5 MPa or 345 bar max (5 000 psi)

S = Material stress = 138 MPa (20 000 psi)

T = Minimum wall thickness, mm (in)

D = Nominal outside diameter, mm (in)

8.3.1.2 Electrically by use of a non-destructive electrical or electronic test capable of detecting defects that would prevent the tubing from passing the hydrostatic pressure proof test.

8.3.2 Calculate the average wall, in the case of tubing being ordered to O.D. and I.D. dimensions, by substracting the mean I.D. from the mean O.D. and dividing by two.

8.3.3 Reject a tube if it leaks during the hydrostatic test or does not meet the nondestructive electrical test.

8.4 Hardness Test

8.4.1 Make one Rockwell B Hardness Test on a tubing specimen from each 1 524 metres (5 000 feet) of tubing or fraction thereof, with a minimum of two tests per production lot.

NOTE: It is not required to make a hardness test on tubes less than 1.65 mm (0.065 in) wall thickness.

8.4.2 Make the hardness test on the inside of a specimen cut from the tube.

NOTE: If the inside diameter of the tube is less than 6.4 mm (0.250 in), the hardness test may be made on the outside of the tube.

8.5 Tensile Test

8.5.1 Make one tension test on a tubing specimen from each 1 524 metres (5 000 feet) of tubing or fraction thereof, with a minimum of two tests per production run.

8.5.2 Determine the yield strength corresponding to a permanent offset of 0.2% of the gage length of the specimen, or to a total extension of 0.5% of the gage length under load.

8.5.3 Allow a retest if the percentage of elongation of any test specimen is less than that specified and any part of the fracture is more than 19 mm (0.750 in) from the center of the gage length, as indicated by scribe marks on the specimen before testing.

8.6 Method of Mechanical Testing

8.6.1 Make specimens and mechanical test for hardness and tensile requirements in accordance with ASTM/A370. In case of conflict, the requirements of this recommended standard prevail over those of ASTM/A370.

8.6.2 Test the specimens at room temperature, $15.5 - 32.2°C$ ($60 - 90°F$).

8.7 Test Specimens

8.7.1 Smooth the ends and remove flaws and burrs from specimens for mechanical test.

8.7.2 Discard any test specimen which shows a flaw or defective marking, and substitute by another specimen.

8.8 Retests

If the result of the mechanical test of any production lot does not conform to the requirements specified, make retests on additional tubes of double the original number from the same production lot, requiring each to conform to the requirements specified.

9 SHIPPING INFORMATION

Mark the following information on a tag securely attached to the bundle or box in which the tubes are shipped:

ANSI/B93.11

9.1 Name or brand of manufacturer.

9.2 Identification statement in accordance with section 10.

9.3 The letters SMLS - hydraulic line to ANSI/B93.11M-1981 - size and wall thickness.

10 INDENTIFICATION STATEMENT

Use the following statement in catalogs and sales literature when electing to comply with this voluntary standard:

"Specifications for seamless low carbon steel line tubing conform to American National Standard, ANSI/B93.11M-1981, **Hydraulic fluid power - Line tubing - Seamless low carbon steel.**"

APPENDIX

to ANSI/B93.11M-1981

2 REFERENCES

ISO 5598[1], Fluid power systems and components - Vocabulary.

ISO 1000, SI units and recommendations for the use of their multiples and of certain other units.

NFPA/T3.8.11, A Bibliography of Fluid Power Tube Fittings and Conductors.

---

[1] At present at the stage of draft.

ANSI/B93.42M-1977 (R1983)

AN INDUSTRY STANDARD FOR FLUID POWER

American National Standard

Method for Testing Hydraulic

Fluid Power Quick Action Couplings

Approved as an ANSI Standard
18 July 1977

---

published by
NATIONAL FLUID POWER ASSOCIATION, INC.
3333 N. Mayfair Road / Milwaukee, WI 53222 / 414-778-3344 / TLX 26898

ANSI/B93.42

FOREWORD

This Foreword is not part of American National Standard Method for Testing Hydraulic Fluid Power Quick Disconnect Couplings, ANSI/B93.42-1977.

The original project, NFPA/T3.20.2-1970, was initiated in 1966 and completed in 1970. The approved NFPA Recommended Standard was then submitted to ANSI/B93 for promulgation as an American National Standard. As a result of comments received during the B93 ballot, a revision of the NFPA document was initiated.

Although the revised document was submitted in sections, with the intent of approving individual sections and then combining the sections into a larger document, this approach was abandoned in the interests of economy and efficiency. On 28 March 1975, the NFPA Technical Staff edited the document for General Review. The individual sections were rejoined and the document was brought into conformity with the NFPA Style Guide. It was then sent to Project Group Chairman Ken Koch for his review.

On 16 April 1975, the Quick Disconnent Section reviewed T3.20.2 R1-19xx, and directed that it be circulated to Project Group members for their comments prior to General Review. This was done by NFPA Headquarters. All comments were resolved and consensus was reached on 28 May 1975. On 25 June 1975, the Final Working Draft, incorporating the comments received during the Project Group's review, was sent to Section members for their comments prior to General Review. No further comments were received from the Section.

The NFPA Technical Staff prepared the General Review Draft of T3.20.2 R1 on 18 July 1975. One comment was received during General Review; it was resolved by the Project Group Chairman on 6 September 1975.

Approval to Ballot was granted by the Technical Board on 5 November 1975. The NFPA Technical Staff prepared the Ballot Draft on 13 November 1975. No negative ballots were received on the proposed recommended standard.

On 4 February 1976, the Technical Board voted to recommend to the Board of Directors that this document be approved as an NFPA Recommended Standard. On 25 February 1976, the Board of Directors granted final approval to NFPA/T3.20.2 R1-1976.

Members of the NFPA Project Group which developed this standard are listed on page 4.

On 27 June 1976, the NFPA Standard was submitted to ANSI Standards Committee B93 for promulgation as an ANSI Standard. Ballot was concluded on 5 October 1976, and resulted in one negative vote and two affirmative comments. Changes made to the standard caused the negative vote and affirmative comments to be withdrawn on 21 April 1977.

The document was then submitted for ANSI Public Review on 19 May 1977. Approval by ANSI Board of Standards Review was granted on 18 July 1977.

The membership roster for ANSI Standards Committee B93 at the time of ballot is listed on pages 4 and 5.

ANSI/B93.42

Members of the NFPA Project Group responsible for the development of this standard included:

Koch, Kenneth	Project Chairman	Componetrol/ Bruning Co.
Saloum, Edward	Section Chairman	Snap-Tite, Inc.
Blickley, George	Technical Auditor	SOR, Inc.
Luecke, John R.	Director of National Technical Services	National Fluid Power Association

Hammond, H.	The Hansen Mfg. Co.
Karcher, T.	The Hansen Mfg. Co.
Lamb, T.	Parker Hannifin
Rogers, R.	Aeroquip Corporation

On 27 July 1977, ANSI Standards Committee B93 was composed of the following: Melvin E. Long, Chairman; John R. Luecke, Co-Secretary; William Toth, Co-Secretary.

AMERICAN SOCIETY OF AGRICULTURAL ENGINEERS
    E. H. Fletcher
AMERICAN SOCIETY OF LUBRICATION ENGINEERS
    M. M. Gurgo
AMERICAN SOCIETY OF MECHANICAL ENGINEERS
    R. Hildebrandt
    T. R. Curran (alternate)
    F. Yeaple (alternate)
AMERICAN SOCIETY FOR TESTING AND MATERIALS
    J. D. Lykins
    J. J. Rothrock (alternate)
CONSTRUCTION INDUSTRY MANUFACTURERS ASSOCIATION
    G. Stewart
    H. T. Larmore (alternate)
FLUID CONTROLS INSTITUTE
    H. H. Kaemmer
    F. Bianchi (alternate)

FLUID POWER SOCIETY
    M. Allen
    R. D. Burgess, Sr.
    W. H. Dreher
    R. W. Hanpeter
    A. Hehn
    R. Read
    A. Tucker
FLUID SEALING ASSOCIATION
    R. Prachel
    J. Scannell (alternate)
INDUSTRIAL TRUCK ASSOCIATION
    C. D. Gibson
INSTRUMENT SOCIETY OF AMERICA
    A. I. Kutz
JOINT INDUSTRY COUNCIL
    R. Muhl
MATERIAL HANDLING INSTITUTE
    J. C. McPherson
    W. Chichester (alternate)
MOTOR VEHICLE MANUFACTURERS ASSOCIATION
    J. Phillipson

NATIONAL FLUID POWER
ASSOCIATION
- J. L. Fisher, Jr.
- W. Forster
- Z. J. Lansky
- M. E. Long
- O. J. Maha
- J. J. Pippenger
- A. O. Roberts

NATIONAL MACHINE TOOL
BUILDERS ASSOCIATION
- E. Loeffler

POWER CRANE AND SHOVEL
ASSOCIATION
- (to be named)

RUBBER MANUFACTURERS
ASSOCIATION
- W. J. Atwell
- N. J. Cyphers (alternate)
- E. J. McCarthy (alternate)

SOCIETY OF AUTOMOTIVE
ENGINEERS
- W. A. Hertel
- E. L. Falendysz
- H. Schultz
- D. B. Shore
- W. L. Snyder
- D. Prevallet
- R. W. White

SOCIETY OF MANUFACTURING
ENGINEERS
- J. Wood

U.S. COAST GUARD
- G. Hicks

U.S. DEPARTMENT OF
DEFENSE
- H. Y. Smith
- P. Hopler (alternate)

INDIVIDUAL MEMBERS
- E. C. Fitch, Jr.
- J. Johnson

ANSI/B93.42

REFERENCES

1. American National Standard Glossary of Terms for Fluid Power, ANSI/B93.2-1971, and Supplements thereto. (DP/5598)

2. SI units and recommendations for the use of their multiples and of certain other units, ISO 1000-1973.

3. Society of Automotive Engineers Recommended Practice for the Determination of Hydraulic Pressure Drop, SAE/ARP 24B-1968.

4. National Fluid Power Association Recommended Standard for Verifying the Fatigue and Static Pressure Ratings of the Pressure Containing Envelope of a Metal Fluid Power Component, NFPA/T2.6.1-1974, and Supplement No. 5 for Quick Disconnect Couplings (NFPA/T3.20.8-1975).

5. National Fluid Power Association Recommended Standard Glossary for Fluid Power Quick Disconnect Couplings, NFPA/T3.20.1-1973.

ANSI/B93.42

METHOD FOR TESTING HYDRAULIC FLUID POWER QUICK DISCONNECT COUPLINGS

INTRODUCTION

In hydraulic fluid power systems, power is transmitted and controlled thru a liquid under pressure within an enclosed circuit. Quick disconnect couplings are used to quickly join or separate fluid conducting lines without the use of tools or special devices.

1. SCOPE

    1.1 To include identical coupling halves, male and female coupling halves, and coupling assemblies used in hydraulic fluid power systems.

    1.2 To include couplings with and without fluid sealing means, when uncoupled.

    1.3 To include only quick disconnect couplings that are connected and disconnected by a linear or rotational motion, or both.

2. PURPOSE

    2.1 To further the understanding and use of quick disconnect couplings.

    2.2 To promote personnel safety.

    2.3 To provide uniform test conditions and procedures.

    2.4 To promote component reliability.

    2.5 To facilitate accurate communications.

    2.6 To provide a basis for component selection and application.

3. TERMS AND DEFINITIONS

    For definition of terms used, see Reference Nos. 1 and 5.

4. UNITS OF MEASUREMENT

    4.1 Units of measurement are used in accordance with ISO/1000-1973 (Reference No. 2).

    4.2 Approximate conversions to Customary U.S. units are given in parentheses following their ISO 1000 counterparts.

ANSI/B93.42

5. **SELECTION AND EXAMINATION OF TEST SAMPLES**

    5.1 Select coupling assemblies representing a production lot in all respects pertaining to design, material, surface treatment, process, etc.

    5.2 Apply adequate identification to the coupling when more than one sample is used.

    5.3 Permanently mark each coupling or coupling half in a manner suitable for identification with the required test procedures and reports.

6. **TEST EQUIPMENT**

    6.1 Use test equipment described in Figures 1 thru 8.

    6.2 Conduct tests at an ambient temperature of 20°C (68°F) to 35°C (95°F).

    6.3 Use either MIL-H-5606 or MIL-H-6083 hydraulic oil as the test fluid.

7. **TEST CONDITIONS ACCURACY**

    TABLE 1 - Test conditions accuracy (unless otherwise specified)

Test Condition	ISO 1000 Unit	Customary US Unit	Maintain Within ($\pm$) of Actual Gage Value
Force	N	lb(f)	3%
Frequency	Hz	cps	10%
Flow rate	L/min	USGPM	3%
Length	mm	in	3%
Mass	kg	lb(m)	3%
Pressure (above atm)	bar	psi	3%
Pressure (below atm)	bar, abs	in Hg, abs	2%
Temperature	°C	°F	2.8°C (5°F)
Time	min	min	3%
Torque	N·m	lb(f) - in	3%

8. PROCEDURES FOR CONNECT FORCE TEST

　　8.1　Lubricate the coupling interfaces with the test fluid.

　　8.2　Insert the coupling in a test fixture.

　　8.3　Maintain the specified internal pressure.

　　8.4　Apply a linear force or torque, or both, to the coupling half until complete connection occurs. During this operation, the locking mechanism may be operated, manually if necessary to permit normal coupling of the halves.

　　8.5　Measure the connecting force or torque, or both.

　　8.6　Repeat the test a total of five times on the test coupling.

　　8.7　Average the results of the five tests to determine the connect force or torque.

　　8.8　Report the average connect force or torque on the test report per Section 21.

　　8.9　Report any conditions of damage or malfunction on the test report per Section 21.

9. PROCEDURES FOR DISCONNECT FORCE TEST

　　9.1　Lubricate the coupling interfaces with the test fluid.

　　9.2　Insert the coupling in a test fixture.

　　9.3　Maintain the specified internal pressure of flow conditions, or both.

　　9.4　Apply a linear force or torque, or both, to the coupling until disconnection occurs.

　　9.5　Measure the disconnect force or torque, or both.

　　9.6　Repeat the test for five disconnections.

　　9.7　Average the test results of the five tests to determine the disconnect force or torque, or both.

　　9.8　Report the average results in the test report per Section 21.

　　9.9　Report any condition of damage or malfunction in the test report per Section 21.

ANSI/B93.42

10. PROCEDURES FOR LEAKAGE TEST

    10.1   Low Pressure, Coupled (Open valve conditions on valved couplings).

        10.1.1   Insert coupling assembly in a test setup as illustrated in Figure 1.

        10.1.2   Apply side-load as illustrated in Figure 2.

        10.1.3   Record the drop in column height during a 30 minute minimum test period.

        10.1.4   Compute the leakage rate in millilitres per hour.

        10.1.5   Report the leakage rate in the test report per Section 21.

    10.2   Low Pressure, Uncoupled (Valved only)

        10.2.1   Install each coupling half in a test setup as illustrated in Figure 3.

        10.2.2   Record the drop in column height during a 30 minute test period.

        10.2.3   Compute the leakage rate in millilitres per hour.

        10.2.4   Report the leakage rate in the test report per Section 21.

    10.3   Maximum Operating Pressure, Coupled (Open valve conditions on valved couplings)

        10.3.1   Pressurize the coupling assembly with test fluid at maximum operating pressure.

        10.3.2   Observe for leakage for 30 minute test period while maintaining maximum operating pressure.

        10.3.3   Collect and measure the leakage in a graduate.

        10.3.4   Compute the leakage rate in millilitres per hour.

        10.3.5   Report the leakage rate in the test report per Section 21.

    10.4   Maximum Operating Pressure, Uncoupled (Valved only)

        10.4.1   Pressurize both coupling halves with the test fluid at maximum operating pressure.

        10.4.2   Observe leakage for a 30 minute test period while maintaining maximum operating pressure.

        10.4.3   Collect and measure the leakage for each coupling half in a graduate.

ANSI/B93.42

    10.4.4   Compute the leakage rate in millilitres per hour.

    10.4.5   Report the leakage rate in the test report per Section 21.

11. PROCEDURES FOR EXTREME TEMPERATURE TEST

    11.1   Maximum Operating Temperature, Coupled (Open valve condition on valved couplings)

        11.1.1   Fill the coupling assembly with test fluid and subject the assembly to the maximum operating temperature for at least six hours.

        NOTE: Coupling must be internally vented to atmosphere during temperature adjustment.

        11.1.2   Allow the coupling to cool to ambient temperature and determine the leakage rate in accordance with Section 10.

        11.1.3   Report the leakage rate in the test report per section 21.

    11.2   Maximum Operating Temperature, Uncoupled (Valved only)

        11.2.1   Fill the coupling halves with test fluid and subject the halves to the maximum operating temperature for at least six hours.

        11.2.2   Allow the coupling half to cool to ambient temperature and manually actuate the valves five times to separate valve seal from sealing surface.

        11.2.3   Determine the leakage rate in accordance with Section 10.

        11.2.4   Report the leakage rate in the test report per Section 21.

    11.3   Minimum Operating Temperature, Coupled (Open valve condition on valved couplings)

        11.3.1   Fill the coupling assembly with test fluid and subject the assembly to the minimum operating temperature for at least four hours.

        11.3.2   Determine the leakage rate at minimum operating temperature in accordance with Section 10.

        11.3.3   Report the leakage rate in the test report per Section 21.

    11.4   Minimum Operating Temperature, Uncoupled (Valved only)

        11.4.1   Fill the coupling halves with the test fluid and subject each half to the minimum operating temperature for at least four hours.

ANSI/B93.42

11.4.2 Manually actuate valves five times to separate valve seal from sealing surface.

11.4.3 Determine the leakage rate at minimum operating temperature in accordance with Section 10.

11.4.4 Report the leakage rate in the test report per Section 21.

12. PROCEDURES FOR PRESSURE IMPLUSE TEST

12.1 Because the pressure impluse test is a terminal test, do not use the test coupling in any further testing.

12.2 Coupled Test

12.2.1 Connect the coupling assembly to a test setup capable of producing pressure impluses as illustrated in Figure 4.

12.2.2 Adjust the test setup to a pressure-time curve within the shaded area of Figure 4.

12.2.3 Conduct 10,000 test cycles at a maximum rate of 1.66 Hz (100 cycles per minute).

12.2.4 Uncouple and couple the test coupling assembly 10 times.

12.2.5 Repeat clauses 12.2.3 and 12.2.4 nine times for 100,000 total test cycles.

12.2.6 Record any evidence of binding or malfunction.

12.2.7 Determine the leakage rate in accordance with Section 10.

12.2.8 Report the leakage rate in the test report per Section 21.

12.3 Uncoupled Test (Valved only)

12.3.1 Connect each coupling half to a test setup capable of producing pressure impluses as illustrated in Figure 4.

12.3.2 Adjust the test setup to a pressure-time cycle within the shaded area of Figure 4.

12.3.3 Conduct 100,000 cycles.

12.3.4 Determine the leakage rate in accordance with Section 10.

12.3.5 Report the leakage rate in the test report per Section 21.

ANSI/B93.42

13. ## PROCEDURES FOR ENDURANCE TEST

    13.1 Because the endurance test is a terminal test, do not use the test coupling for any further testing.

    13.2 Connect the coupling assembly to a pressure source capable of providing 1 bar (15 psi) internal pressure.

        13.2.1 Compressed air may be used.

        13.2.2 Record the type of test media used.

    13.3 Couple and uncouple the assembly 25,000 times for coupling sizes up to and including 1/2 inch, and 5,000 times for coupling sizes greater than 1/2 inch.

    13.4 Do not exceed a coupling rate of 600 connect/disconnects per hour.

    13.5 Record any evidence of binding or malfunction.

    13.6 Determine the leakage rate in accordance with Section 10.

    13.7 Report the leakage rate in the test report per Section 21.

14. ## PROCEDURES FOR PRESSURE DROP TEST

    14.1 This test is applicable only to couplings with tube fittings. A pressure drop test for couplings with pipe thread fittings will be covered in a subsequent document.

    14.2 Install the coupling assembly in a test setup as illustrated in Figure 5.

    14.3 Maintain fluid temperature at $38^{\circ}C \pm 11^{\circ}C$ ($100^{\circ}F \pm 20^{\circ}F$) and record the temperature.

    14.4 Select at least six flow rates from 25% to 150% of the rated flow, including 100% of rated flow.

    NOTE: If rated flow is not specified, use 6.1 m/s (20 feet per second) in an equivalent tube size as rated flow.

    14.5 Determine and record the pressure drop of the coupling assembly in male half-to-female half and female half-to-male half directions at the selected flow rates.

    14.6 Remove the coupling assembly from the test setup and install an equal length of tubing and, if used, fittings of the corresponding size.

    14.7 Determine and record the pressure drop at the same flow rates used in clause 14.5.

14.8 Subtract the values obtained in clause 14.7 from those obtained in clause 14.5. The difference is the net pressure drop of the coupling assembly.

14.9 Plot the net pressure drop on graph paper.

14.10 Draw a smooth line thru the points for each flow direction.

NOTE: Full logarithmic graph paper is recommended to permit using a straight line. The line may not pass through points, but should represent a common value between the points.

14.11 If the pressure drop values at any one flow rate, in one direction of flow through the coupling, differ by less than 10 percent of the pressure drop in the other direction of flow through the coupling, use the higher of the two values.

15. PROCEDURES FOR VACUUM TEST

15.1 This procedure is recommended only for vacuum tests when leakage rate measurement is not required.

15.2 Coupled Test

15.2.1 Install the coupling in a test setup as illustrated in Figure 6.

15.2.2 Apply side-load to the coupling assembly as illustrated in Figure 2.

15.2.3 Start vacuum pump and pull a vacuum to a specified value.

15.2.4 Allow 10 minutes for stabilization at the specified vacuum.

15.2.5 Observe the vacuum gage for loss of vacuum during a specified time period.

15.2.6 Report any loss of vacuum during the time period.

15.3 Uncoupled Test (Valved only)

15.3.1 Install the coupling half in a test setup as illustrated in Figure 6.

15.3.2 Start the vacuum pump and pull a vacuum to a specified value.

15.3.3 Allow 10 minutes for stabilization at the specified vacuum.

15.3.4 Observe the vacuum gage for loss of vacuum during a specified time period.

15.3.5 Report any loss of vacuum during the time period.

ANSI/B93.42

16. PROCEDURES FOR AIR INCLUSION TEST

    16.1   Install the coupling assembly in a test setup as illustrated in Figure 7.

    16.2   Record the fluid level of the graduate cylinder.

    16.3   Uncouple and couple the coupling assembly.

    16.4   After each uncouple/couple cycle, bump or tap the coupling assembly to clear all air bubbles from the interior of the assembly.

    16.5   Repeat clause 16.3 and 16.4 until the fluid displaced by air in the graduate cylinder exceeds 10 minor divisions on the graduate scale.

    16.6   Adjust the graduate vertically so the fluid levels are coincident.

    16.7   Record the fluid level of the graduate cylinder.

    16.8   Subtract the value recorded in clause 16.7 from the value recorded in clause 16.2, and divide the difference by the number of uncouple/couple cycles.

    16.9   Report the air inclusion in standard millilitres per uncouple/couple cycle in the test report per Section 21.

17. PROCEDURES FOR SPILLAGE TEST

    17.1   Install the coupling assembly in a test setup as illustrated in Figure 8.

    17.2   Maintain a fluid pressure of 1 bar (15 psi).

    17.3   Record the fluid level of the graduate cylinder.

    17.4   Couple and uncouple the assembly.

    17.5   After each uncouple, dump the spillage from the assembly.

    17.6   After each couple, tap or bump the assembly to clear all air bubbles from the coupling interior.

    17.7   Repeat clauses 17.4, 17.5 and 17.6 until the fluid level of the graduate cylinder has dropped a minimum of 10 minor divisions on the scale.

    17.8   Record the fluid level of the graduate.

    17.9   Subtract the value recorded in clause 17.8 from the value recorded in clause 17.3, and divide the difference by the number of couple/uncouple cycles.

ANSI/B93.42

17.10 Report the spillage in millilitres per couple/uncouple cycle in the test report per Section 21.

NOTE: Use a low viscosity fluid if the viscosity of the standard test fluid prevents the prompt clearing of bubbles. Record the fluid type if a substitute fluid is used.

18. PROCEDURES FOR BURST TEST

   18.1 Provide suitable personnel protection when conducting burst tests.

   18.2 Burst pressure, uncoupled (Valved only)

   18.2.1 Pressurize the coupling halves at a rate not exceeding 1725 bar (25,000 psi) per minute.

   18.2.2 Report the burst pressure in the test report per Section 21.

   18.3 Burst pressure, coupled

   18.3.1 Pressurize the coupling assembly at a rate not exceeding 1725 bar (25,000 psi) per minute.

   18.3.2 Report the burst pressure in the test report per Section 21.

19. PROCEDURES FOR RATED FATIGUE AND STATIC PRESSURE TEST

   19.1 Coupled (Open valve condition on valved couplings)

   Verify the rated fatigue and static pressure by testing in accordance with Reference No. 4.

   19.2 Uncoupled (Valved only)

   Verify the rated fatigue and static pressure by testing in accordance with Reference No. 4.

ANSI/B93.42

20. DATA ACCURACY

TABLE 2 - Data Accuracy

Quantity	ISO 1000 Unit	Customary US Unit	Maintain Within ($\pm$) of Actual Gage Value
Force	N	lb(f)	3%
Pressure	bar	psi	3%
Pressure drop	bar	psi	3%
Temperature	$^{\circ}$C	$^{\circ}$F	2.8°C (5°F)
Torque	N·m	lb(f) - in	3%
Volume (leakage)	mL		1%

ANSI/B93.42

21. DATA PRESENTATION

TABLE 3 - Typical form for presentation of test data

Coupling Manufacturer: _____

Coupling P/N: _____ S/N or Ident.: _____

Date Tested: _____ Test By: _____

Name of Test	Test Results		Remarks
	ISO 1000 Unit	Customary US	
Connect Force			
Force	_____ N	_____ lb(f)	_____ Test Pressure
Torque	_____ N·m	_____ lb(f) - in	
Disconnect Force			
Force	_____ N	_____ lb(f)	_____ Test Pressure
Torque	_____ N·m	_____ lb(f) - in	_____ Flow Rate
Leakage			
Low pressure, coupled	_____ mL/hr		_____ Test Pressure
Low pressure, uncoupled	_____ mL/hr		_____ Test Pressure
Max op pressure, coupled	_____ mL/hr		_____ Test Pressure
Max op pressure, uncoupled	_____ mL/hr		_____ Test Pressure

TABLE 3 continued

TABLE 3 - Typical form for presentation of test data (continued)

Name of Test	Test Results		Remarks
	ISO 1000 Unit	Customary US	
Extreme Temperature			
Max op temp, coupled	____ °C	____ °F	
	____ mL/hr	____ mL/hr	
Max op temp, uncoupled	____ °C	____ °F	
	____ mL/hr	____ mL/hr	
Min op temp, coupled	____ °C	____ °F	
	____ mL/hr	____ mL/hr	
Min op temp, uncoupled	____ °C	____ °F	
	____ mL/hr	____ mL/hr	
Pressure Impluse			
Coupled	____ mL/hr		_____ Test Pressure
Uncoupled	____ mL/hr		_____ Test Pressure
Endurance	____ mL/hr		_____ No. of Cycles
			_____ Media
Pressure Drop	See graph.		_____ Rated Flow
			_____ Fluid Temp.
Vacuum	____ bar, abs	____ in Hg	

TABLE 3 continued

ANSI/B93.42

TABLE 3 - Typical form for presentation of test data (continued)

Name of Test	Test Results		Remarks
	ISO 1000 Unit	Customary US	
Air Inclusion	_____ mL /couple-uncouple cycle		
Spillage	_____ mL /couple uncouple cycle		
Burst			
Uncoupled	_____ bar	_____ psi	
Coupled	_____ bar	_____ psi	

22. SUMMARY OF DESIGNATED INFORMATION

The following designated information is needed when applying this recommended standard to a particular application or use:

22.1　Rated Flow

22.2　Rated Pressure

22.3　Maximum Operating Pressure

22.4　Maximum Operating Temperature

22.5　Minimum Operating Temperature

22.6　Vacuum Test

22.7　Rated Static Pressure

22.8　Rated Fatigue Pressure

23. JUSTIFICATION STATEMENT

These recommended standard test procedures are based on the combined expert experiences of those who have participated in the development and review of the original NFPA Recommended Standard, its review in ANSI/B93, and its subsequent revision as an NFPA Recommended Standard.

24. TEST/PRODUCTION SIMILARITY

Utilize managerial controls necessary to maintain substantial similarity between test and production components or elements.

25. IDENTIFICATION STATEMENT

Use the following statement in catalogs and sales literature when electing to comply with this voluntary standard:

> "Method of obtaining and presenting performance data conforms to American National Standard, ANSI/B93.42-1977."

ANSI/B93.42

26. KEY WORDS

The following Key Words, useful in indexes and information retrieval systems, are suggested for this recommended standard:

fluid power

quick disconnect coupling

testing, air inclusion

testing, burst

testing, connect force

testing, disconnect force

testing, endurance

testing, extreme temperature

testing, leakage

testing, pressure drop

testing, pressure impulse

testing, quick disconnect coupling

testing, spillage

testing, vacuum

ANSI/B93.42

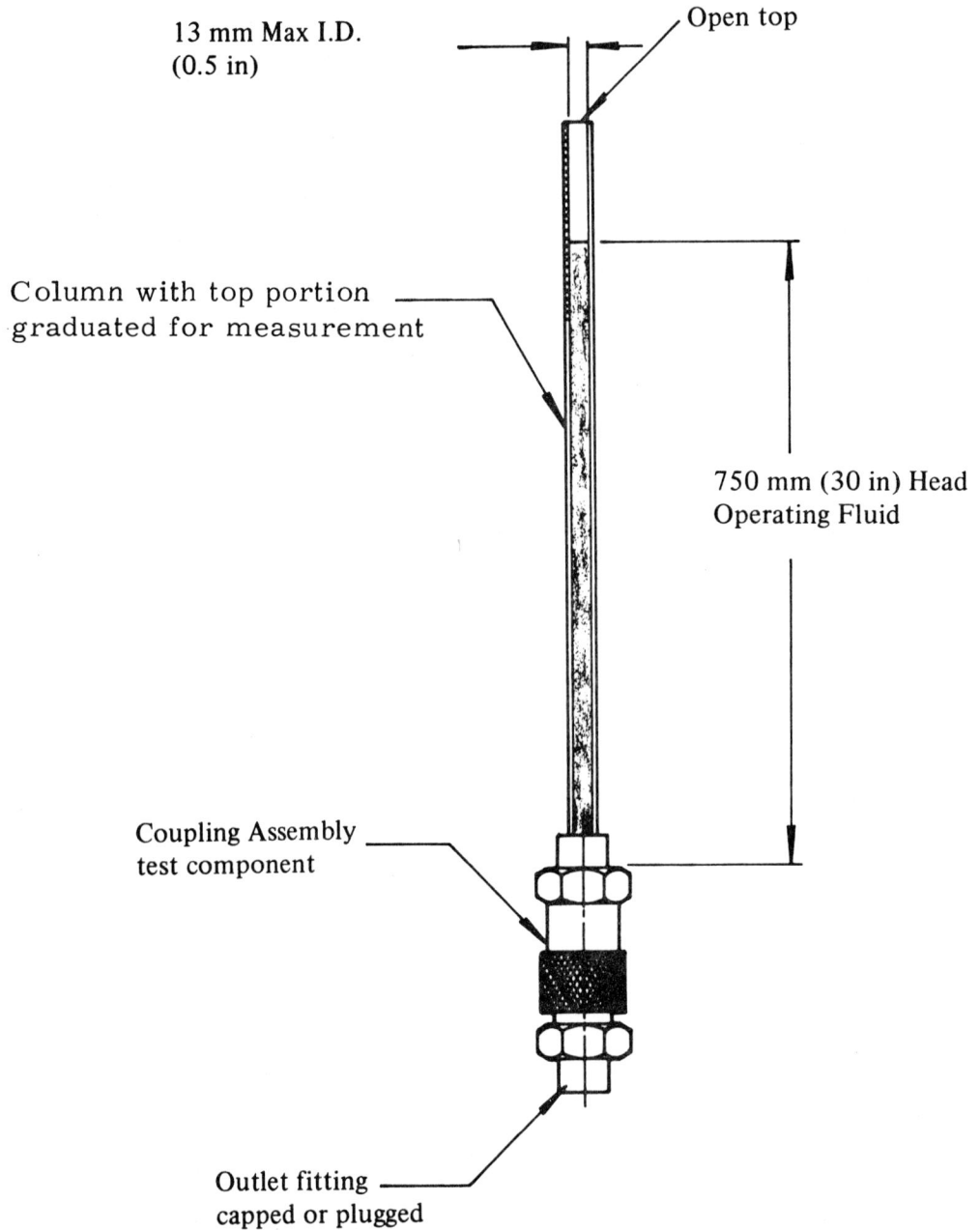

FIGURE 1 - Low pressure leak test setup (Coupled)

- 132 -

ANSI/B93.42

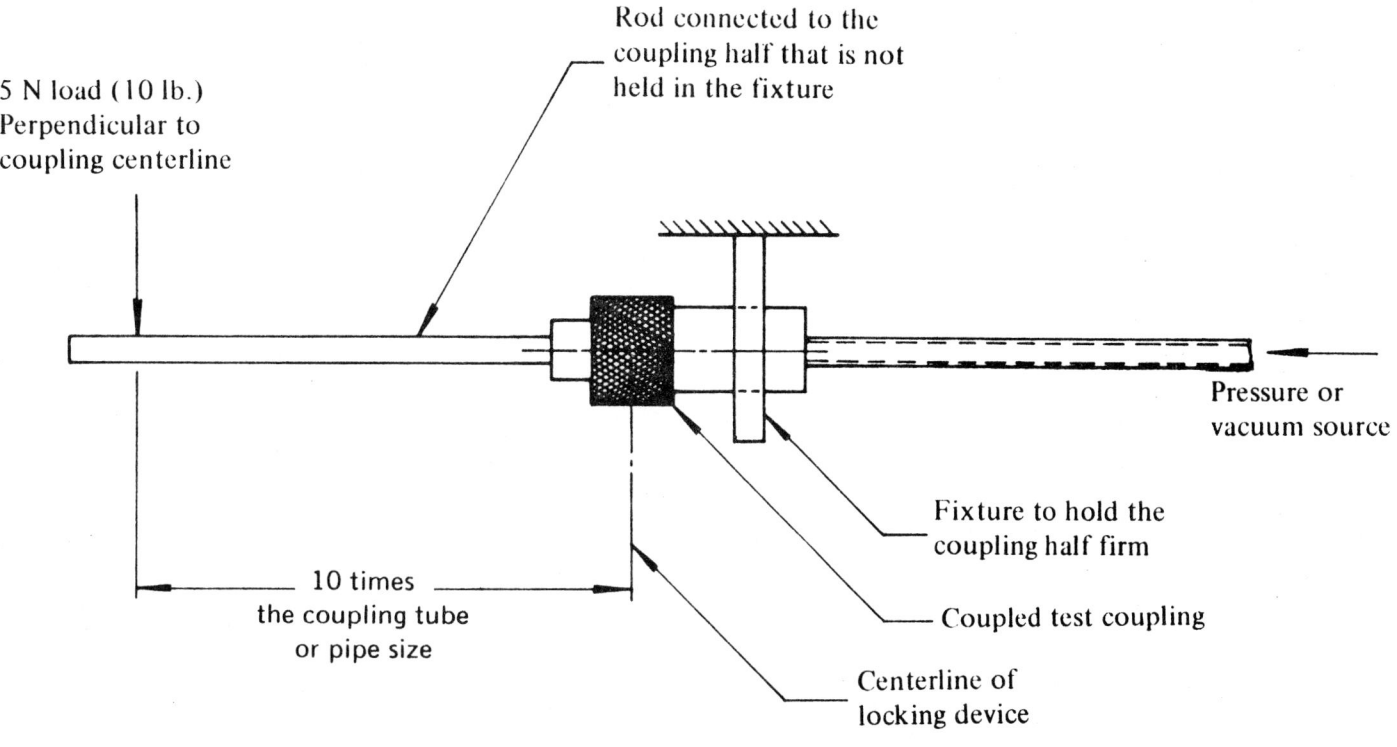

FIGURE 2 - Sideload fixture

ANSI/B93.42

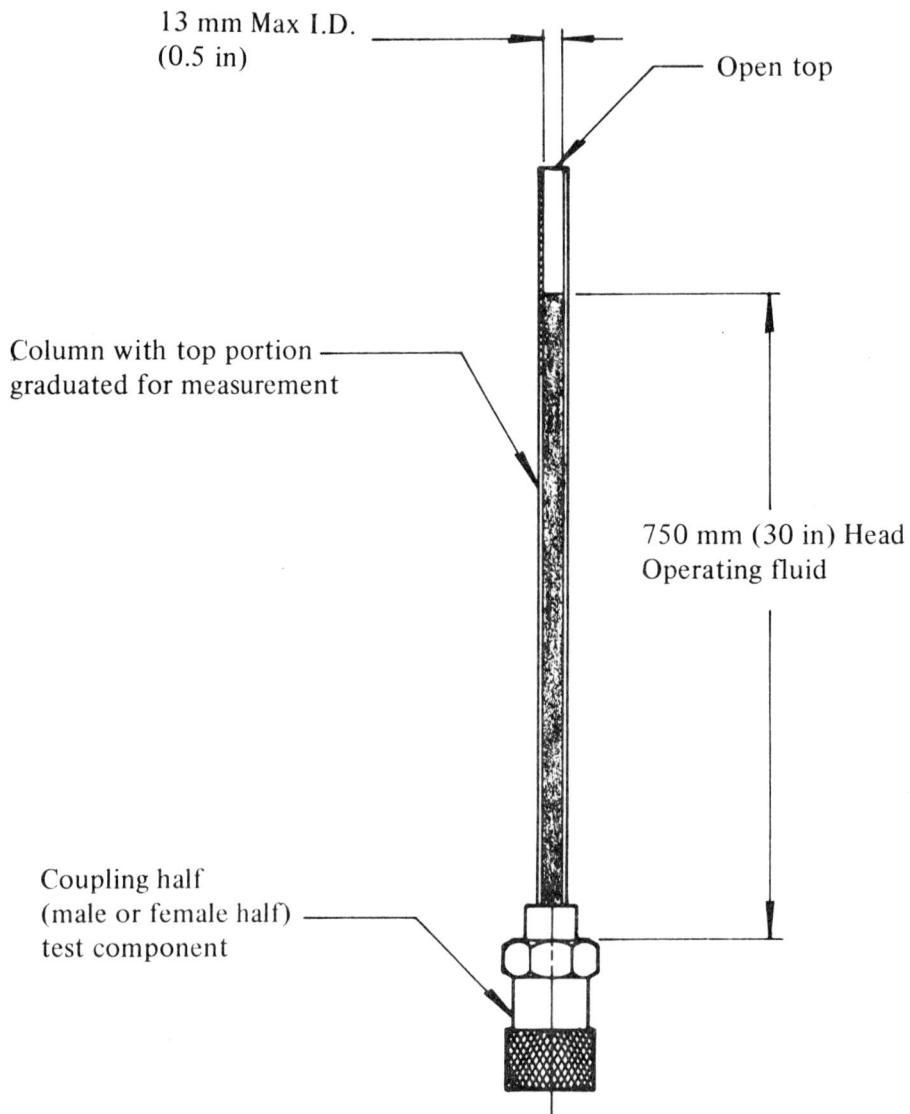

FIGURE 3 - Low pressure leak test setup (Uncoupled half)

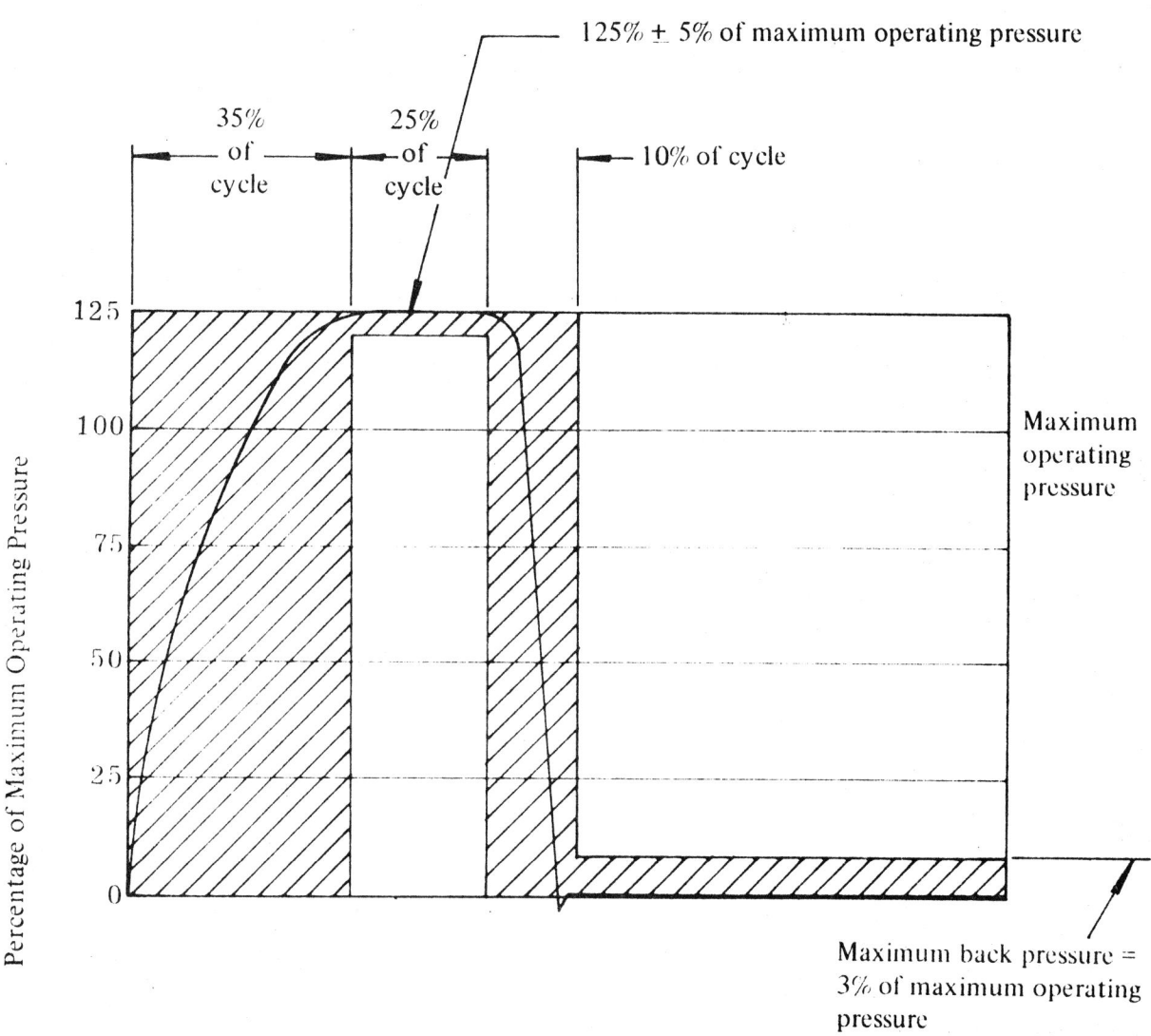

FIGURE 4 - Pressure impluse test

ANSI/B93.42

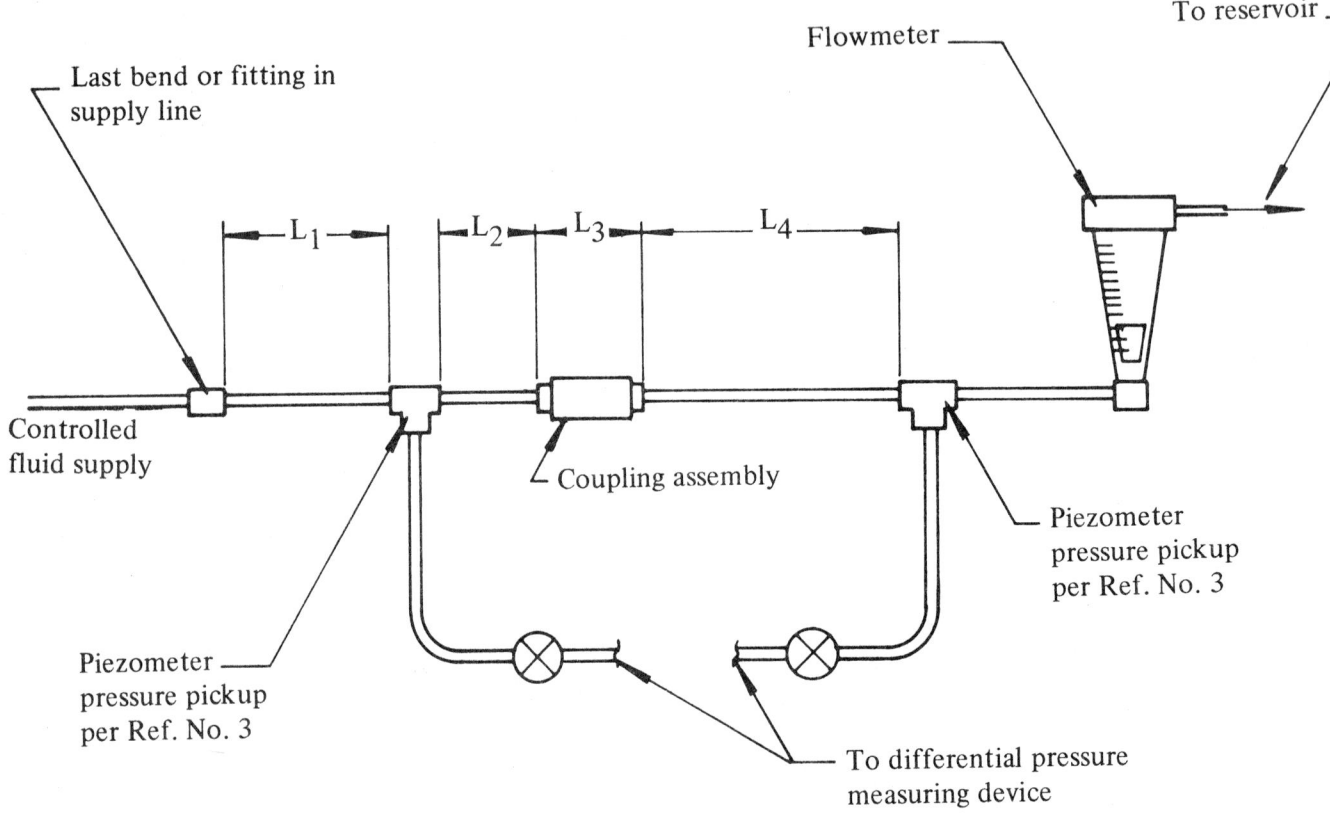

L_1 = Coupling inlet tube or pipe size, 15 diameters minimum length

L_2 = Coupling inlet tube or pipe size, 4 diameters minimum length

L_3 = Coupling assembly plus end fittings

L_4 = Coupling outlet tube or pipe size, 15 diameters minimum length

FIGURE 5 - Pressure drop test

ANSI/B93.42

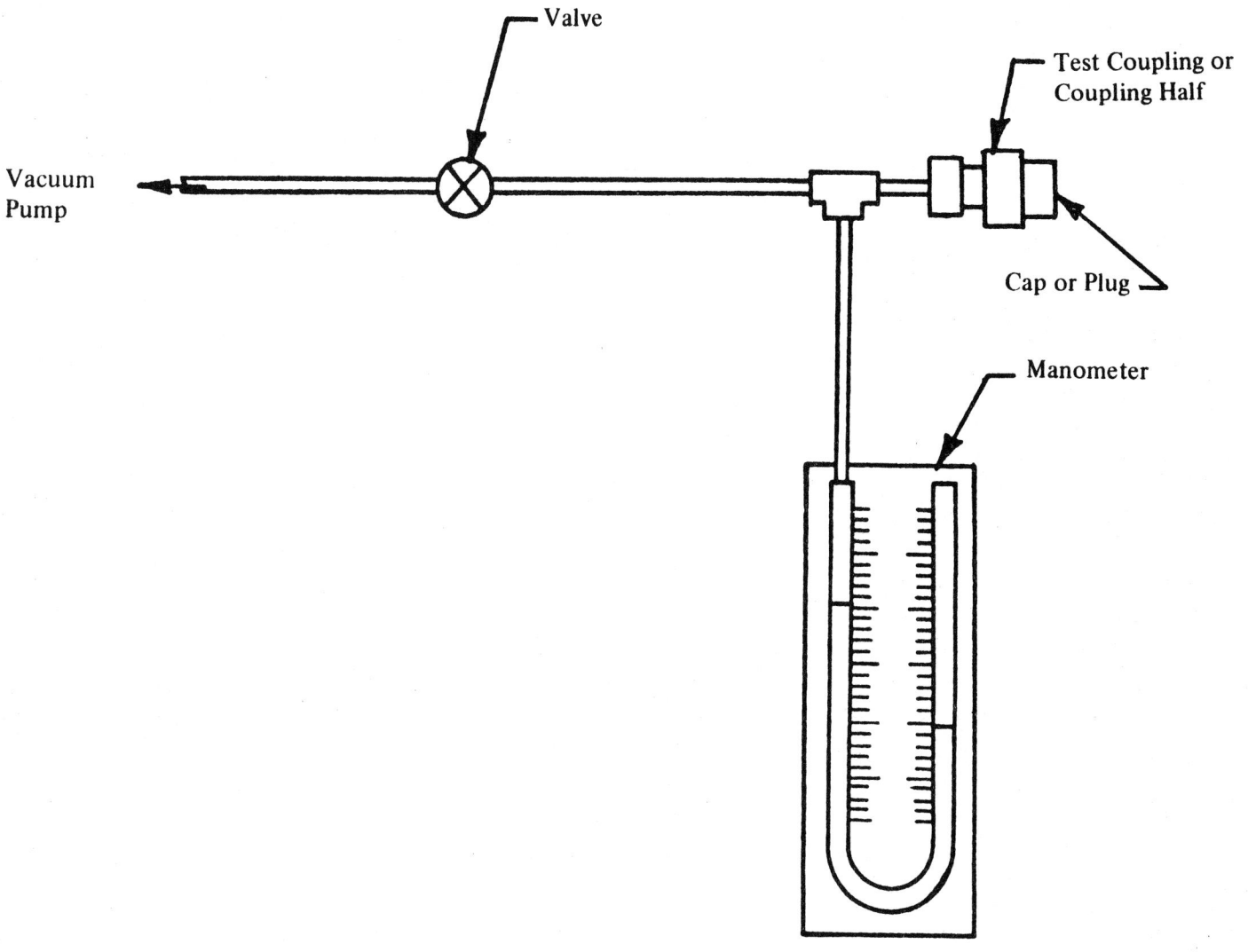

FIGURE 6 - Vacuum leakage test

ANSI/B93.42

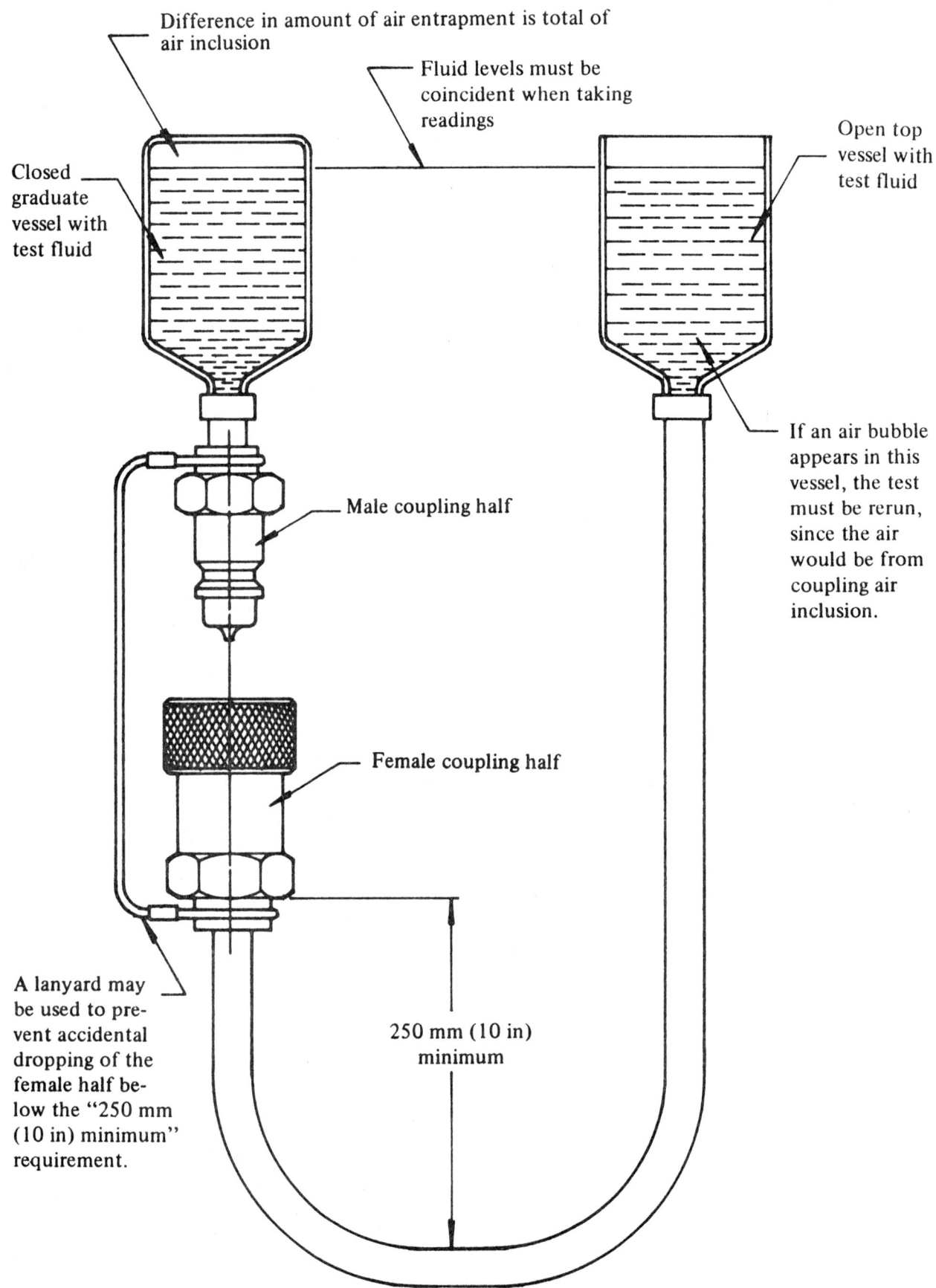

FIGURE 7 - Air inclusion

ANSI/B93.42

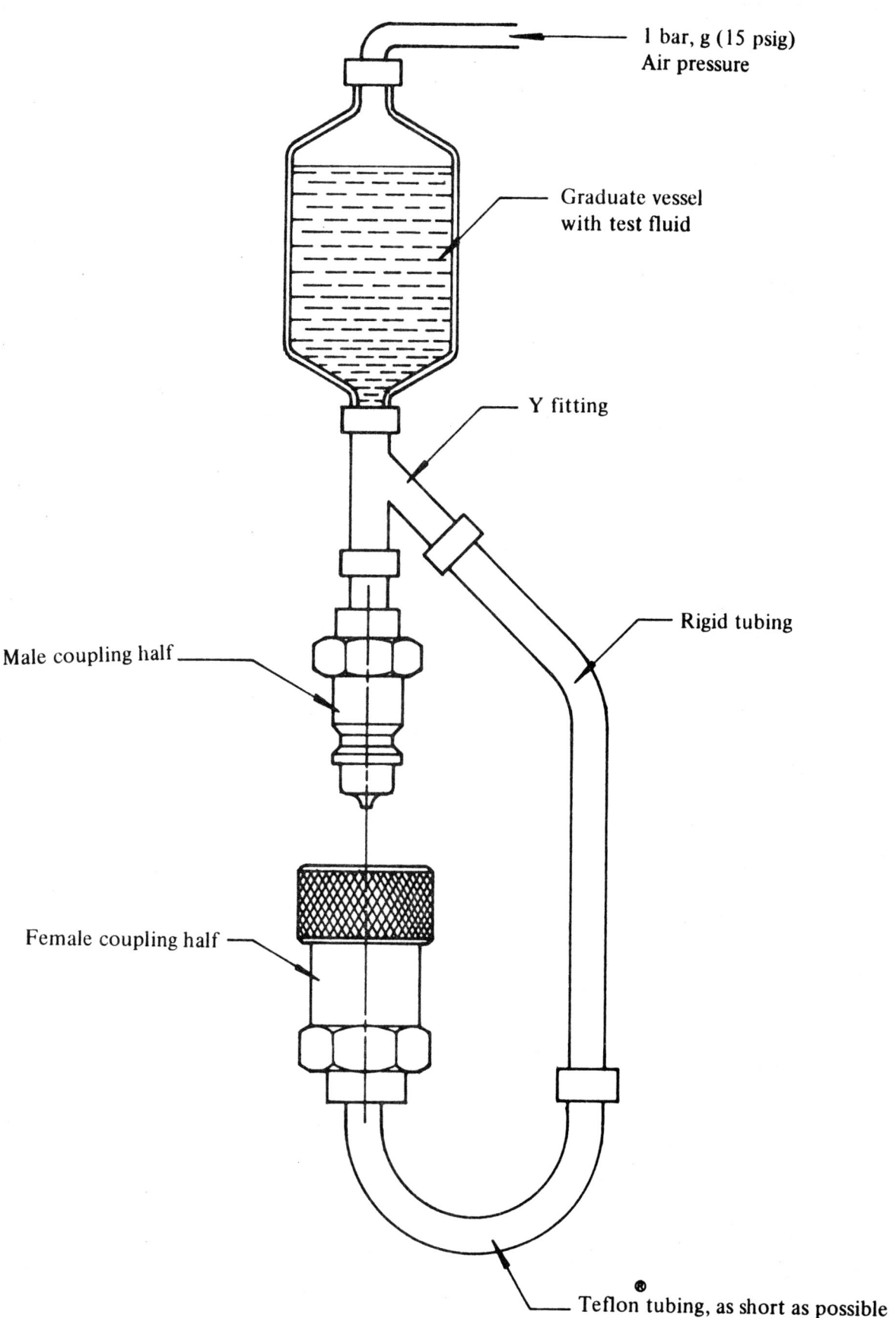

FIGURE 8 - Spillage

NOTES

ANSI/B93.51M-1980

AN INDUSTRY STANDARD FOR FLUID POWER

American National Standard

Pneumatic fluid power -

Quick action couplings -

Test conditions and procedures

Approved as an ANSI Standard
13 June 1980

---

published by
NATIONAL FLUID POWER ASSOCIATION, INC.

3333 N. Mayfair Road / Milwaukee, WI 53222 / 414-778-3344 / TLX 26898

ANSI/B93.51

FOREWORD

This Foreword is not part of American National Standard Pneumatic fluid power - Quick action couplings - Test conditions and procedures, ANSI/B93.51M-1980.

NFPA Recommended Standard T3.20.3 was originally approved by the Board of Directors on 11 November 1973.

The Quick Action Coupling Section reviewed this standard at their 6 March 1974 meeting and agreed that a revision to the document was required to reflect the current state of the art. A TSP was prepared and received Technical Board approval on 5 June 1974. Draft No. 1 was prepared on 6 March 1977 and discussed at the 6 April 1977 Section meeting. Chairman Tom Karcher discussed the comments and recommendations and included them in Draft No. 3 on 1 July 1977. Draft No. 4 was prepared on 1 November 1978. Headquarters prepared the General Review Draft on 13 January 1978. Three comments were received when the document was circulated for General Review. All three of these were resolved through correspondence and were reported to the Quick Action Coupling Section at its 18 April 1978 meeting.

The NFPA Technical Board granted approval to ballot at its 4 May 1978 meeting. NFPA Headquarters prepared the Ballot Draft on 26 May 1978.

The Ballot closed with one negative comment, which was resolved by changing the metric units to agree with NFPA preferred metric units standard. At its 17 February 1979 meeting, the NFPA Technical Board reviewed this proposed standard and voted to recommend to the Board of Directors that this document be approved as an NFPA Recommended Standard.

The NFPA Board of Directors granted final approval to NFPA/T3.20.3 R1 on 3 March 1979.

Project Group Members who developed this standard:

Karcher, Tom
Project Chairman
Hansen Mfg. Co.

Saloum, Ed
Section Chairman
Snap-Tite, Inc.

Koch, Ken
Section Vice Chairman
Gould, Inc./Bruning

Lytle, John*
Technical Auditor
Ross Operating Valve

White, Jim
Director of Technical Services
National Fluid Power Association

Fischer, D.
Aeroquip

Hammond, H.
Hansen Mfg. Co.

Herzan, G.
Parker Hannifin Corp.

Royer, W.
Perfecting Division

---

* Company affiliation has changed.

Favorable ANSI/B93 ballot and Public Review were completed and in May 1980 ANSI/B93.51M was submitted to ANSI Board of Standards Review. Approval was granted on 13 June 1980.

The Membership roster for Standard Committee B93 at the time of approval was comprised of:

Jack McPherson
Chairman
Eaton Corp.

Robert Uhl
Co-Secretary
Society of Automotive Engineers

James C. White
Co-Secretary
National Fluid Power Association

American Society of Agricultural Engineers
Ed Fletcher

American Society for Engineering Education
William R. Smith

American Society of Lubrication Engineers
M. M. Gurgo

American Society of Mechanical Engineers
Robert Hildebrandt
Thomas R. Curran (alternate)
Frank Yeaple (alternate)

Compressed Air and Gas Institute
D. E. Bonn

Construction Industry Manufacturers Association
Glen Stewart

Fluid Controls Institute
Herbert H. Kaemmer
Eric Bianchi (alternate)

Fluid Power Distributors Association
Thomas H. Neff

Fluid Power Society
Marsh Allen
Edward Briggs
R. D. Burgess, Sr.
Ray Fiedler
William Dreher
Robert W. Hanpeter
Anton Hehn
Richard Read
Allen Tucker

Fluid Sealing Association
Ronald Prachel
John Scannell (alternate)

Industrial Truck Association
C. D. Gibson

Instrument Society of America
Aaron I. Kutz

Joint Industry Council
Robert Muhl

Material Handling Institute
Jack McPherson
William Chichester (alternate)

Motor Vehicle Manufacturers
Jim Phillipson

National Fluid Power Association
James L. Fisher, Jr.
Walter Forster
Z. J. Lansky
Otto Maha
John Pippenger
A. O. Roberts

National Machine Tool Builders
John B. Deam

Power Crane and Shovel Association
(to be named)

Rubber Manufacturers Association
William Hertel
John Loder
E. J. McCarthy (alternate)

Society of Automotive Engineers
William Hertel
Eugene Falendysz
Henry Schultz
D. B. Shore
W. L. Snyder
David Prevallet
Robert White

Society of Manufacturing Engineers
Raymond Grisdale

Individual Members
Dr. E. C. Fitch, Jr.
Jack Johnson

cg

mm

ANSI/B93.51

Pneumatic fluid power - Quick action couplings - Test conditions and procedures

0 INTRODUCTION

In pneumatic fluid power systems, power is transmitted and controlled thru a gas under pressure within an enclosed circuit. Quick action couplings are used to quickly join or separate fluid conducting lines without the use of tools or special devices.

1 SCOPE

This National Standard includes:

— identical coupling halves, male or female coupling halves, and coupling assemblies used in pneumatic fluid power systems.

— couplings with and without fluid sealing means, when uncoupled.

— only quick action couplings that are connected and disconnected by a linear or rotary motion.

2 FIELD OF APPLICATION

This National Standard is intended to:

— further the understanding and use of quick action couplings.

— provide uniform test conditions and procedures.

— facilitate accurate communications.

— provide a basis for component selection and application.

3 REFERENCES

ANSI/B93.2M-1971 and Supplement ANSI/B93.2A-1978, **American National Standard, Glossary of Terms for Fluid Power.**

NFPA/T3.20.1-1973, **National Fluid Power Association, Recommended Standard Glossary of Terms for Fluid Power Quick Action Couplings.**

NFPA/T2.10.1M-1978, **National Fluid Power Association, Recommended Standard Metric Units for Fluid Power Applications.**

ISO 554-1976, **Standard atmospheres for conditioning or testing - Standard reference atmosphere - Specifications.**

NFPA/T2.6.1M-1974 and Supplement No. 5 for Quick Action couplings (NFPA/T3.20.8M-1975), **National Fluid Power Association, Recommended Standard for Verifying the Fatigue and Static Pressure Ratings of the Pressure Containing Envelope of a Metal Fluid Power Component.**

4 TERMS AND DEFINITIONS

For definitions of terms used, see ANSI/B93.2M and Supplement ANSI/B93.2A, **American National Standard, Glossary of Terms for Fluid Power.**

5 UNITS OF MEASUREMENT

5.1 Units of measurement are used in accordance with NFPA/T2.10.1M. This document agrees with ISO 1000.

5.2 Conversions to Customary U. S. units are shown in parenthesis after their metric counterparts and are made in accordance with NFPA/T2.10.1M.

5.3 Gas, under standard temperature, humidity, and pressure conditions in accordance with ISO 554.

ANSI/B93.51

6 LETTER SYMBOLS

The following letter symbols apply to this document:

6.1 Metric units

cm^3/s (ANR)	=	cubic centimetres per second
$°C$	=	degrees Celsius
mm	=	millimetre
ρ	=	bar
N	=	newton
dm^3/s (ANR)	=	cubic decimetres per second
s	=	seconds
N·m	=	newton metre

6.2 "Customary US" units

SCIM	=	standard cubic inches per minute
$°F$	=	degrees Farenheit
in	=	inches
in Hg	=	inches of mercury
lb(f)	=	pounds
psi	=	pounds per square inch
SCFM	=	standard cubic feet per minute
min	=	minutes
lb·ft	=	pound feet

7 TEST EQUIPMENT

7.1 Use filter air or nitrogen. [Recommended 40 micrometre (40 micron) or better filtration.]

7.2 For bubble test immersion fluids, use water with an optional corrosion inhibitor.

7.3 Conduct all tests at manufacturer's specified test pressure unless otherwise specified.

7.4 Conduct all tests at 24°C (75°F) and maintain temperature within the limits in table 1.

8 TEST CONDITIONS ACCURACY

Set up and maintain equipment accuracy within the limits in table 1.

TABLE 1 - Test conditions accuracy

Test Condition	Metric Unit	Customary US Unit	Maintain Within (±) of Actual Measured Value
Temperature	°C	°F	5.5°C (10°F)
Leakage Rate	cm^3/s (ANR)	SCIM	2%
Sideload	N	lb	5%
Internal Pressure	bar	psi	2%
Pressure Drop (ΔP)	bar	psi	2%
Flow Rate	dm^3/s (ANR)	SCFM	2%
Vacuum	bar	in Hg	2%
Time	s	min	3%
Torque	N·m	lb·ft, lb·in	3%

9 TEST PROCEDURES

9.1 Selection and examination of test samples.

9.1.1 Select coupling assemblies representing a production lot in all respects pertaining to design, material, surface treatment, process, etc.

9.1.2 Apply adequate identification to the coupling when more than one sample is used.

9.1.3 Permanently mark each coupling or coupling half in a manner suitable for identification with the required test procedures and reports.

9.2 Operational Tests

9.2.1 Disconnect force.

a) Insert the Coupling Assembly in a test fixture similar to that in figure 1 or figure 2.

NOTE: If Coupling Assembly requires rotation of sleeve, see figure 2.

b) Maintain manufacturer's specified test pressure. [Maximum 6.9 bar (100 psi)].

c) Apply force or torque to the locking mechanism until disconnection occurs.

d) Measure the disconnection force or torque.

e) Repeat the test a total of five (5) times.

f) Use the median of the five (5) tests to determine the disconnect force or torque.

g) Record any damage or malfunction.

9.2.2 Connect force.

a) Insert the coupling in a test fixture similar to that in figure 3 and figure 4.

NOTE: If Coupling Assembly requires rotation of sleeve, see figure 4.

b) Maintain the manufacturer's specified internal test pressure. [Maximum 6.9 bar (100 psi).

c) Apply a force or torque to the coupling half until complete connection occurs.

d) Measure the connecting force or torque.

e) Repeat the test a total of five (5) times.

f) Use the median of the five (5) tests to determine the connect force or torque.

g) Record any damage or malfunction.

9.2.3 Leakage test

9.2.3.1 Low pressure, coupled. [Less than 6.9 bar (100 psi)]

a) Install the coupling in a test container as illustrated in figure 5.

b) Apply a side load of 44.5N (10lb) as illustrated in figure 6.

c) Maintain the specified internal test pressure.

d) Place an inverted graduated cylinder filled with the bubble fluid over the couplings so as to collect the escaping gas for a minimum of five (5) minutes.

e) Adjust the cylinder vertically until the level of the fluid within the graduated cylinder is coincident with the fluid level in the test vessel.

f) Record the volume of gas within the graduated cylinder and the time interval.

9.2.3.2 Low pressure, uncoupled (valved only). [Less than 6.9 bar (100 psi)];

a) Install valved coupling halves in a test container as illustrated in figure 5.

b) Maintain the specified internal test pressure for a minimum of five (5) minutes.

c) Measure and record leakage in accordance with 9.2.3.1 (d through f).

9.2.3.3 Maximum operating pressure coupled.

NOTE: Conduct tests in a burst chamber or other suitable location for safety reasons.

a) Install the coupling in a test container as illustrated in figure 5.

b) Maintain the manufacturer's specified test pressure for a minimum of five (5) minutes.

c) Measure and record leakage in accordance with 9.2.3.1 (d through f).

9.2.3.4 Maximum operating pressure, uncoupled (valved only).

NOTE: Conduct tests in a burst chamber or other suitable location for safety reasons.

a) Install valved coupling halves in test container as illustrated in figure 5.

b) Maintain the manufacturer's specified test pressure for a minimum of five (5) minutes.

c) Measure and record leakage in accordance with 9.2.3.1 (d through f).

9.2.4 Extreme temperature test.

9.2.4.1 Maximum operating temperature, coupled.

a) Use a test set-up similar to that in figure 7.

b) Subject coupling to the manufacturer's recommended maximum operating temperature for six (6) hours minimum.

NOTE: Reduce pressure to zero and return temperature to ambient.

c) Determine leakage in accordance with steps under 9.2.3.1 and 9.2.3.3.

d) Couple and uncouple the coupling.

e) Record any evidence of binding or malfunction.

9.2.4.2 Maximum operating temperature uncoupled (valved only).

a) Use a test set-up similar to that in figure 7.

b) Subject valved coupling halves to the manufacturer's recommended maximum operating temperature for six (6) hours minimum.

NOTE: Reduce pressure to zero and return temperature to ambient.

c) Determine leakage in accordance with steps under 9.2.3.2 and 9.2.3.4.

9.2.4.3 Minimum operating temperature, coupled.

a) Use a test set-up similar to that in figure 7.

b) Subject coupling to the manufacturer's recommended minimum operating temperature for six (6) hours minimum.

NOTE: Reduce pressure to zero and return temperature to ambient.

c) Determine leakage in accordance with steps under 9.2.3.1 and 9.2.3.3.

d) Couple and uncouple the couplings.

e) Record any evidence of binding or malfunction.

9.2.4.4 Minimum operating temperature, uncoupled (valved only).

a) Use test set-up similar to that in figure 7.

b) Subject valved coupling halves to the manufacturer's recommended minimum operating temperature for six (6) hours minimum.

NOTE: Reduce pressure to zero and return temperature to ambient.

c) Determine leakage in accordance with steps under 9.2.3.2. and 9.2.3.4.

9.2.5 Endurance Test.

a) Pressurize the coupling assembly using test fluid 1 bar (14.5 psi) internal pressure (valved only).

b) Couple and uncouple the coupling assembly 25,000 times for coupling sizes up to and including 12.7 mm (1/2 inch); 5,000 times for greater than 12.7 mm (1/2 inch). Do not exceed a coupling rate of 3600/h (3600/hr).

c) Record any evidence of binding or malfunction.

d) Determine leakage in accordance with steps under 9.2.3.

NOTE: This is a terminal test and the couplings used in this test should not be used for further testing.

9.2.6 Pressure drop test.

a) Install the coupling assembly in a test set-up illustrated in figure 8.

b) Maintain specified internal test pressure during the entire test, using the test media.

c) Select at least six (6) proportional pressure drop levels.

d) Determine and record the flow within the limits of table 1 at each selected pressure drop reading. (Reference paragraph 12 for data presentation.)

e) Remove the coupling assembly from the test set-up and install an equal length of tare tube with constant internal diameter and end attachments the same as the test coupling ends of the corresponding size.

f) Determine and record flow versus pressure drop as in d (9.2.6).

g) Refer to 12.1 and 12.2 for data presentation.

9.2.7 Vacuum leakage test, coupled.

a) Connect the coupling to test set-up illustrated in figure 9.

b) Apply side load to the coupling assembly as shown in figure 6.

c) Start vacuum pump and pull a vacuum to the specified test presssure.

d) Allow ten (10) minutes for stablization at the specified test pressure.

e) Observe the vacuum gage for loss of vacuum during a specified time period and report any loss of vacuum during the time period.

9.2.8 Vacuum leakage test, uncoupled (valved only).

NOTE: Applicable to self-sealing halves only.

a) Connect the coupling half to test set-up illustrated in figure 9.

b) Start vacuum pump and pull a vacuum to a specified test pressure.

c) Allow ten (10) minutes for stabilization at the specified vacuum.

d) Observe the vacuum gage for loss of vacuum during a specified time period and report any loss of vacuum during the time period.

10 PROCEDURES FOR RATED AND STATIC PRESSURE TEST

10.1 Coupled (open valved condition on valved couplings).

Verify the rated fatigue and static pressure by testing in accordance with NFPA/T2.6.1M.

11 DATA ACCURACY

Select and maintain instumentation so that Data Accuracy is within the limits in table 2.

TABLE 2 - **Data accuracy**

Quantity	Metric Unit	Customary US Unit	Maintain Within (±) of Actual Measured Value
Pressure	bar	psi	2%
Pressure Drop (ΔP)	bar	psi	2%
Flow	cm^3/s (ANR)	SCIM	2%
Length	mm	in	1%

12 DATA PRESENTATION

12.1 Net pressure drops may be presented graphically or numerically using the values obtained in 9.2.6.

12.2 To obtain net pressure drops subtract tare pressure drop from gross pressure drop.

12.3 Note all deviations or modifications of the test procedure within the test data.

12.4 Have available a record of all the following minimum test data in all test reports referencing this recommended standard:

12.4.1 All physical values pertaining to the test as required by the user of this document.

12.4.2 All additional provisions or modifications pertaining to the test.

13 JUSTIFICATION STATEMENT

This recommended standard verification procedure is based on the combined expert experiences of those who have participated in its preparation and review.

14 TEST/PRODUCTION SIMILARITY

Utilize managerial controls necessary to maintain substantial similarity between test and production components or elements.

15 IDENTIFICATION STATEMENT

Use the following statement in catalogs and sales literature when electing to comply with this voluntary standard:

"Performance data has been measured in accordance with American National Standard, ANSI/B93.51M-1980, **Pneumatic fluid power - Quick action couplings - Test conditions and procedures.**"

ANSI/B93.51

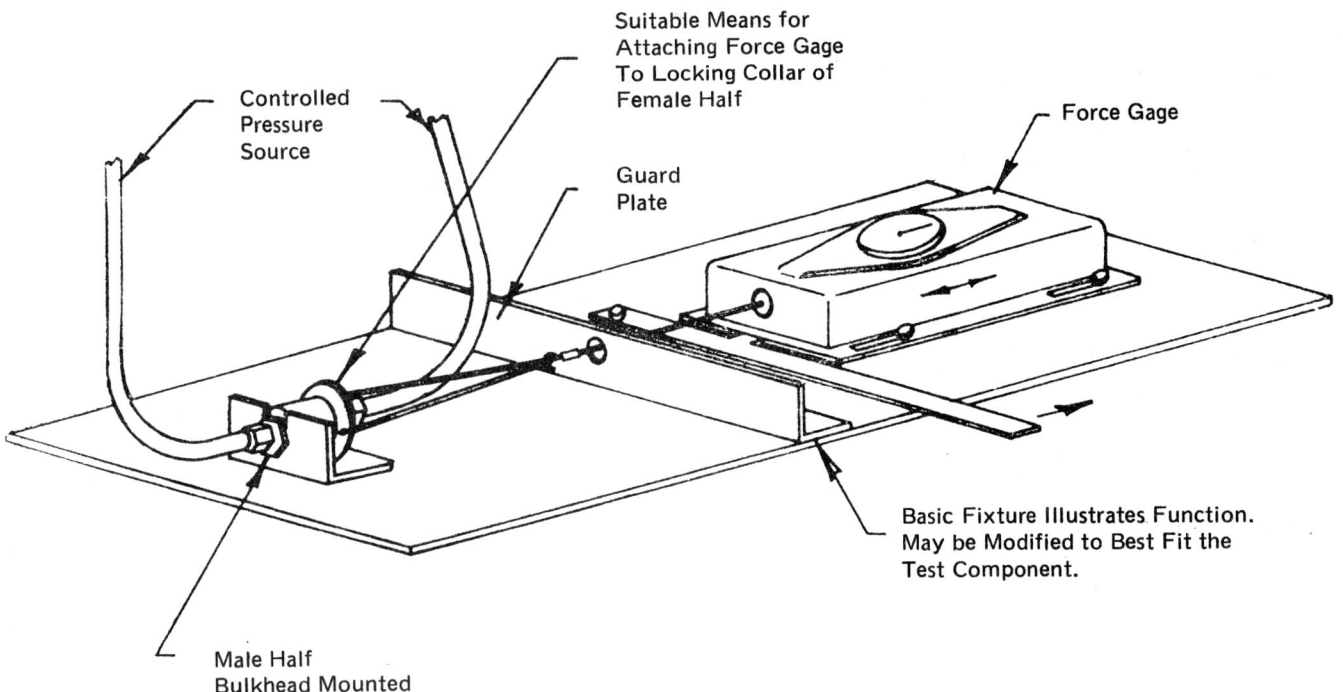

FIGURE 1 - Disconnect force test set-up

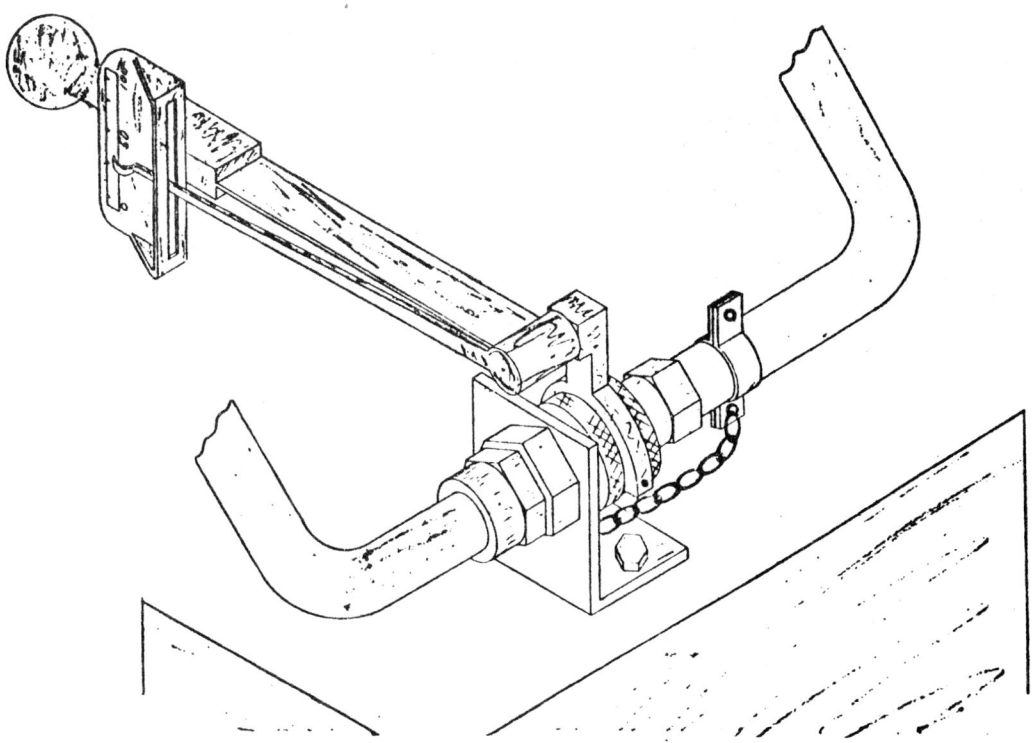

FIGURE 2 - Disconnect torque test set-up

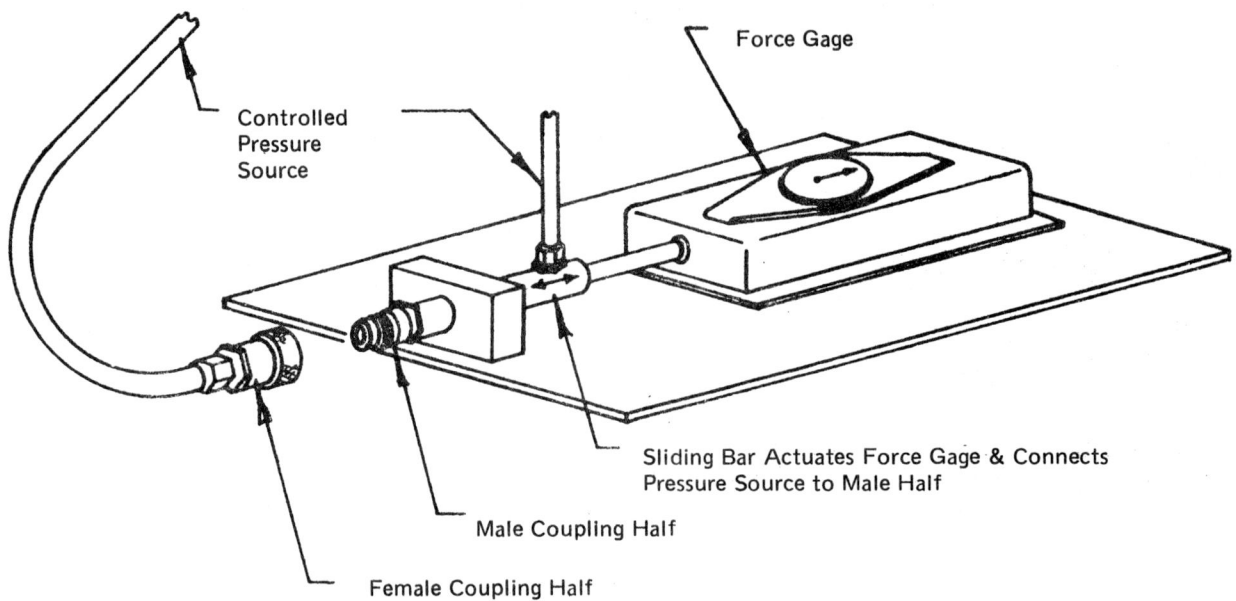

FIGURE 3 - **Connect force test set-up**

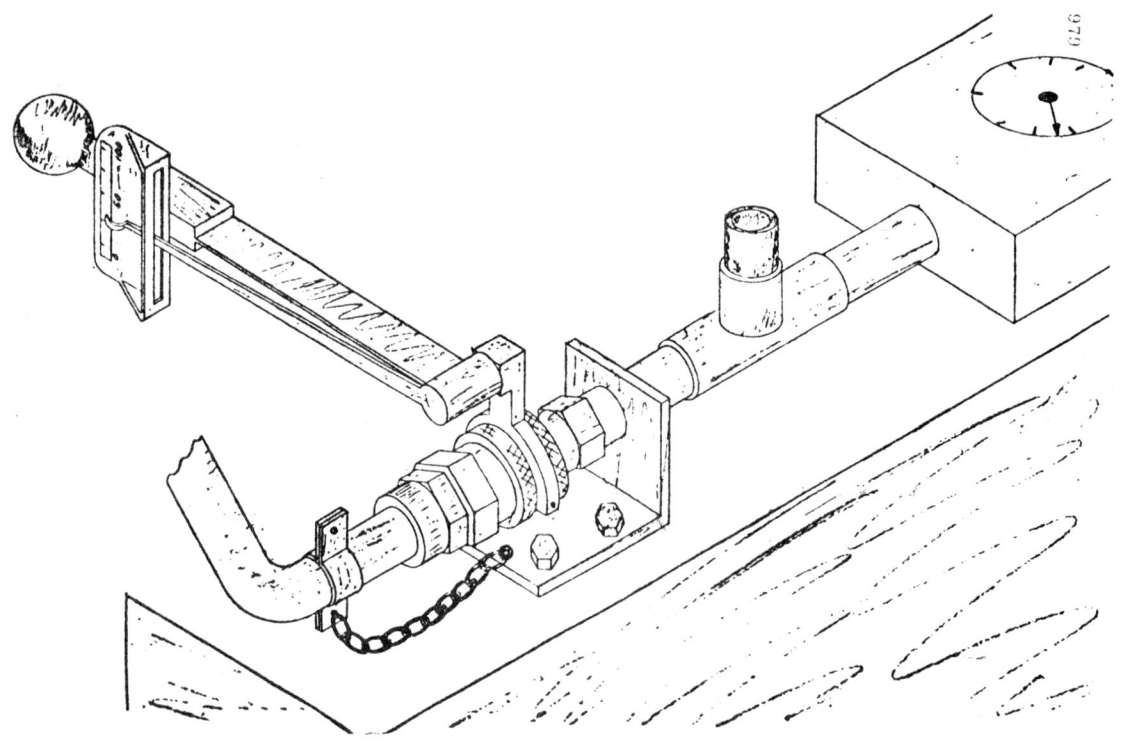

FIGURE 4 - **Connect torque test set-up**

ANSI/B93.51

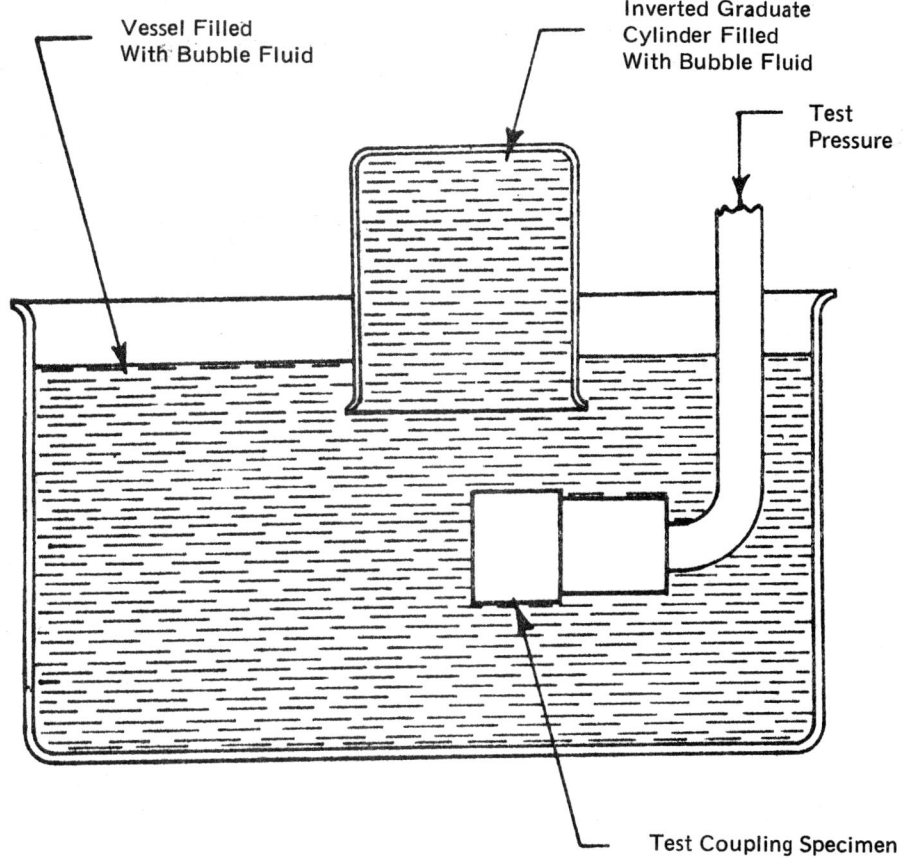

FIGURE 5 - Leakage test set-up

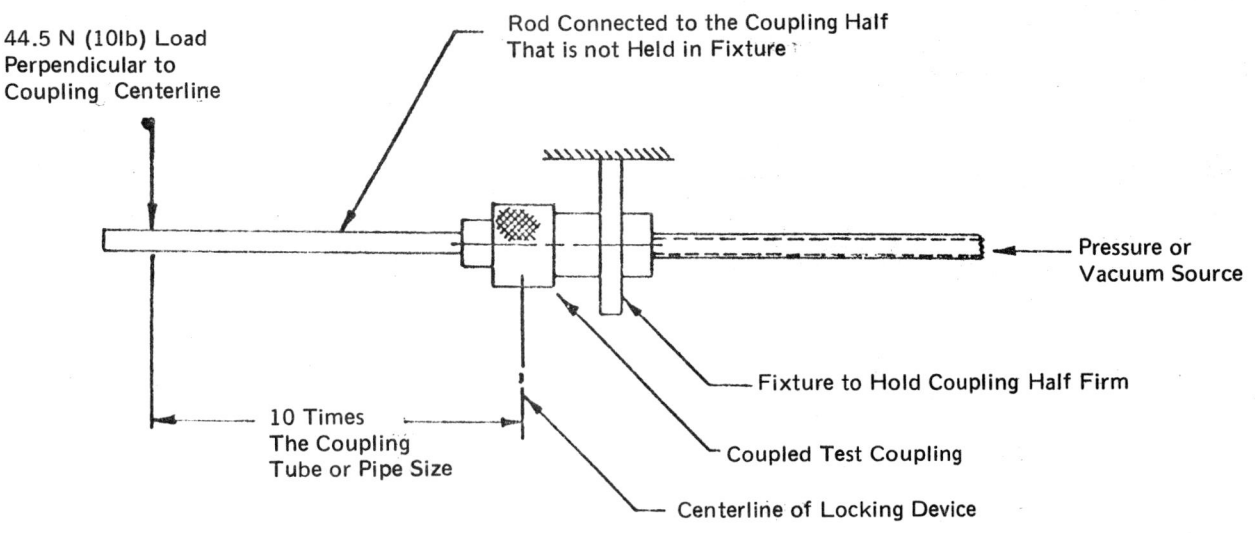

FIGURE 6 - Sideload fixture

- 151 -

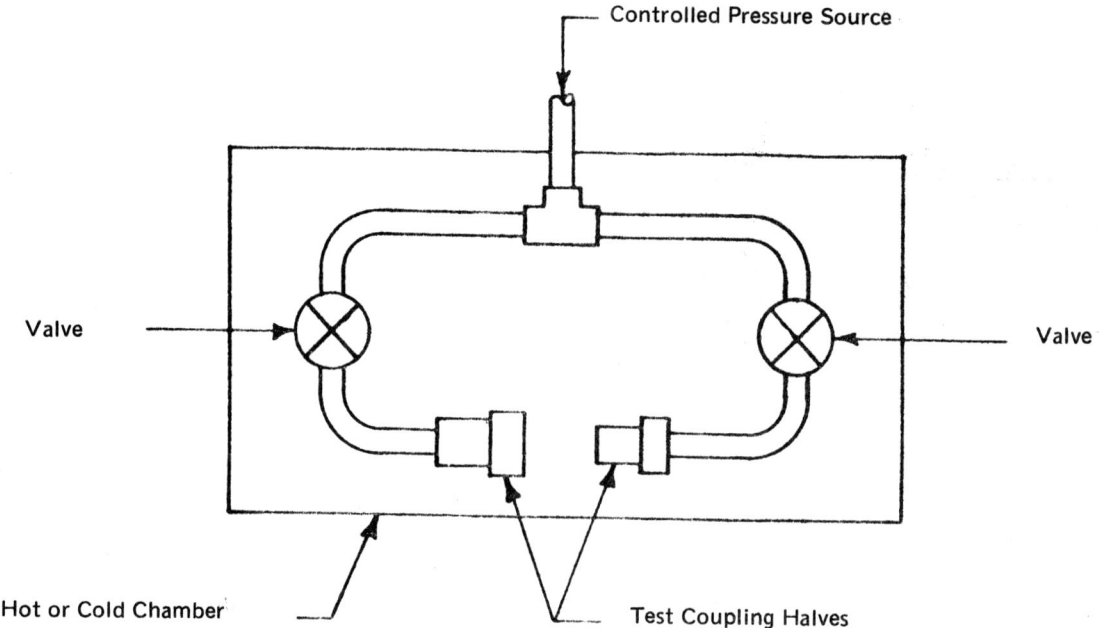

FIGURE 7 - Extreme temperature test set-up

ANSI/B93.51

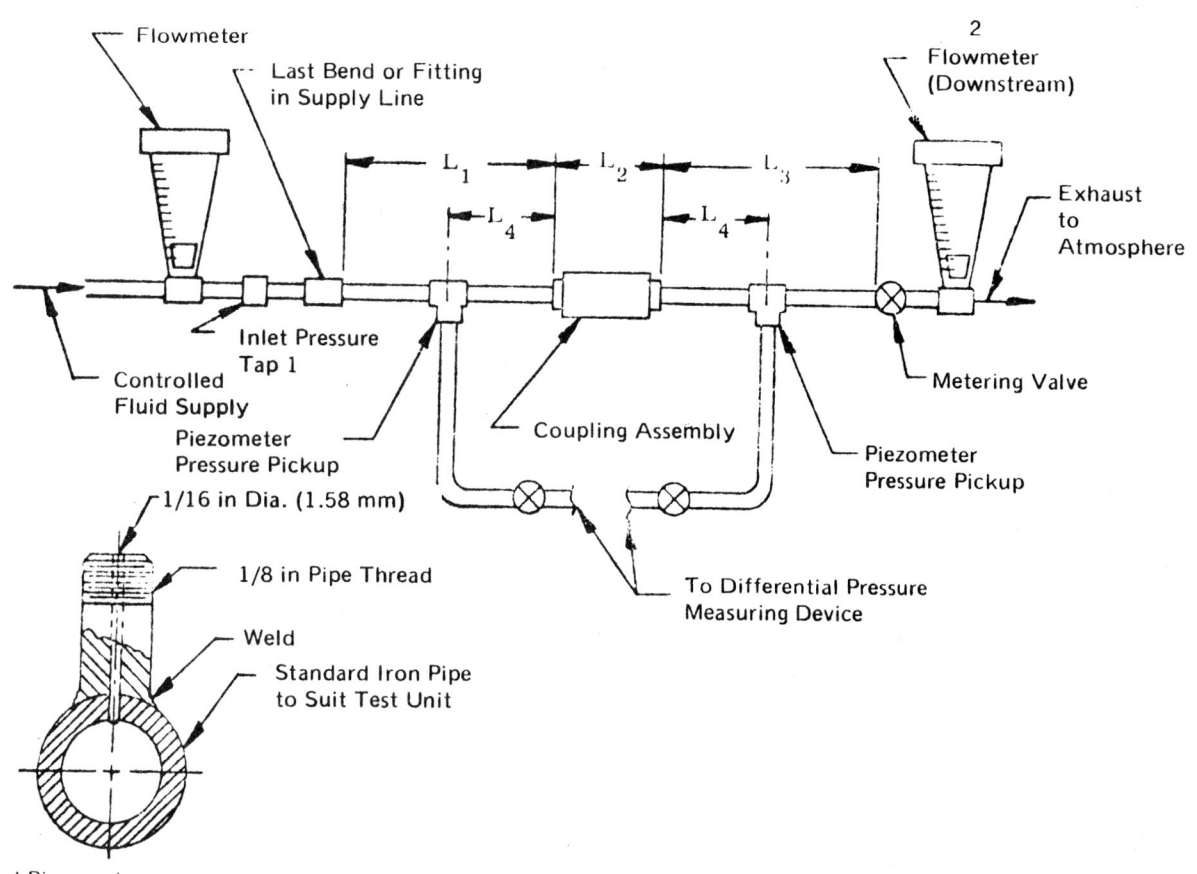

L_1 = Coupling inlet pipe size, 20 diameters length
L_2 = Coupling assembly plus end fittings
L_3 = Coupling inlet pipe size, 20 diameters length
L_4 = Coupling inlet pipe size, 10 diameters length

NOTE 1: Position inlet pressure tap as close as practical to last fitting of supply line.

NOTE 2: Optional flowmeter on downstream side or discharge is acceptable.

FIGURE 8 - Pressure drop test set-up

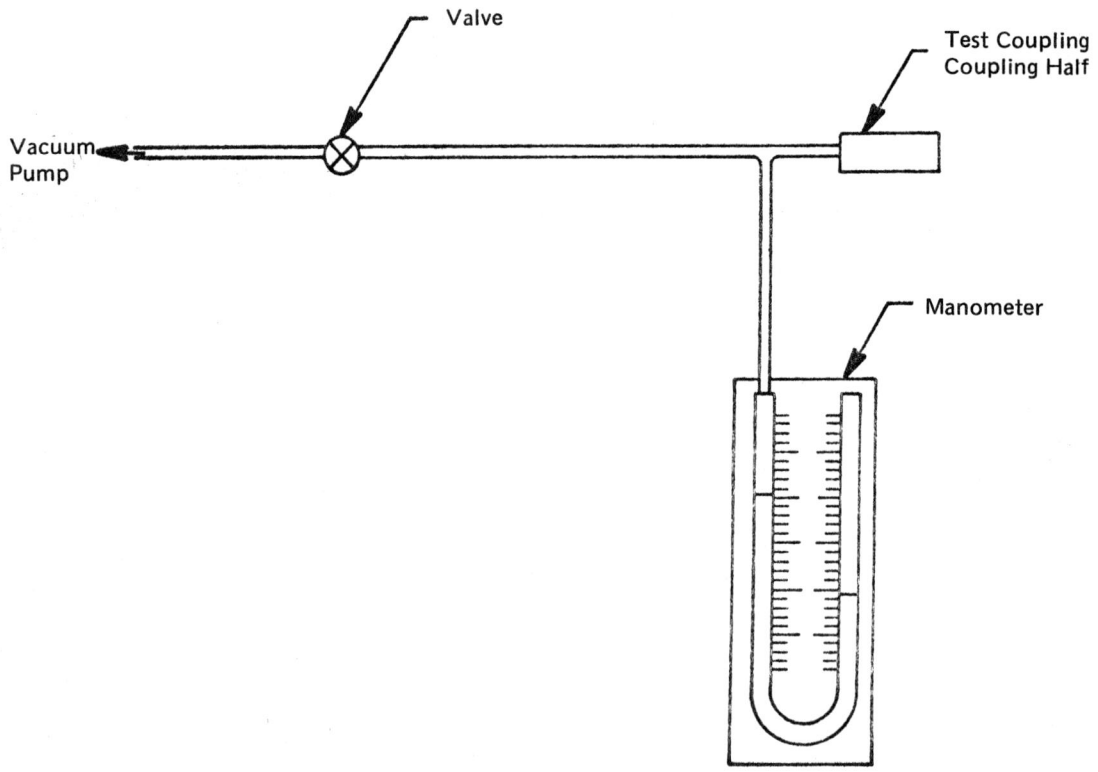

FIGURE 9 - **Vacuum leakage test set-up**

APPENDIX

to ANSI/B93.51M-1980

3 REFERENCES

ISO 5598 [1], Fluid power systems and components - Vocabulary.

---

[1] At present at the stage of draft.

NFPA Recommended Standard
T3.20.1-1973 (R1981)

AN INDUSTRY STANDARD FOR FLUID POWER

Glossary For Fluid Power

Quick Action Couplings

Approved as an NFPA Recommended Standard
11 November 1973

---

published by
NATIONAL FLUID POWER ASSOCIATION, INC.
3333 N. Mayfair Road / Milwaukee, WI 53222 / 414-778-3344 / TLX 26898

NFPA/T3.20.1

FOREWORD

(This Foreword is not part of NFPA Recommended Standard Glossary of Terms for Fluid Power Quick Disconnect Couplings, T3.20.1-1973.)

This project was recommended by a coupling manufacturer to the National Fluid Power Association on 2 February 1966. The project was assigned to the Fittings and Connectors Section of the NFPA Technical Board.

On 8 December 1966 the Fittings and Connectors Section Committee voted to recommend to the Technical Board that a new component section be initiated for quick disconnect couplings.

The Technical Board approved the Quick Disconnect Coupling Section under the chairmanship of Wayne Barmore of the Bruning Company.

The first project meeting was held 22 June 1967. After a series of five draft writings and reviews, a General Review Draft was prepared on 1 November 1971.

The General Review period closed on 22 December 1971 and the comments received were resolved on 11 April 1972. The Ballot Draft was prepared on 12 October 1972. Balloting closed on 28 February 1973, and resulted in unanimous approval of the document. The NFPA Technical Board recommended the document to the Board of Directors on 4 October 1973.

The NFPA Board of Directors approved the document as NFPA Recommended Standard T3.20.1-1973 on 11 November 1973. Terms contained in this document will be incorporated into the American National Standard Glossary of Terms for Fluid Power, ANSI/B93.2-1970 at its next revision.

NFPA/T3.20.1

MEMBERS OF THE PROJECT GROUP DEVELOPING THIS STANDARD

Saloum, Edward   Project Chairman   Snap-Tite, Inc.

Barmore, Wayne   Section Chairman   Bruning Company

Ritchie, Rex   Section Vice Chairman   Aeroquip Corporation

Morgan, James I. Secretariat   National Fluid Power Assn.

Currie, W.   Parker Hannifin Corporation
Dekker, D.   Michigan Seamless Tube Co.
Dellis, R.   Stratoflex, Inc.
Fletcher, E.   ASAE
Hofmann, J.   Hofmann Engineering Co.
Koch, K.   Bruning Company
Kozak, T.   Michigan Seamless Tube Co.
Lamb, T.   Parker Hannifin Corporation
Lessen, C.   Massey-Ferguson
Malpass, W.   Scovill Fluid Power Division
Rogers, R.   Aeroquip Corporation
Smith, C.   Roylyn, Inc.
Stratman, P.   Pioneer Div. - Parker Hannifin
Struck, L.   Imperial Eastman
Thorson, C.   Char-Lynn Company
Warren, J.   Stratoflex, Inc.
Zopf, K.   The Hansen Manufacturing Co.

REFERENCES

1. American National Standard Glossary of Terms for Fluid Power, ANSI/B93.2-1971. (ISO/TC 131/SC 1 (USA 2) 3).

NFPA/T3.20.1

GLOSSARY OF TERMS FOR
FLUID POWER QUICK DISCONNECT COUPLINGS

INTRODUCTION

In fluid power systems, power is transmitted and controlled thru a fluid (liquid or gas) under pressure within an enclosed circuit.

A Quick Disconnect Coupling is used to quickly join or separate fluid conducting lines without the use of tools or special devices.

1. SCOPE

   To include identical coupling halves (hermaphrodite), male or female coupling halves, and coupling assemblies used in fluid power systems.

2. PURPOSE

   To establish a basic source for, and common understanding of, terms related to quick disconnect couplings used in fluid power systems.

3. TERMS AND DEFINITIONS

   (For definition of other terms, see Reference No. 1)

   (Terms set forth are intended for insertion in Section 55 of Reference No. 1.)

   3.1 Air Inclusion. The ambient atmosphere forced or trapped into the system during the connection of the quick disconnect halves.

   3.2 Break-Away. Automatic separation of a mounted coupling when a force is applied axially to the unmounted coupling half.

   3.3 Connect Under Pressure. Ability to connect coupling halves with internal line pressure applied to either both sides or one si

NFPA/T3.20.1

3.4 Coupling, Quick Disconnect. A component which can quickly join or separate a fluid line without the use of tools or special devices.

3.5 Coupling, Quick Disconnect, One Valve. A quick disconnect coupling with a shut-off valve in one half.

3.6 Coupling, Quick Disconnect, Valved. A quick disconnect coupling with a shut-off valve in each half.

3.7 Spillage. The fluid removed from the system during disconnection of a coupling assembly.

4. IDENTIFICATION STATEMENT

Use the following statement in catalogs and sales literature when electing to comply with this voluntary standard:

"Definitions of terms conform to NFPA Recommended Standard, NFPA/T3.20.1-1973."

- NOTES -

**NFPA Information Report
T3.20.7 R1-1983**

AN INDUSTRY STANDARD FOR FLUID POWER

NFPA Information Report
A Bibliography of
Fluid Power Quick Action Coupling Standards

Published 1984

published by
NATIONAL FLUID POWER ASSOCIATION, INC.
3333 N. Mayfair Road / Milwaukee, WI 53222 / 414-778-3344 / TLX 26898

NFPA/T3.20.7 R1

SOURCE OF DOCUMENTS

AFNOR	Association fancaise de normalisation Tour Europe Cedex 7 92080 Paris - La Defense France
ANSI	American National Standards Institute 1430 Broadway New York, NY 10018
ASAE	American Society of Agricultural Engineers 2950 Niles Road St. Joseph, MI 49085
BSI	British Standards Institution 2 Park Street London W1A 2BS United Kingdom
CETOP	European Committee on Oil Hydraulic and Pneumatic Control c/o AHEM 192-198 Vauxhall Bridge Road London, SW1V England
DIN	Deutsches Institut fur Normung Burggrafenstrasse 4-10 Postfach 1107 D-1000 Berlin 30 West Germany
GOST	USSR/URSS USSR State Committee for Standards Leninsky Prospekt 9 Moskva 117049 USSR
ISO	International Standards Organization 1, rue de Varumbe Case postale 56 CH - 1211 Geneva 20 Switzerland
NFPA	National Fluid Power Association, Inc. 3333 North Mayfair Road Milwaukee, WI 53222
NPFC	Naval Publications and Forms Center 5801 Tabor Avenue Philadelphia, PA 19120 Attn: NPFC 105
SAE	Society of Automotive Engineers 400 Commonwealth Drive Warrendale, PA 15096
VDMA	Verein Deutscher Maschinenbau-Ansralten e.V. Normengruppe Maschinenbau Fachbereich Fluidechnik Lyoner Strasse 18 Postfach 710109 6000 Frankfurt 71 West Germany

NOTE: This Bibliography provides a list of standards related to fluid power quick action couplings which have been issued by the NFPA and other standards-writing organizations. Listing in this Bibliography is only for reference and does not imply Association endorsement.

NFPA/T3.20.7 R1

Identification Number	Title	Source	Date of Issue
AFNOR E 48-058	Hydraulic and Pneumatic System Quick Acting Coupling Nominal Pressure 40 Connecting Dimensions for Plugs	AFNOR	1976
AFNOR N 49-053	Pneumatic Fluid Power Systems Quick Acting Coupling Nominal Pressure NP16 - Connecting Dimensions for Plugs	AFNOR	1976
ANSI/B93.42M-1977 (R1983)	Method for Testing Hydraulic Fluid Power Quick Action Couplings	NFPA (ANSI)	1976
ANSI/B93.51M-1980	Pneumatic fluid power - Quick action couplings - Test conditions and procedures	NFPA (ANSI)	1979
ANSI/B93.64M-1983	Hydraulic fluid power - Quick action couplings (with pipe thread or SAE straight thread connections) - Method of measuring and reporting pressure drop	NFPA (ANSI)	1981
ANSI/B93.68M-1983	Hydraulic fluid power - Quick action couplings - Surge flow test (short duration flow)	NFPA (ANSI)	1982
ASAE S 366	Dimensions for Cylindrical Hydraulic Couplers for Agricultural Tractors(SAE-J-1036)	ASAE	1982
BS 2464	Quick Action Couplings: Part 3	BSI	1968
CETOP RP29P	Pneumatic Quick Action Couplings	CETOP	1969
CETOP RP59P	Quick Action Couplings - Plug Dimensions (10 bar)	CETOP	1974
DIN 24328	Olhydraulik und Pneumatik Hydraulik-Steekkupplungen mit Aussengewinde	DIN	1974
GOST 12853	Hydraulic and Pneumatic Drives and Lubricating Systems - Connecting Couplings	GOST	1967
ISO 5675	Agricultural tractors and machinery - Hydraulic couplers for general purposes - Specifications	ISO	1981
MIL-C-3486B	Coupling and Coupling Halves, Quick-disconnect, Air Hose, Bowes Type	NPFC	1966
MIL-C-4109E	Coupling Assembly, Low Pressure, Air Hose, Quick-disconnect	NPFC	1976
MIL-C-25427A (1)	Coupling Assembly, Hydraulic Self-sealing, Quick-disconnect	NPFC	1963
MIL-H-8775D (1)	Hydraulic System Components, Aircraft and Missiles, General Specification	NPFC	1976
MS-24333D	Coupling Assembly, Hydraulic Self-sealing, Quick-disconnect Flared Fitting to Internal Thread Boss (Asg)	NPFC	1968
MS-24334C	Coupling Assembly, Hydraulic, Self-sealing Flareless Fitting to Internal Thread Boss (Asg)	NPFC	1968
NFPA/T3.20.1-(R1981)	Glossary for Fluid Power Quick Action Couplings	NFPA	1973
NFPA/T3.20.8M-(R1981)	Quick Action Pressure Rating Supplement (to be used with NFPA/T2.6.1-1974)	NFPA	1975

NFPA/T3.20.7 R1

Identification Number	Title	Source	Date of Issue
NFPA/T3.20.12M-	Hydraulic fluid power - Quick action couplings - Surge flow test (long duration flow)	NFPA	1983
SAE-AIR-737C	Aerospace Hydraulic and Pneumatic Specifications and Standards	SAE	1971
SAE-AIR-1047A	A Guide for the Selection of Quick-Disconnect Couplings for Aerospace Fluid Systems	SAE	1978
SAE-ARP-24B	Determination of Hydraulic Pressure Drop	SAE	1968
SAE-ARP-219	Procedure and Method for Conducting Test of Hydraulic Components in Contamination Controlled Systems	SAE	1961
SAE-ARP-243B	Nomenclature, Aircraft Hydraulic and Pneumatics Systems	SAE	1965
SAE-ARP-603E	Impulse Testing of Hydraulic Hose Assemblies, Tubings, and Fittings	SAE	1979
SAE-ARP-868	Pressure Drop Test for Fuel System Components	SAE	1966
SAE-J-1036	Dimensional Standard for Cylindrical Hydraulic Couplers for Agricultural Tractors	SAE	1973
VDMA-24-328	Oilhydraulic and Pneumatic Hydraulic-plug Coupling	VDMA	1969

NFPA Recommended Standard
T2.6.1M S5 (T3.20.8M)-1975
(R1981)

AN INDUSTRY STANDARD FOR FLUID POWER

Quick Action Couplings Pressure Rating

Supplement No. 5 to

NFPA Recommended Standard for Verifying the
Fatigue and Static Pressure Ratings of the
Pressure Containing Envelope of a Metal Fluid Power Component

Approved as an NFPA Recommended Standard
25 July 1975

---

published by
NATIONAL FLUID POWER ASSOCIATION, INC.
3333 N. Mayfair Road / Milwaukee, WI 53222 / 414-778-3344 / TLX 26898

NFPA/T2.6.1 S5 (T3.20.8)

FOREWORD

(This Foreword is not part of NFPA Recommended Standard Quick Disconnect Pressure Rating Supplement No. 5 to NFPA Recommended Standard for Verifying the Fatigue and Static Pressure Ratings of the Pressure Containing Envelope of a Metal Fluid Power Component, NFPA/T2.6.1 S5 (T3.20.8)-1975.)

Early in 1974, the approval of NFPA/T2.6.1-1974 established a group of common requirements intended to provide an industry-wide philosophy and basic standard, providing a rationale for judging a component's ability as a pressure containing envelope. Although the specific applicability of NFPA/T2.6.1-1974 is limited, it immediately established a uniform base for subsequent, more specific proposed NFPA Recommended Standards for individual components.

The Quick Disconnect Section met on 20 November 1973 to discuss the preliminary requirements and plans for developing Project T3.20.8.

The first draft was completed by the NFPA Technical Staff on 17 December 1973, to include recent information from the Project Group and changes in the basic document, T2.6.1-1974. The T3.20.8 Project Group reached consensus on 6 March 1974 after making several modifications to the draft document.

The General Review Draft was completed by Headquarters' Technical Staff on 20 March 1974. Comments resulting from general industry review were resolved on 4 November 1974, and the project was approved for ballot by the Technical Board on 5 December 1974. Headquarters prepared the Ballot Draft on 15 January 1975. No negative ballots were received on this standard; however, two general suggestions were received, one of which was incorporated in the document on 3 July 1975.

On 22 May 1975, the Technical Board unanimously recommended approval of NFPA/T2.6.1 S5 (T3.20.8)-19xx.

The NFPA Board of Directors granted approval to this standard on 25 July 1975.

NFPA/T2.6.1 S5 (T3.20.8)

PROJECT GROUP MEMBERS WHO DEVELOPED THIS STANDARD

Koch, Ken	Project Chairman	Bruning Company
Karcher, T. D.	Project Vice Chairman	The Hansen Mfg. Co.
Saloum, Edward	Section Chairman	Snap-Tite, Inc.
Moyer, Don	Technical Auditor	International Harvester Co.
Luecke, John R.	Director of National Technical Services	National Fluid Power Association

Hammond, H.   The Hansen Mfg. Co.
Lamb, T.   Parker Hannifin Corp.

REFERENCES

1. American National Standard Glossary of Terms for Fluid Power, ANSI/B93.2-1971, and Supplements thereto. (ISO/TC 131/SC 1 (USA-2) 3)

2. SI units and recommendations for the use of their multiples and of certain other units, ISO 1000-1973.

3. National Fluid Power Association Recommended Standard Method for Verifying the Fatigue and Static Pressure Ratings of the Pressure Containing Envelope of a Metal Fluid Power Component, NFPA/T2.6.1-1974.

NFPA/T2.6.1 S5 (T3.20.8)

QUICK DISCONNECT PRESSURE RATING

SUPPLEMENT NO. 5 to

NFPA Recommended Standard for Verifying the

Fatigue and Static Pressure Ratings of the

Pressure Containing Envelope of a Metal Fluid Power Component

> HEADQUARTERS NOTE - The Project Group which developed this Supplement intended it for use only with the basic pressure rating document, NFPA/T2.6.1-1974. This Supplement provides a list of additions, deletions and changes to the basic document which are necessary for the establishment of pressure ratings for metal quick disconnect couplings. All clauses which have been modified are so indicated in this Supplement. Read all other clauses as they appear in the basic pressure rating document, NFPA/T2.6.1-1974. Since revisions of the basic pressure rating document will apply to this Supplement, the reader is cautioned to always use the most recent edition of the basic document. When reading this Supplement with the basic document, <u>use the phrase "quick disconnect coupling(s)" whenever the term "components" appears in the basic document.</u>

INTRODUCTION

CHANGE the first two paragraphs of the basic document to read:

In fluid power systems, power is transmitted and controlled thru a fluid (liquid or gas) under pressure within an enclosed circuit. Quick disconnect couplings are used to quickly join or separate fluid conducting lines without the use of tools or special devices. A basic requirement of fluid power quick disconnect couplings is that they should be capable of adequately containing the pressurized fluid.

The basic pressure rating document, NFPA/T2.6.1-1974, established a group of common requirements intended to provide an industry-wide philosophy and basic standard, providing a rationale for judging a component's ability as a pressure containing envelope. Although the specific applicability of NFPA/T2.6.1-1974 is limited, it immediately established a uniform base for subsequent, more specific proposed NFPA Recommended Standards for individual fluid power components. This Recommended Standard implements NFPA/T2.6.1-1974 and specifically applies to metal fluid power quick disconnect couplings.

1. SCOPE

   ADD the following clauses:

   1.1.5 Quick disconnect couplings that are connected and disconnected by a linear or rotational motion, or both.

   1.1.6 Quick disconnect couplings with or without sealing means, when coupled; and couplings with sealing means, when uncoupled.

   DELETE clause 1.2.

2. PURPOSE

   NO CHANGE from the basic document.

3. TERMS AND DEFINITIONS

   ADD the following NOTE to clause 3.9:

   NOTE: It is the responsibility of the quick disconnect coupling manufacturer to define "excessive seal leakage".

   CHANGE clause 3.12 to read:

   3.12 Elements. The integral pieces which make up a quick disconnect coupling (e.g., seal, poppet, end fitting, etc.)

4. UNITS OF MEASUREMENT

   NO CHANGE from the basic document.

NFPA/T2.6.1 S5 (T3.20.8)

5. LETTER SYMBOLS

   NO CHANGE from the basic document.

6. OUTLINE OF PROCEDURES FOR RFP VERIFICATION BY TEST

   NO CHANGE from the basic document.

7. RFP VERIFICATION BY SIMILARITY

   CHANGE clause 7.1 to read:

   7.1 The rated fatigue pressure of quick disconnect couplings can be verified analytically by geometric similarity based upon tests conducted in accordance with this standard on similar quick disconnect couplings. This practice is acceptable only when the pressure containing capability of the quick disconnect coupling is not affected by modification such as seal compounds, end fitting style, sleeve lock, etc.

8. OUTLINE OF PROCEDURES FOR RSP VERIFICATION BY TEST

   NO CHANGE from the basic document.

9. PREPARATIONS FOR TESTING

   CHANGE clause 9.4 to read:

   9.4 Apply different ratings to the coupled unit and the uncoupled halves of the pressure containing envelope as required.

   CHANGE clauses 9.8 up to and including 9.11 to read:

   9.8 Perform cyclic and static tests on separate pressure containing envelopes when a sufficient number of test pieces is available.

   9.9 Perform coupled and uncoupled tests on separate pressure containing envelopes.

9.10 Perform cyclic tests first, when both RFP and RSP verification tests must be made on the same pressure containing envelopes.

9.11 Complete machining and processing of test quick disconnect couplings to the degree necessary for duplicating operating stress distributions and final strength in the pressure containing envelope.

ADD the following clauses:

9.12 It is permissible to make modifications to the test piece to facilitate cyclic or static tests providing that such modifications do not increase the pressure capabilities of the pressure containing envelope being tested. This would include substitution of materials on non-metallic gaskets and seals to facilitate testing.

9.13 Examine the test quick disconnect couplings for quality defects such as material cracks, dimensions outside of tolerance limits, etc. It is recommended that the pressure containing elements of each test quick disconnect coupling be inspected for compliance with the drawing requirements prior to testing. If a quality defect is found, replace the element or quick disconnect coupling with a new part.

10. TEST EQUIPMENT

CHANGE clauses 10.1.3 and 10.1.4 to read:

10.1.3 Mount the pressure measuring instrument directly into the pressure containing envelope, thru a pressurized port which is not being used to supply the test fluid.

10.1.4 When no such port (clause 10.1.3) is available, minimize restrictions between the instrument and the pressure containing envelope by mounting the pressure measuring instrument on the pressure supply line.

NOTE: It is recommended that it be verified that the pressure generated in the pressure containing envelope at the actual cycling rate be the intended value.

NFPA/T2.6.1 S5 (T3.20.8)

11. TEST CONDITIONS ACCURACY

    NO CHANGE from the basic document.

12. PROCEDURES FOR CYCLIC TEST

    ADD the following clause:

    12.1.6 Do not rotate the male half with respect to the female half during the cyclic testing of the coupled unit.

13. PROCEDURES FOR STATIC TEST

    NO CHANGE from the basic document.

14. ADDITIONAL PROVISIONS

    CHANGE the entire Section to read:

    14.1   During cyclic tests, re-tighten only those coupling halves that are coupled by a threaded member and are normally tightened by hand in service. Re-tightening of such couplings does not constitute a failure.

    14.2   During cyclic tests, do not re-torque those threaded connections that are not normally re-tightened in service, e.g., threaded elements torqued by the manufacturer.

    14.3   For coupled tests, ensure that the stressing of the quick disconnect couplings is not encumbered by the mounting fixture.

15. CRITERIA FOR RFP VERIFICATION

    ADD the following clause:

    15.4   Consider permanent metal deformation, which interferes with the proper functioning of the quick disconnect coupling, a failure. This includes connect and disconnect function and operation of the valves.

    NOTE: Functional verification other than specified by clause 15.4 is beyond the scope of this document; and when required, must be dealt with by separate agreement.

16. CRITERIA FOR RSP VERIFICATION

    CHANGE clause 16.4 to read:

    16.4   Consider permanent deformation, which interferes in any way with the proper functioning of the quick disconnect coupling, a failure. This includes connect and disconnect function and operation of valves.

    NOTE: Functional verification other than specified by clause 16.4 is beyond the scope of this document; and when required, must be dealt with by separate agreement.

17. DATA PRESENTATION

    ADD the following clause:

    17.2.3   Failure leakage requirement as required by clause 3.9.

18. SUMMARY OF DESIGNATED INFORMATION

    NO CHANGE from the basic document.

NFPA/T2.6.1 S5 (T3.20.8)

19. JUSTIFICATION STATEMENT

    CHANGE the first two paragraphs to read:

    This recommended standard verification procedure is based upon the combined expert experiences of those who have participated in its preparation and review, and in the preparation and review of the basic document, NFPA/T2.6.1-1974, and its tutorial reference.

20. TEST/PRODUCTION SIMILARITY

    NO CHANGE from the basic document.

21. IDENTIFICATION STATEMENT

    CHANGE the quoted statement to read:

    "Method of verifying rated fatigue and static pressure of quick disconnect couplings conforms to NFPA Recommended Standard, NFPA/T2.6.1 S5 (T3.20.8)-19xx, category _____.*"

ANSI/B93.68M-1983

AN INDUSTRY STANDARD FOR FLUID POWER

American National Standard

Hydraulic fluid power -

Quick action couplings -

Surge flow test (short duration flow)
(NFPA/T3.20.11M-1982)

Approved as an ANSI Standard
16 May 1983

Descriptors: hydraulic fluid power; quick action couplings; rated flow; surge, flow; surge, pressure; testing, leakage; testing, pressure drop; testing, quick action couplings.

---

published by
NATIONAL FLUID POWER ASSOCIATION, INC.
3333 N. Mayfair Road / Milwaukee, WI 53222 / 414-778-3344 / TLX 26898

ANSI/B93.68

FOREWORD

This Foreword is not part of American National Standard Hydraulic fluid power - Quick action couplings - Surge flow test, (short duration flow) ANSI/B93.68M-1983.

ANSI/B93.42M-1977, Hydraulic Quick Disconnect Test Procedure included a surge flow test. The test was impractical to conduct on larger size couplings due to the high flow rates required (five times rated flow). In addition, it did not simulate most applications which have short duration high surge flows.

The general revision to B93.42M-1977 was processed without this test since lengthy development time was anticipated for the surge flow test.

After lab tests were conducted by participating member companies, Draft No. 1 was prepared on 1 May 1979.

Comments received were reviewed and answered. As a result, Draft No. 2 was prepared 16 October 1979.

Headquarters Technical Staff prepared the document for General Review on 12 September 1980. As a result of the General Review, metric corrections were noted; the title was changed; a data presentation section added; and a note to the 9.2. A TSP will be initiated to reactivate the five times rated flow document as an alternative to this surge document.

T3.20.11M was presented to the Technical Board for approval to ballot on 11 February 1981. Approval was granted contingent on the Technical Auditors letter of resolution. Said letter was received by Headquarters on 17 February 1981.

Headquarters Technical Staff prepared the document for Ballot on 24 July 1981. Ballot closed 24 August 1981 with one comment.

The comment was reviewed and resolved at the 14 October 1981, T3.20 Section meeting. The document was presented and granted approval by the Technical Board on 24 February, 1982.

The Board of Directors granted final approval on 20 October 1982.

Project Group members who developed this standard:

Ken Koch
Project Chairman &
Section Vice Chairman
Imperial Clevite Inc./Bruning Operations

Ed Saloum
Section Chairman
Snap-Tite, Inc.

Don Fischer
Section Secretary
Aeroquip Corp.

David Sallberg
Technical Auditor
Koehring Company/Pegasus

James C. White*
Director of Technical Services
National Fluid Power Association

C. Grigsby
Schrader Bellows Div.

G. Herzan
Parker Hannifin Corp.

T. Karcher
Hansen Manufacturing Co.

O. Maldavs
Imperial Clevite Inc./Bruning Operations

S. Meisinger
Safe Way Hydraulics Inc.

S. Patin
John Deere Product Engineering Center

R. Rogers
Aeroquip Corp.

It was intended that this NFPA Recommended Standard, upon approval, be submitted to ANSI for promulgation as an American National Standard.

On 21 January 1983, ANSI/B93.68M was submitted to ANSI Committee B93 for ballot. Balloting closed 1 April 1983 with unanimous approval.

ANSI/B93.68M was forwarded to ANSI Board of Standards Review on 19 April 1983 and granted approval 16 May 1983.

The membership roster for the Standards Committee B93 at the time of ballot:

Jack C. McPherson
Chairman

Daniel B. Shore
Vice Chairman

Allen E. Tucker
Co-Secretary

William G. Wagner
Co-Secretary

American Society of Agricultural Engineers
Ed Fletcher

American Society for Engineering Education
Wm. R. Smith

---

*Company affiliation has changed

American Society of Lubrication Engineers
(To be named)

Compressed Air & Gas Institute
David E. Bonn
John Addington (alternate)

Construction Industry Manufacturers Association
Glenn Stewart

Fluid Controls Institute
Jude Pauli
Eric Bianchi (alternate)

Fluid Power Distributor Association
Thomas Neff

Fluid Power Society
Robert L. Firth
Carroll Grigsby
Robert W. Hanpeter
Marty King
Richard Read
Ronald Smith
Alan Tiedman

Fluid Sealing Association
John Scannell
Alex Pilecki (alternate)

Instrument Society of America
(To be named)

Joint Industry Council
Robert Muhl

Material Handling Institute
Jack C. McPherson
Willard Chichester (alternate)

National Fluid Power Association
Richard N. Bailey
John Bowbin
Walter Forster
Z. J. Lansky
Paul Schacht

National Machine Tool Builders Association
John B. Deam

Rubber Manufacturers Association
William A. Hertel
John R. Loder
E. J. McCarthy (alternate)

Society of Automotive Engineers
William A. Hertel
John T. Parrett
Henry Schultz
Daniel B. Shore
W. L. Snyder
Robert W. White

U. S. Dept. of Defense
William P. Coyne

Company Members
L. L. Schmaltz
Don McGeachy
John Welker (alternate)

Individual Members
Dr. E. C. Fitch, Jr.
Otto Maha
John J. Pippenger
A. O. Roberts
Jack Walrad
Tom Wanke
Frank Yeaple

ksg

ANSI/B93.68

**Hydraulic fluid power -
Quick action couplings -
Surge flow test (short duration flow)**

0 INTRODUCTION

In hydraulic fluid power systems, power is transmitted and controlled through a liquid under pressure within an enclosed circuit. Quick action couplings are used to quickly join or separate fluid conduction lines without the use of tools or special devices.

It is recognized that hydraulic quick action couplings may be exposed to surge flow conditions above the normal rated flow. Depending on the application, the surge flows may be short duration and/or long duration. Testing couplings under the two conditions generally requires two different test methods. The user may select either method or both methods as appropriate for his application or needs.

NFPA/T3.20.12M, Hydraulic Surge Flow Test, is recommended for testing hydraulic quick action couplings under long duration surge flow conditions. A continuous flow test method is used.

This document (ANSI/B93.68M) is recommended for short duration surges. An Accumulator-Discharge Test method is used to generate this short duration surge flows.

1 SCOPE AND FIELD OF APPLICATION

1.1 This National Standard:

— Applies to identical coupling halves, male or female coupling halves, and coupling assemblies used in hydraulic fluid power systems;

— Includes couplings with and without fluid sealing means, when uncoupled.

1.2 This National Standard is intended to:

— Promote the understanding and use of quick action couplings;

— Provide uniform test conditions and procedures;

— Promote product reliability;

— Facilitate accurate communications;

— Provide a basis for product selection and application.

2 REFERENCES

NFPA/T3.20.1-1971 (R1981), **National Fluid Power Association Recommended Standard Glossary of Terms for Fluid Power Quick Action Couplings.**

NFPA/T3.20.6M-1982, **National Fluid Power Association Recommended Standard Hydraulic fluid power - Quick action couplings (with pipe thread or SAE straight thread connections) - Method of measuring and reporting pressure drop.**

NFPA/T2.10.1M-1978, **National Fluid Power Association Recommended Standard Metric Units for Fluid Power Applications.**

ANSI/B93.2-1971 and Supplement ANSI/B93.2A-1978, **American National Association Glossary of Terms for Fluid Power.**

ANSI/B93.42M-1981, **American National Standard Test Conditions and Procedures for Hydraulic Fluid Power Quick Action Couplings.**

ISO 1000-1981, **International Standard SI units and recommendations for the use of their multiples and of certain other units.**

3 TERMS AND DEFINITIONS

For definition of terms used, see ANSI/B93.2 and Supplement ANSI/B93.2A, and NFPA/T3.20.1.

4 UNITS OF MEASUREMENT

4.1 Units of measurement are used in accordance with NFPA/T2.10.1M. This document is in agreement with ISO 1000.

4.2 Approximate conversions to Customary US units are shown in parentheses after their metric counterparts and are made in accordance with NFPA/T2.10.1M.

5 SELECTION AND EXAMINATION OF TEST SAMPLES

5.1 Select coupling assemblies representing a production lot in all respects pertaining to design, material, surface treatment process, etc.

5.2 Apply adequate identification to the coupling when more than one sample is used.

5.3 Permanently mark each coupling or coupling half in a manner suitable for identification with the required test procedures and reports.

6 TEST EQUIPMENT

6.1 Use test equipment described in figure 1.

6.2 Conduct tests at an ambient temperature of 20°C (68°F) to 35°C (95°F).

6.3 Specify the fluid type, viscosity and temperature used for this test.

7 TEST CONDITIONS ACCURACY

Table 1 - Test conditions accuracy (unless otherwise specified)

Test Condition	SI Units	Customary US Unit	Maintain Within (±) of Actual Gage Value
Force	N	lbf	3%
Frequency	Hz	cps	10%
Flow rate	L/min	gpm	3%
Length	mm	in	3%
Mass	kg	lb	3%
Pressure (above atm)	bar	psi	3%
Pressure (below atm)	bar, abs	in Hg, abs	2%
Temperature	°C	°F	2.8°C (5°F)
Time	min	min	3%
Torque	N·m	lb·in	3%

8 TEST PROCEDURE

8.1 THIS PROCEDURE INVOLVES HIGH OIL VELOCITIES. PRECAUTIONS SHOULD BE EXERCISED IN SETUP METHODS, TEST PROCEDURES AND EQUIPMENT USED TO AVOID SAFETY HAZARDS AND EQUIPMENT DAMAGE.

8.2 Leakage test the coupling in accordance with ANSI/B93.42M.

8.3 Pressure drop test the coupling in accordance with ANSI/B93.42M or NFPA/T3.20.6M.

8.4 Multiply the gross pressure drop at rated flow by 25 to establish the surge test pressure. (Gross pressure drop as obtained in accordance with ANSI/B93.42M or NFPA/T3.20.6M).

NOTE: If rated flow is not specified, use the rated flow shown in table 3. See appendix for an explanation on the establishment of the surge test pressure.

8.5 Install the coupling in a test circuit as illustrated in figure 1.

8.6 Adjust the oil supply circuit discharge characteristics to meet the pressure-time curve illustrated in figure 2. The point of pressure-time measurement is shown on figure 1. One oil discharge equals one cycle. Include a copy of the actual pressure-time curve in the test data.

8.7 Conduct 100 cycles.

8.8 Reverse the coupling in the circuit. Adjust the oil discharge characteristics if necessary to meet 8.4 and 8.6. (Adjustment is necessary only if the coupling pressure drop differs by more than 10% between direction at rated flow.)

8.9 Conduct 100 cycles in this flow direction.

8.10 Leakage test the coupling in accordance with ANSI/B93.42M.

8.11 Pressure drop test the coupling in accordance with ANSI/B93.42M.

8.12 Record the following test results.

8.12.1 Leakage before and after the cycle test.

8.12.2 Pressure drop before and after the cycle test.

8.12.3 Visual signs of damage caused by the cycle test.

9 SUMMARY OF DESIGNATED INFORMATION

The following designated information is needed when applying this recommended standard to a particular application or use:

9.1 Rated flow.

9.2 Surge test pressure (see appendix).

9.3 Maximum operating pressure.

10 DATA ACCURACY

Table 2 - Data Accuracy

Quantity	SI Units	Customary US Unit	Maintain Within (±) of Actual Gage Value
Force	N	lbf	3%
Pressure	bar	psi	3%
Pressure drop	bar	psi	3%
Temperature	°C	°F	2.8°C(5°F)
Torque	N·m	lb·in	3%
Volume (leakage)	mL		1%

11 DATA PRESENTATION

Record test data on a form similar to table 4.

12 JUSTIFICATION STATEMENT

This recommended standard is based on the combined expert experiences of those who have participated in the development and review of this standard.

13 TEST/PRODUCTION SIMILARITY

Utilize managerial controls necessary to maintain substantial similarity between test and production components or elements.

14 IDENTIFICATION STATEMENT

Use the following statement in catalogs and sales literature when electing to comply with this National Standard:

"Method of obtaining and presenting performance data conforms to ANSI/B93.68M-1983, Hydraulic fluid power - Quick action couplings - Surge flow test (short duration flow)."

Table 3 - Rated Flows

Coupling Size	Rated Flow	
5 mm (1/8")	3 L/min	(0.8 gpm)
6.3 (1/4")	12	(3)
10 (3/8")	23	(6)
12.5 (1/2")	45	(12)
20 (3/4")	106	(28)
25 (1")	189	(50)
31.5 (1-1/4")	288	(76)
40 (1-1/2")	379	(100)
50 (2")	757	(200)

NOTE: Rated flow based on approximately 6 m/s (20 ft/sec) velocity in a conductor with an inside diameter equal to the coupling size listed above.

ANSI/B93.68

Table 4 - **Typical form for presentation of test data**

Coupling Manufacturer:_____

Coupling P/N:_____ S/N or Ident.:_____

Date Tested:_____ Test By:_____

Name of Test	Test Results		Remarks
	Si Units	Customary US Unit	
Surge Flow:			
Rated Flow	_____L/min	_____gpm	
Surge Test Pressure	_____bar	_____psi	
Max. Oper. Pressure	_____bar	_____psi	
Fluid			
Viscosity	_____mm^2/s	_____cSt	
Fluid Temperature	_____°C	_____°F	
Leakage Before Cycle Test:			
Low Pressure Coupled	_____mL/h		_____Test Press.
Low Pressure Uncoupled	_____mL/h		_____Test Press.
Max. Oper. Pressure, Coupled	_____mL/h		_____Test Press.
Max. Oper. Pressure, Uncoupled	_____mL/h		_____Test Press.
Pressure Drop Before Cycle Test			Attach Graph or Chart
Leakage After Cycle Test:			
Low Pressure, Coupled	_____mL/h		_____Test Press.
Low Pressure, Uncoupled	_____mL/h		_____Test Press.
Max. Oper. Pressure Coupled	_____mL/h		_____Test Press.
Max. Oper. Pressure Uncoupled	_____mL/h		_____Test Press.
Pressure Drop After Cycle Test			Attach Graph or Chart
Damage After Cycle Test			

ANSI/B93.68

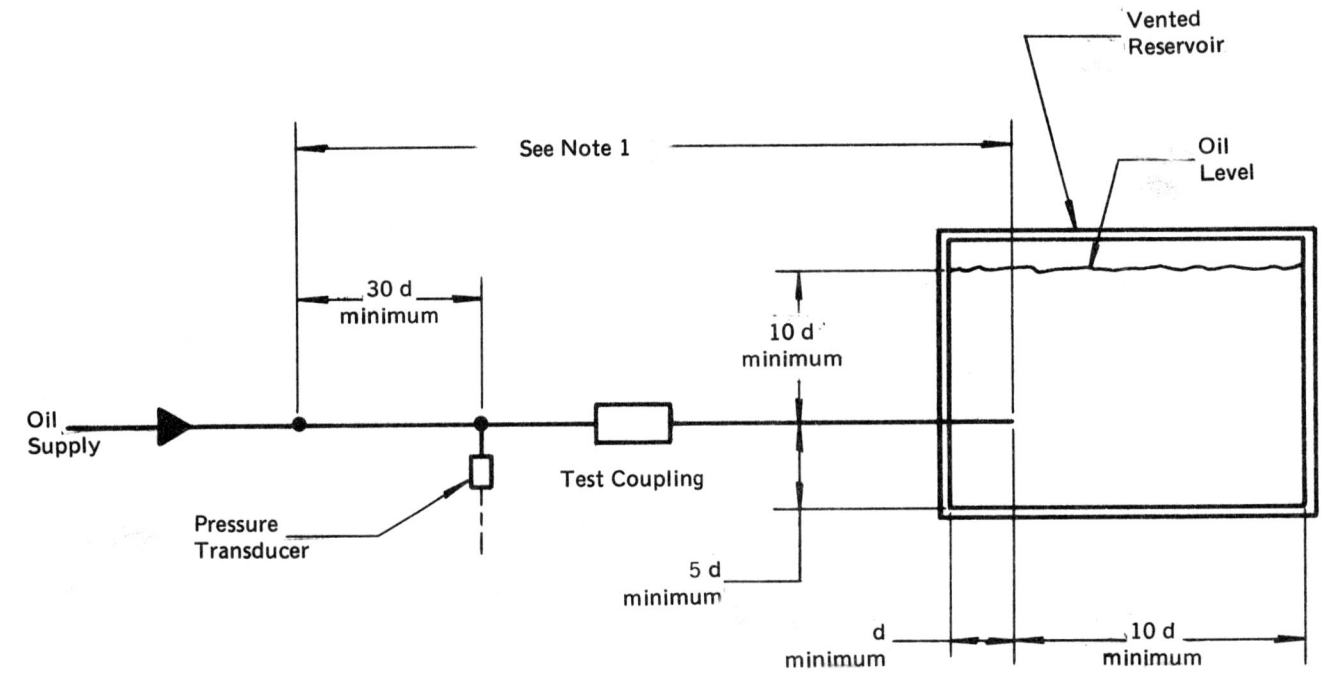

d = Tubing I. D.

FIGURE 1 - Test circuit

NOTES: 1. Inlet flow tubes and test unit are the same as used for the pressure drop test. Outlet flow tubes are the same size and length as used in the pressure drop test.

2. Oil supply to be capable of producing a pressure/time curve at the transducer in accordance with figure 2. Exact details of the oil supply circuit are optional.

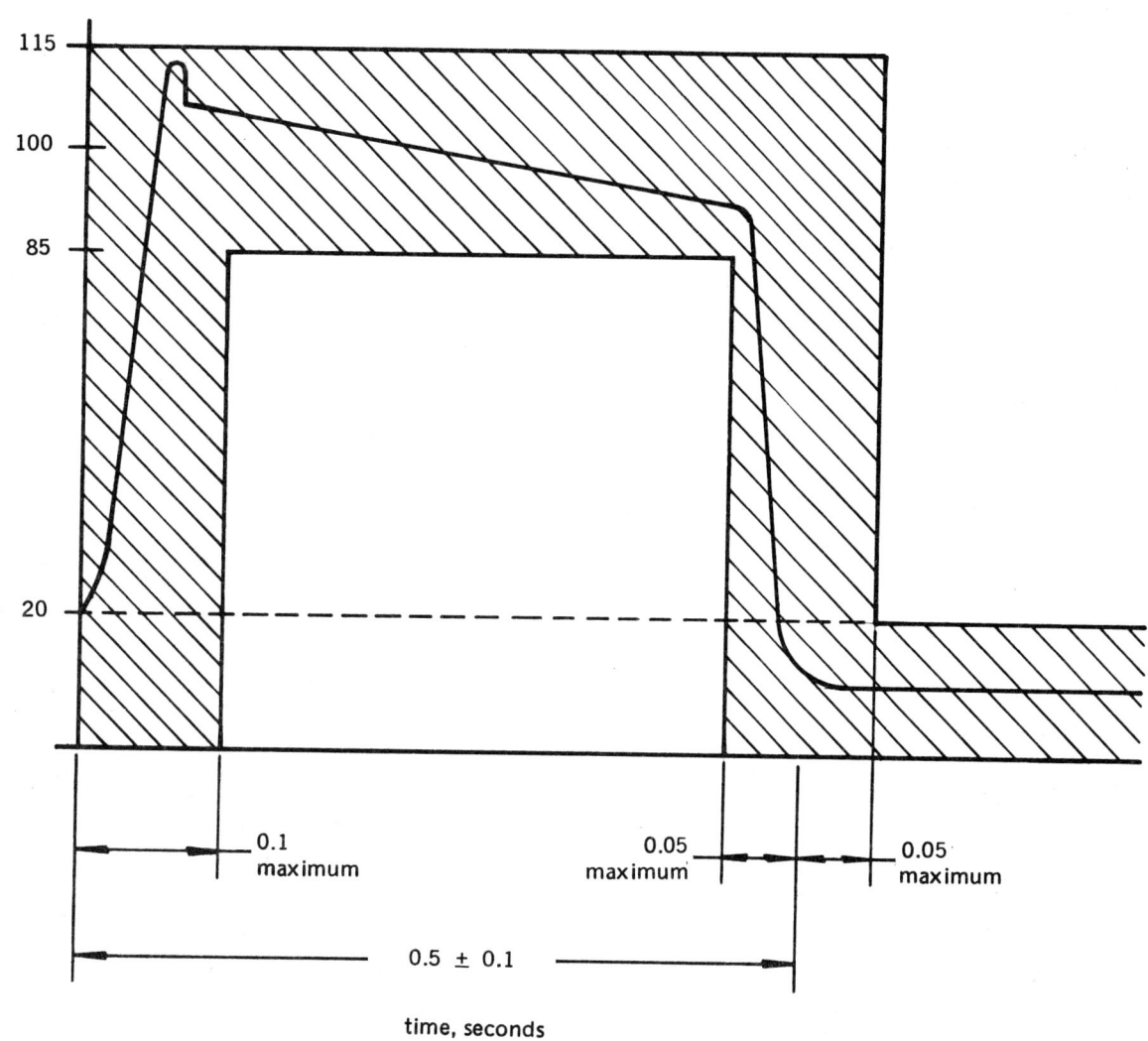

FIGURE 2 - Pressure/time curve

Appendix

to ANSI/B93.68M-1983

2 REFERENCES

ISO 5598[1], Fluid power systems and components - Vocabulary.

NOTE: This test is intended to test the coupling under short durations of flows at approximately 5 times the rated flow. Since the pressure drop increases at approximately the square of the flow rate, 25 times the pressure drop at rated flow was selected for this procedure.

---

[1] Presently at the stage of draft.

ANSI/B93.69M-1983

AN INDUSTRY STANDARD FOR FLUID POWER

American National Standard

Hydraulic fluid power -

Quick action couplings -

Surge flow test (long duration flow)
(NFPA/T3.20.12M-1983)

Approved as an ANSI Standard
30 December 1983

Descriptors: hydraulic fluid power; quick action couplings; rated flow, surge, flow; surge, pressure; testing, leakage; testing, pressure drop; testing.

published by
NATIONAL FLUID POWER ASSOCIATION, INC.
3333 N. Mayfair Road / Milwaukee, WI 53222 / 414-778-3344 / TLX 26898

ANSI/B93.69

FOREWORD

This Foreword is not part of American National Standard Hydraulic fluid power - Quick action couplings - Surge flow test (long duration flow), ANSI/B93.69M-1983 (NFPA/T3.20.12M-1983).

ANSI/B93.62M (NFPA/T3.20.11M) Hydraulic fluid power - Quick action couplings - Surge flow test (short duration flow) was designed to test for short duration flow surges typically found in hydraulic circuits. However, some applications experience longer durations of surge flows. It was recognized that a completely different test procedure and test equipment were required to evaluate couplings under those conditions.

A new project to develop a test procedure for longer duration surge flow was approved by the Technical Board on 11 February 1981.

The Project Group prepared a first Draft on 15 March 1981. This Draft was reviewed and revised at the T3.20 Section meeting of 25 March 1981.

Draft No. 2 was prepared on 6 April 1981 for circulation within Section T3.20. It was then forwarded to NFPA Headquarters for General Review.

Headquarters Technical Staff prepared the document for General Review on 24 April 1981. General Review closed on 25 May 1981. Editorial comments received were incorporated into the draft. The Technical Board granted approval to Ballot on 15 September 1981.

The Ballot Draft was prepared by Headquarters Technical Staff on 17 September 1982. Successful Ballot closed 15 October 1982.

The NFPA Technical Board approved T3.20.12M on 2 February 1983. The document was then forwarded to the Board of Directors who granted final approval on 27 February 1983.

Project Group members who developed this standard:

Ken Koch
Project Chairman &
Section Vice Chairman
Imperial Clevite, Inc./Bruning Operations

Edward Saloum
Section Chairman
Snap-Tite, Inc.

Donald Fischer
Section Secretary
Aeroquip Corp.

Ed Ratkay *
Technical Auditor
Commercial Shearing, Inc.

Allen E. Tucker
Director of Technical Services
National Fluid Power Association

Carroll Grigsby
Schrader Bellows Division

Eugene Herzan
Parker Hannifin Corp.

Thomas Karcher
Hansen Manufacturing Co.

Ojars Maldavs
Imperial Clevite, Inc.

Stanlee Meisinger
Safe Way Hydraulics, Inc.

Steve Patin
John Deere Product Engineering Center

R. Rogers
Aeroquip Corp.

It is the intent of the Project Group and the Quick Action Coupling Section (T3.20) to forward NFPA/T3.20.12M to ANSI Committee B93 for promulgation as an ANSI standard.

On 2 September 1983, ANSI/B93.69M was submitted to ANSI Committee B93 for ballot. Successful balloting concluded 21 November 1983.

ANSI/B93.69M was forwarded to ANSI Board of Standards Review on 29 November 1983 and granted approval on 30 December 1983.

The membership roster of Standards Committee B93 at the time of ballot:

Jack C. McPherson
Chairman

Daniel B. Shore
Vice Chairman

Allen E. Tucker
Co-Secretary

William G. Wagner
Co-Secretary

American Society of Agricultural Engineers
Ed Fletcher

American Society for Engineering Education
(to be named)

American Society of Lubrication Engineers
(to be named)

Compressed Air & Gas Institute
David E. Bonn
John Addington (alternate)

Construction Industry Manufacturers Association
Glenn Stewart

Fluid Controls Institute
Jude Pauli
Eric Bianchi (alternate)

Fluid Power Distributors Association
Thomas Neff

Fluid Power Society
Robert L. Firth
Carroll Grigsby
Robert W. Hanpeter
Marty King
Richard Read
Ronald Smith
Alan Tiedman

Fluid Sealing Association
John Scannell
Alex Pilecki

Instrument Society of America
(to be named)

Joint Industry Council
Robert Muhl

Material Handling Institute
Jack C. McPherson
Willard Chichester (alternate)

National Fluid Power Association
Richard N. Bailey
John Bowbin
Walter Forster
Z. J. Lansky
Robert Kay
Paul Schacht

---

* Retired

National Machine Tool Builders Association
John B. Deam

Rubber Manufacturers Association
William A. Hertel
John R. Loder
E. J. McCarthy (alternate)

Society of Automotive Engineers
William A. Hertel
John T. Parrett
Henry Schultz
Daniel B. Shore
W. L. Snyder
Robert W. White

US Deptartment of Defense
William P. Coyne

Company Members
Lloyd L. Schmalz
Don McGeachy
Logan Mathis
John Welker (alternate)

Individual Member
Dr. E. C. Fitch
Otto Maha
John J. Pippenger
A. O. Roberts
Jack Walrad
Tom Wanke
Frank Yeaple

mkm

ANSI/B93.69

Hydraulic fluid power - Quick action couplings -
Surge flow test (long duration flow)

0 INTRODUCTION

In hydraulic fluid power systems, power is transmitted and controlled thru a liquid under pressure within an enclosed circuit. Quick action couplings are used to quickly join or separate fluid conducting lines without the use of tools or special devices.

It is recognized that hydraulic quick action couplings may be exposed to surge flow conditions above the normal rated flow. Depending on the application, the surge flows may be short duration and/or long duration. Testing couplings under the two conditions generally required two different test methods. The user may select either method or both methods as appropriate for his/her application or needs.

ANSI/B93.68M, (Hydraulic fluid power - Quick action couplings - Surge flow test (short duration flow)) is recommended for short duration flow surges. An accumulator-discharge test method is used to generate the short duration surge flow.

This document, ANSI/B93.69M, is recommended for testing hydraulic quick action couplings under longer duration surge flow conditions. A continuous flow test method is used.

1 SCOPE AND FIELD OF APPLICATION

1.1 This National Standard applies to identical coupling halves, male or female coupling halves, and coupling assemblies used in hydraulic fluid power systems.

1.2 This National Standard includes couplings with and without fluid sealing means, when uncoupled.

1.3 This National Standard intends to:

- Promote the understanding and use of quick action couplings;
- Provide uniform test conditions and procedures;
- Facilitate accurate communications;
- Provide a basis for component selection and application.

2 REFERENCES

ANSI/B93.2-1971 and Supplement ANSI/B93.2A-1978, **American National Standard Glossary of Terms for Fluid Power.**

ANSI/B93.42M-1977 (R1983), **American National Standard Method for Testing Hydraulic Fluid Power Quick Action Couplings.**

ANSI/B93.64M-1983, **American National Standard Hydraulic fluid power - Quick action couplings (with pipe thread or SAE straight thread connections) - Method of measuring and reporting pressure drop.**

ANSI/B93.68M-1983, **American National Standard Hydraulic fluid power - Quick action couplings - Surge flow test (short duration flow).**

NFPA/T2.10.1M-1978, **National Fluid Power Association Recommended Standard Metric Units for Fluid Power Applications.**

NFPA/T3.20.1-1973 (R1981), **National Fluid Power Association Recommended Standard Glossary of Terms for Fluid Power Quick Action Couplings.**

ISO 1000-1981, **International Standard SI units and recommendations for the use of their multiples and of certain other units.**

3 TERMS AND DEFINITIONS

For definitions of terms used, see ANSI/B93.2, ANSI/B93.2A and NFPA/T3.20.1.

4 UNITS OF MEASUREMENT

4.1 Units of measurement are used in accordance with NFPA/T2.10.1M. This document agrees with ISO 1000.

4.2 Approximate conversions to Customary US units are shown in parentheses after their metric counterparts and are made in accordance with NFPA/T2.10.1M.

5 SELECTION AND EXAMINATION OF TEST SAMPLES

5.1 Select coupling assemblies representing a production lot in all respects pertaining to design, material, surface treatment, process, etc.

5.2 Apply adequate identification to the coupling when more than one sample is used.

5.3 Permanently mark each coupling or coupling half in a manner suitable for identification with the required test procedures and reports.

6 TEST CONDITIONS

6.1 Conduct test at an ambient temperature of 20°C (68°F) to 35°C (95°F).

6.2 Specify the fluid type, viscosity and temperature used for this test.

7 TEST CONDITION ACCURACY

TABLE 1 - **Test conditions accuracy**
(unless otherwise specified)

Test Condition	Metric Unit	Customary US Unit	Maintain Within (±) of Actual Gage Value
Force	N	lb	3%
Frequency	Hz	cps	10%
Flow rate	L/min	gpm	3%
Length	mm	in	3%
Mass	kg	lb	3%
Pressure (above atm)	bar	psi	3%
Pressure (below atm)	bar, abs	in Hg, abs	2%
Temperature	°C	°F	2.8°C (5°F)
Time	s/min/h	sec/min/hr	3%
Torque	N·m	lb·in	3%

8 TEST PROCEDURE

8.1 Leak test the coupling in accordance with ANSI/B93.42M.

8.2 Pressure drop test the coupling in accordance with ANSI/B93.42M.

8.3 Subject the coupling to the specified surge flow rate for a duration of five seconds minimum in each direction. Repeat the cycle for a total of 100 times.

NOTES:

1) If the surge flow rate is not specified, use five times the coupling rated flow.

2) If rated flow is not specified, use the rated flows of table 3.

8.4 Leak test the coupling in accordance with ANSI/B93.42M.

8.5 Pressure drop test the couplings in accordance with ANSI/B93.42M.

8.6 Record the following results:

   a) Leakage before and after the surge cycle tests;

   b) Pressure drop before and after the surge cycle test;

   c) Visual signs of damage caused by the cycle test.

9 SUMMARY OF DESIGNATED INFORMATION

9.1 The following designated information is needed when applying this recommended standard to a particular application or use:

   a) Rated flow;

   b) Surge flow test.

10 DATA ACCURACY

Table 2 - **Data accuracy**

Quantity	Metric Unit	Customary US Unit	Maintain Within (±) of Actual Gage Value
Force	N	lb	3%
Pressure	bar	psi	3%
Pressure drop	bar	psi	3%
Temperature	°C	°F	2.8°C (5°F)
Torque	N·m	lb·in	3%
Volume	mL	oz	1%

11 DATA PRESENTATION

Record test data on a form similar to table 4.

12 JUSTIFICATION STATEMENT

This recommended standard is based on the combined expert experiences of those who have participated in the development and review of this standard.

13 TEST/PRODUCTION SIMILARITY

Utilize managerial controls necessary to maintain substantial similarity between test and production components or elements.

ANSI/B93.69

14 IDENTIFICATION STATEMENT

Use the following statement in catalogs and sales literature when electing to comply with this voluntary standard:

"Method of obtaining and presenting performance data conforms to American National Standard, ANSI/B93.69M-1983, **Hydraulic fluid power - Quick action couplings - Surge flow test (long duration flow)."**

Table 3 - **Rated flows**

Coupling Size	Rated Flow	
5. mm (1/8")	3 L/min	(0.8) gpm
6.3 mm (1/4")	12 L/min	(3) gpm
10. mm (3/8")	23 L/min	(6) gpm
12.5 mm (1/2")	45 L/min	(12) gpm
20. mm (3/4")	106 L/min	(28) gpm
25. mm (1 ")	189 L/min	(50) gpm
31.5 mm (1-1/4")	288 L/min	(76) gpm
40. mm (1-1/2")	379 L/min	(100) gpm
50. mm (2 ")	757 L/min	(200) gpm

NOTE: Rated flow based on approximately 6 m/s (20ft/sec) velocity in a conductor with an inside diameter equal to the coupling size listed above.

Table 4 -**Typical form for presentation of test data**

Coupling Manufacturer: _____
Coupling P/N: _____ S/N or Identification: _____
Test By: _____ Date Tested: _____

Name of Test	Test Results		Remarks
	Metric Unit	Customary US Unit	
Surge Flow			
Rated Flow	...................L/min	...................gpm	
Surge Test Flow	...................L/min	...................gpm	
Fluid			
Viscosity	...................mm²/s	...................cSt	
Fluid Temperature	...................°C	...................°F	
Leakage Before Cycle Test			
Low Pressure Coupled	...................mL/h		...................Test Pressure
Low Pressure Uncoupled	...................mL/h		...................Test Pressure
Maximum Operating Pressure, Coupled	...................mL/h		...................Test Pressure
Maximum Operating Pressure, Uncoupled	...................mL/h		...................Test Pressure
Pressure Drop Before Cycle Test			Attach Graph or Chart
Leakage After Cycle Test:			
Low Pressure, Coupled	...................mL/h		...................Test Pressure
Low Pressure, Uncoupled	...................mL/h		...................Test Pressure
Maximum Operating Pressure, Coupled	...................mL/h		...................Test Pressure
Maximum Operating Pressure, Uncoupled	...................mL/h		...................Test Pressure
Pressure Drop After Cycle Test			Attach Graph or Chart
Damage After Cycle Test			

APPENDIX to

ANSI/B93.69M-1983

This Appendix is not part of the ANSI/B93.69M-1983 standard, but is included for information purposes only.

2 REFERENCES

ISO 5598[1], **International Standard Fluid power systems and components - Vocabulary.**

---
[1] **Presently at the stage of draft.**

NOTES

NFPA Bibliography
T3.26.1 R1-1977

AN INDUSTRY STANDARD FOR FLUID POWER

A Bibliography of

Fluid Power Hose, Hose Fittings and Hose Assemblies

published by
NATIONAL FLUID POWER ASSOCIATION, INC.

3333 N. Mayfair Road / Milwaukee, WI 53222 / 414-778-3344 / TLX 26898

SOURCE OF DOCUMENTS

AAR	Association of American Railroads 1920 L Street, N.W. Washington, DC 20036
API	American Petroleum Institute 1801 K Street, N.W. Washington, DC 20006
ASTM	American Society for Testing and Materials 1916 Race Street Philadelphia, PA 19103
DOT	Department of Transportation National Highway Traffic Safety Administration 400 Seventh Street, S.W. Washington, DC 20590
FAA	Federal Aviation Administration 400 Seventh Street, S.W. Washington, DC 20590
ISO	International Standards Organization 1, Rue de Varembe 1211 Geneva 20 SWITZERLAND
JIC	Joint Industrial Council 7901 Westpark Drive McLean, VA 22101
NPFC	Commanding Officer, Naval Publications and Forms Center 5801 Tabor Avenue Philadelphia, PA 19120 Attn: NPFC 105
SAE	Society of Automotive Engineers 400 Commonwealth Drive Warrendale, PA 15096

NOTE: This bibliography provides a list of standards related to fluid power hose, hose fittings and hose assemblies which have been issued by standards-writing organizations other than NFPA, ANSI/B93 and ISO/TC 131. Listing in this bibliography is only for reference and does not imply Association endorsement.

NFPA/T3.26.1 R1

Identification Number	Title	Source	Date of Issue
AAR/M-601A	Hose, Air Brake and Train Air Signal	AAR	1968
AAR/M-608	Hose, Air Pneumatic Tool (Wrapped or Braided)	AAR	1966
AAR/M-609	Hose, Lubrication Oil (Wrapped or Braided)	AAR	1949
AAR/M-618	Hose, Air, Wire Reinforced	AAR	1968
API/7	Spec. for Rotary Drilling Equipment	API	1969
ASTM D-380	Standard Method of Testing Rubber Hose	ASTM	1965
ASTM D-571	Standard Method of Testing Auto Hydraulic Brake Hose	ASTM	1955
ASTM D-622	Standard Method of Testing Auto Air Brake and Vacuum Brake Hose	ASTM	1965
FAA/PPE-3	Standard Fire Test Apparatus and Procedures	FAA	1962
FAA/TSO-C75	Hydraulic Hose Assemblies-Part 514	FAA	1962
FMVSS 571.106	Federal Motor Vehicle Safety Standard, Brake Hoses	DOT	
ISO/1307	Rubber hose - Bore sizes, tolerances on length, and test pressures	ISO	1975
ISO/1402	Rubber hose - Hydrostatic testing	ISO	1974
ISO/1436	Wire reinforced, rubber covered hydraulic hose	ISO	1972

NFPA/T3.26.1 R1

Identification Number	Title	Source	Date of Issue
ISO/1746	Rubber hoses- Bending test	ISO	1976
ISO/1823	Rubber hoses for oil suction and discharge	ISO	1975
ISO/2398	Industrial rubber hose for compressed air (up to 2,5 MPa)	ISO	1975
JIC/H4	Hydraulic Standard for Industrial Equipment	JIC	
MIL-A-52525B	Adapters, Elbows and Clamp Halves, for Hose, Hydraulic, Field Attachable, Reusable, Pressure Type	NPFC	1976
MIL-H-775C	Hose, Rubber, Plastic Fabric, or Metal; and Fittings, Nozzles and Strainers	NPFC	1973
MIL-H-3992C	Hose and Hose Assembly, Rubber, Air and Vacuum Brake, Automotive	NPFC	1968
MIL-H-5593B	Hose, Aircraft, Low Pressure, Flexible	NPFC	1968
MIL-H-8788B	Hose, Hydraulic and Pneumatic, High Pressure	NPFC	1977
MIL-H-8790	Hose Assemblies, Hydraulic High Pressure (3000 psi)	NPFC	1970
MIL-H-8794D	Hose, Rubber, Hydraulic and Oil Resistant	NPFC	1971
MIL-H-13531B	Hose, Rubber and Hose Assembly Rubber (Hydraulic, Flexible)	NPFC	1975
MIL-H-13719C	Hose, Assembly, Rubber: Hydraulic Brake	NPFC	1972

NFPA/T3.26.1 R1

Identification Number	Title	Source	Date of Issue
MIL-H-24135	Hose, Reinforced, Water and Oil Resistant, and End Fittings, Reusable, For Flexible Hose Connections	NPFC	1968
MIL-H-24136	Hose, Synthetic Rubber, Polyester Reinforced and End Fittings, Reusable, for Flexible Hose Connections	NPFC	1970
MIL-H-24239	Hose Assemblies, Hydraulic Nonmagnetic, for Minesweeping Tensiometers	NPFC	1967
MIL-H-25579C	Hose Assembly, Tetrafluoroethylene, High Temperature, Medium Pressure, General Requirements.	NPFC	1971
MIL-H-26666D	Hose Assembly, Pneumatic High Pressure	NPFC	1968
MIL-H-27516	Hose, Rubber and Hose Assembly	NPFC	1974
MIL-H-28523	Hose Assembly, Rubber, Hydraulic and Pneumatic Jetting (400 psi) Working Pressure	NPFC	1969
MIL-H-38209B	Hose Reel Assembly, Fuel Servicing, General Requirements for	NPFC	1971
MIL-H-38360A	Hose Assembly, Tetrafluoroethylene, High Temperature, High Pressure Hydraulic and Pneumatic	NPFC	1973
MIL-H-38390A	Hose Assembly, Tetrafluoroethylene, Pneumatic, High Pressure	NPFC	1965

NFPA/T3.26.1 R1

Identification Number	Title	Source	Date of Issue
MIL-H-52471B	Hose and Hose Assemblies, Rubber Hydraulic, Pressure-type	NPFC	1976
MIL-H-52544	Hose, Rubber; Oil Suction, Wire Reinforced	NPFC	1974
MS 8004B	Hose Assembly, Detachable Fittings Tetrafluoroethylene, High Temperature, Medium Pressure, Flange to Flange	NPFC	1976
MS 27364D	Hose Assembly, Tetrafluoroethylene, Hydraulic, Pneumatic (3000 psi), Straight to Elbow 45°, Flared Tube	NPFC	1972
MS 27365D	Hose Assembly, Tetrafluoroethylene, Hydraulic, Pneumatic (3000 psi), Straight to Elbow 90°, Flared Tube	NPFC	1972
MS 27366D	Hose Assembly, Tetrafluoroethylene, Hydraulic, Pneumatic (3000 psi), Elbow 45° to Elbow 45°, Flared Tube	NPFC	1972
MS 27367D	Hose Assembly, Tetrafluoroethylene, Hydraulic, Pneumatic (3000 psi), Elbow 45° to Elbow 90°, Flared Tube	NPFC	1972
MS 27368D	Hose Assembly, Tetrafluoroethylene, Hydraulic, Pneumatic (3000 psi), Elbow 90° to Elbow 90°, Flared Tube	NPFC	1972
MS 27369C	Hose Assembly, Tetrafluorethylene, Hydraulic, Pneumatic (3000 psi), Flareless.	NPFC	1972

NFPA/T3.26.1 R1

Identification Number	Title	Source	Date of Issue
MS 27370D	Hose Assembly, Tetrafluoroethylene, Hydraulic, Pneumatic, (3000 psi), Straight to Elbow 45°, Flareless Tube	NPFC	1972
MS 27371D	Hose Assembly, Tetrafluoroethylene, Hydraulic, Pneumatic (3000 psi), Straight to Elbow 90°, Flareless Tube	NPFC	1972
MS 27372D	Hose Assembly, Tetrafluoroethylene, Hydraulic, Pneumatic (3000 psi), Elbow 45° to Elbow 45°, Flareless Tube	NPFC	1972
MS 27374D	Hose Assembly, Tetrafluoroethylene, Hydraulic, Pneumatic (3000 psi), Flareless Tube	NPFC	1972
MS 28741G	Hose Assembly, Detachable End Fitting, Medium Pressure	NPFC	1971
MS 28759B	Hose Assembly, Rubber, Hydraulic and Pneumatic (3000 psi), Flared Tube	NPFC	1969
MS 28920A	Hose Assembly, Rubber, Hydraulic, (3000 psi), Fitting End to Elbow 45°, Flared Tube	NPFC	1969
MS 28921A	Hose Assembly Rubber, Hydraulic, (3000 psi), Fitting End to Elbow 90°, Flared Tube	NPFC	1969
MS 28924A	Hose Assembly, Rubber, Hydraulic, (3000 psi), Elbow 90° to Elbow 90°, Flared Tube	NPFC	1969

NFPA/T3.26.1 R1

Identification Number	Title	Source	Date of Issue
MS 39262A	Hose, Rubber: Hydraulic, Pressure Type Single Wire Braid Reinforcement (for Engineer Equipment)	NPFC	1970
MS 39263A	Hose, Rubber: Hydraulic, Pressure Type, Double Wire Braid Reinforcement	NPFC	1970
MS 39264A	Hose, Rubber: Hydraulic, Pressure Type, Four Spiral Wrap Reinforcement	NPFC	1970
MS 39265A	Hose Rubber: Hydraulic Pressure Type, Six Spiral Wrap Reinforcement (for Engineer Equipment)	NPFC	1970
MS 39266A	Hose Assembly, Rubber: Hydraulic Pressure Type, Length Measurement	NPFC	1970
MS 39267A	Hose Assembly, Rubber: Hydraulic, Pressure Type, Minimum Bend Radius (for Engineer Equipment)	NPFC	1970
MS 39306	Hose Assembly, Rubber: Hydraulic, Oil Suction, with Reusable Couplings	NPFC	1976
MS 39316B	Hose Assemblies: Hydraulic, Single Wire Braid Reinforcement, with Reusable Couplings	NPFC	1976
MS 39317B	Hose Assemblies, Hydraulic, Double Wire Braid Reinforcement, with Reusable Couplings	NPFC	1976
MS 39318B	Hose Assemblies, Hydraulic, 4-Spiral Wrap - Reinforcement, with Reusable Couplings	NPFC	1976
MS 39319B	Hose Assemblies: Hydraulic 6-Spiral Wrap Reinforcement, with Reusable Coupling	NPFC	1976

Identification Number	Title	Source	Date of Issue
MS 39320A	Hose Assembly: Measurement of Coupling Orientation Angle	NPFC	1970
MS 39325A	Hose Assemblies, Air Brake	NPFC	1974
MS 500077C	Hose Assembly, Rubber: Hydraulic and Pneumatic, Medium Pressure, Flared Tube	NPFC	1976
MS 500083C	Hose Assembly, Rubber: Hydraulic and Pneumatic, High Pressure, Flared Tube	NPFC	1973
SAE AIR 797A	Hose Characteristics and Selection Chart	SAE	1968
SAE ARP 24B	Determination of Hydraulic Pressure Drop	SAE	1968
SAE ARP 600	Torque Determination, Method of, For Tube or Hose End Fittings	SAE	1968
SAE ARP 601	Hose Assemblies, Flexible Metal, Aeronautical, Low Pressure	SAE	1960
SAE ARP 602	Hose Assemblies, Flexible Metal, Aeronautical, Medium Pressure	SAE	1960
SAE ARP 603A	Impulse Test Equipment for Testing Hydraulic System Components	SAE	1963
SAE ARP 604D	Hose Assemblies, Aircraft and Missiles, High Temperature, High Pressure	SAE	1967

NFPA/T3.26.1 R1

Identification Number	Title	Source	Date of Issue
SAE ARP 611	Tetrafluoroethylene Hose Assembly Cleaning Methods	SAE	1963
SAE ARP 614A	Hose Assemblies, Aircraft and Missile, High Temperature (450°F) High Pressure (4000 psi), Aircraft and Missile Fluid Systems	SAE	1965
SAE ARP 620	High Temperature Hose Assembly, Convoluted Tetrafluoroethylene, for Aircraft	SAE	1965
SAE ARP 683	Installation Procedures and Torques for Fluid Connections	SAE	1964
SAE ARP 824	Hose Assemblies, Flexible Metal High Pressure and High Temperature	SAE	1966
SAE ARP 907	Missle and Rocket Applications, Hose Assemblies, Flexible Metal, High Pressure and High Temperature	SAE	1966
SAE ARP 908	Hose Fitting - Installation and Qualification Test Torque Requirements	SAE	1968
SAE AS 756	Fitting End Assembly, Universal	SAE	1963
SAE AS 758	Fittings, Installation in Straight Threaded Boss	SAE	1963
SAE AS 759	Fittings, Universal or Internally Threaded, Flared and Flareless	SAE	1963
SAE AS 1011	Fitting End; Attachable, Male for 37° Flared Tube Connection, Flexible Metal Hose - Straight, 45° Elbow and 90° Elbow Styles	SAE	1969

- 204 -

NFPA/T3.26.1 R1

Identification Number	Title	Source	Date of Issue
SAE AS 1012	Fitting End; Attachable, Swivel Female 37° Machined Flare, Flexible Metal Hose - Straight, 45° Elbow and 90° Elbow Styles	SAE	1969
SAE AS 1013	Fitting End; Attachable, Swivel Female 37° Flared Tube, Flexible Metal Hose - Straight, 45° Elbow and 90° Elbow Styles	SAE	1969
SAE AS 1014	Fitting End; Attachable, Swivel Flange, Flexible Metal Hose - Straight, 45° Elbow and 90° Elbow Styles	SAE	1969
SAE AS 1015	Fitting End; Attachable, Female Flareless, Machined Nipple, Flexible Metal Hose - Straight, 45° Elbow and 90° Elbow Styles	SAE	1969
SAE AS 1016	Fitting End; Attachable, Flareless Male, Flexible Metal Hose	SAE	1969
SAE J246b	Spherical Sleeve (Compression) Tube Fittings	SAE	1974
SAE J343c	Tests and Procedures for SAE 100R Series Hydraulic Hose and Hose Assemblies	SAE	1975
SAE J514g	Hydraulic Tube Fittings	SAE	1974
SAE J515	Hydraulic "O" Ring	SAE	1967
SAE J516b	Hydraulic Hose Fittings	SAE	1975
SAE J517c	Hydraulic Hose	SAE	1976
SAE J518c	Hydraulic Flanged Tube, Pipe and Hose Connections, 4 Bolt Split Flange Type	SAE	1972

NFPA/T3.26.1 R1

Identification Number	Title	Source	Date of Issue
SAE J844c	Air Brake Tubing and Pipe	SAE	1970
SAE J962a	Formed Tube Ends for Hose Connections	SAE	1976
SAE J1131	Performance Requirements for SAE J844D Nonmetallic Tubing and Fitting Assemblies used in Automotive Air Brake Systems	SAE	1976
SAE J1401	Hydraulic Brake Hose	SAE	1967
SAE J1402b	Air Brake Hose	SAE	1971

1987 FLUID POWER STANDARDS

VOLUME A
COMMUNICATIONS

Shirley C. Seal, Editor

National Fluid Power Association, Inc.
Milwaukee, Wisconsin

Copyright 1987 by the

NATIONAL FLUID POWER ASSOCIATION, INC.

Printed in the USA

All technical reports, citations, references and related data including standards and practices approved and/or recommended are advisory only. Use thereof by anyone for any purpose is entirely voluntary and in any event without risk of any nature to the National Fluid Power Association, Inc., its officers, directors or authors of such work. There is no agreement by or between anyone to adhere to any NFPA Recommended Standard, policy or practice, and related matters. In formulating and approving technical reports, the Technical Board, its councils and committees and/or the National Fluid Power Association, Inc. will not investigate or consider citations, references or patents which may or may not apply to such subject matter since prospective users of such reports and data alone are responsible for establishing necessary safeguards in connection with utilization of such matters, including technical data, proprietary rights or patentable materials.

Recommended standards and/or policies and procedures are subject to periodic review and may be changed without notice. Recommended standards, after publication, may be revised or withdrawn at any time and current information on all approved recommended standards may be received by calling or writing the National Fluid Power Association, Inc.

An approved NFPA Recommended Standard implies a consensus of those substantially concerned with its scope and provisions and is intended as a guide to aid the manufacturer, the consumer and the general public. The publication of the NFPA Recommended Standard does not in any respect preclude anyone, whether they have participated in the development of or approved the recommended standard or not, from manufacturing, marketing purchasing or using products, processes or procedures not conforming to the recommended standard.

Participation by federal agency representative(s) or person(s) affiliated with industry is not to be interpreted as government or industry endorsement of this standard and/or policy and procedure.

This publication may not be reproduced in whole or in part without the written permission of the National Fluid Power Association, Inc.

```
Sci Ref
TJ
840
.N55
1987
v.A
```

ISBN 0-942220-82-X (Set)
ISBN 0-942220-83-8 (Volume A)

NATIONAL FLUID POWER ASSOCIATION, INC., 1987

Contents

NFPA — Providing Services for the U.S. Fluid Power Industry v

American National Standard Fluid power
systems and products - Glossary, ANSI/B93.2-1986 1

Metric Units for Fluid Power Applications, NFPA/T2.10.1M-1978. 105

Survey of Metric Language Usage by the U.S. Fluid Power Industry,
NFPA/T2.10.2M-1977 119

Hydraulic fluid power - Pumps and motors - Glossary,
NFPA/T3.9.13-1982. 145

SI units and recommendations for the use of their multiples and of
certain other units, ISO 1000-1981. 157

Fluid power systems and components - Graphic symbols, ISO 1219-1976 . . .173

Fluid power systems and components - Nominal pressures, ISO 2944-1974 . .199

Foreword

FLUID POWER STANDARDS, Seventh Edition

Fluid Power Standards is comprised of ten separate volumes available as a set or separately. The titles of the volumes are:

Volume A	Communication including graphic symbols and metric units
Volume B	Pressure Rating
Volume C	Pumps, Motors, Power Units and Reservoirs
Volume D	Filtration and Contamination
Volume E	Conductors and Associated Products
Volume F	Control Products
Volume G	Cylinders and Accumulators
Volume H	Fluids, Lubricants and Sealing Devices
Volume I	Testing
Volume J	Bibliographies

ORDERING INFORMATION

The volumes of Fluid Power Standards can be ordered by contacting the National Fluid Power Associations Publications Department, 3333 N. Mayfair Rd., Milwaukee, Wisconsin 53222 USA Telephone (414) 778-3344 Telex 26-898.

WHAT IS CONTAINED IN FLUID POWER STANDARDS

NFPA's ten volume set, Fluid Power Standards contains both NFPA and ANSI (American National Standards Institute) Fluid Power standards.

HOW FLUID POWER STANDARDS ARE DEVELOPED

The National Fluid Power Association coordinates development of Fluid Power standards on industry, national and international levels. The association holds the Secretariat for the Fluid Power Committee work of the American National Standards Institute (ANSI) and the International Organization for Standardization (ISO).

On the Industry Level (NFPA)

NFPA standards originate in Product Sections and Technology Committees, and are submitted to the NFPA Technical Board and Board of Directors for approval. Product Sections are generally composed of individuals involved in the design, manufacture, performance and application of specific Fluid Power Products. Technology Committees are comprised of experts in a single broad area of Fluid Power technology, applying to many products.

More than 350 persons currently contribute to the standards writing work of NFPA's technical committees and are helping to develop approximately 140 projects as proposed NFPA Recommended Standards, Information Reports and Recommended Practices. Qualified engineers from NFPA member companies are eligible for participation in NFPA technical committees.

On the National Level (ANSI)

After a standard has been approved by the NFPA, it may be submitted to Standards Committee B93, Fluid Power Systems and Products, accredited by the American National Standards Institute. NFPA serves as secretariat for this committee, which ballots the standard for approval as an American National Standard. To date, more than 60 NFPA Recommended Standards have advanced to become American National Standards.

On the International Level (ISO)

Under authority granted by the International Organization for Standardization and the American National Standards Institute, NFPA administers the Secretariat of ISO Technical Committee 131 (TC 131) Fluid power systems. More than 35 nations participate in standards writing through TC 131. The U. S. Fluid Power Industry is represented through USA TAG to ISO/TC 131, a committee also administered by the NFPA.

On an average, the NFPA coordinates 44 deliberating sessions per year in various parts of the industrialized world.

Participating in Standards Development

Participation in the NFPA, ANSI/B93 and ISO/TC 131 committees is open to qualified engineers. For details on how to become involved, contact the Director of Technical Services at National Fluid Power Association, 3333 North Mayfair Rd., Milwaukee, Wisconsin 53222 USA Telephone (414) 778-3344 Telex 26-898.

Five Year Review of Fluid Power Standards

An NFPA Recommended Standard is subject to revision at anytime by the appropriate technical committee. They are reviewed every five years and if not revised, either reaffirmed or withdrawn. Comments are invited either for revision of any standard or for additional standards. Comments should be addressed to the Director of Technical Services at NFPA Headquarters. Comments will receive careful consideration at a meeting of the appropriate technical committee. Commentators are welcome to attend the meeting.

Obsolete Fluid Power Standards

The 1987 edition of these Fluid Power Standards volumes makes the 1984 edition obsolete. For practical purposes, it is not wise to use obsolete volumes. However, for teaching purposes, the outdated volumes might be useful. Standards contained in each volume have an NFPA or ANSI reference number that includes the year the standard was first approved and the year it was reaffirmed.

Disclaimer

See complete disclaimer on Copyright page (reverse of Title Page).

NATIONAL FLUID POWER ASSOCIATION

The National Fluid Power Association is the trade association for manufacturers of hydraulic and pneumatic products and systems.

Founded in 1953, NFPA has 190 corporate members across the U.S. NFPA members produce more than 75 percent of all U.S. Fluid Power production.

NFPA was founded to pursue activities that advance the performance and application of Fluid Power products, better materials, designs and standards, organize committees to further the art and science of Fluid Power, support activities to advance knowledge and understanding of Fluid Power, and represent the members to the Federal Government, industry and user organizations.

Association activities in the areas of technical standards and services, marketing research and statistics, public affairs, industry promotion and management are directed by volunteer Boards and supported by full-time professional headquarters staff.

NFPA provides administrative services to these organizations in the Fluid Power Industry:

- Fluid Power Committee B93, the committee for national fluid power standards, accredited by American National Standards Institute,
- ISO/TC 131 Fluid power systems, the committee for international fluid power standards, of the International Organization for Standardization,
- USA Technical Advisory Group (USA TAG) to ISO/TC 131, the national committee for international Fluid Power standards,
- Fluid Power Appeals Boards,
- Fluid Power Coordinating Council,
- Fluid Power Educational Foundation,
- International Fluid Power Exposition
- International Fluid Power Exposition Applications Conference.

For information on membership in NFPA, or participation of any of the Fluid Power technical committees, contact the National Fluid Power Association, 3333 North Mayfair Rd., Milwaukee, Wisconsin 53222 USA (414) 778-3344 Telex 26-898.

- NOTES -

ANSI/B93.2-1986

AN INDUSTRY STANDARD FOR FLUID POWER

Fluid power systems and products -

Glossary

Approved 16 June 1986

---

published by
NATIONAL FLUID POWER ASSOCIATION, INC.
3333 N. Mayfair Road / Milwaukee, WI 53222 / 414-778-3344 / TLX 26898

ANSI/B93.2

FOREWORD

This Foreword is not part of American National Standard Fluid power systems and products - Glossary, ANSI/B93.2-1986.

The National Fluid Power Association initiated the compilation of a Glossary of Terms for Fluid Power in 1953. Further improvements and expansion were essential to meet the growing needs of the fluid power industry, users and educational programs

In 1958, the project was reactivated under the direction of the NFPA Technical Board. A Terminology Coordinatng Committee was established consisting of representatives from component manufacturers, technical publishers, technical librarians, educators, users, general interests, representatives from associations, and societies, etc., which had a related interest. The Scope and Purpose (SCOPE AND FIELD OF APPLICATION) were defined. New editions were published in 1960, 1962 and 1964.

The Fourth Edition published in 1964 was approved as ANSI/B93.2-1965 on 29 April 1965. A multi-ligual translation of ANSI/B93.2 was prepared by Italian chapters of the Fluid Power Society and was published in 1969.

On 18 January 1971, the Fifth Edition of the NFPA standard was submitted to the American National Standards Committee B93 for promulgation as an ANSI standard. After successful balloting was concluded, approval by the ANSI Board of Standards Review was granted on 14 July 1971.

ANSI/B93.2-1971 formed the basis for ISO work as part of TC 131, Fluid Power activity. Under the Chairmanship of of the Japanese Industrial Standards Committee for ISO/TC 131/SC 1 (Terms, classifications & symbols) with significant contributions from CETOP (European Committee for Oilhydraulic and Pneumatic Control) a draft document of ISO 5598 was issued.

In January 1980, the need for a revision of the NFPA Glossary was recognized to assist in worldwide uniformity and benefit to the individual user of the document.

Headquarters assigned the Project Group number NFPA/T2.1.1 R1. This document is intended to replace ANSI/B93.2-1971 and assimilate ANSI/B93.2A-1978 into a single document. The bilingual draft (ISO 5598), is the main basis for revisions and additions to develop NFPA/T2.1.1 R1.

The Title, Scope and Purpose was submitted and granted unanimous approval by the NFPA Technical Board on 6 February 1980.

The proposed document was forwarded to Headquarters whereupon, after preparation by the Technical Staff, it was circulated as a General Review Draft on 12 September 1980. Numerous comments were received, reviewed and incorporated into the document.

The re-edited document was forwarded to the Technical Board for approval to Ballot. Unanimous approval was granted on 13 May 1981 with the proviso Headquarters receive a favorable letter from the Technical Auditor prior to Ballot. The favorable letter was received by Headquarters 8 October 1981.

All voting members of the NFPA Coordinating Committees and Product Sections were Balloted on 2 April 1982. Balloting ended 3 May 1982 with multiple comments. Comments were reviewed and resolved.

Because of Project Chairman, J. Fisher's retirement and no new Coordinating Committee Chairperson to recommend approval, final approval of NFPA/T2.1.1 R1 was tabled by the Technical Board on 16 September 1982 and 18 May 1983.

On 15 September 1983, NFPA/T2.1.1 R1 was presented to the Technical Board for final approval by newly-appointed Project Chairman, J. White and Technical Auditor, R. Mathias. Unanimous approval was granted with the recommendation the document be forwarded to ANSC/B93 for promulgation as an American National Standard.

Consequently, NFPA/T2.1.1 R1 was forwarded to the Board of Directors whereupon, it received final approval 5 December 1983.

Project Group members who developed this standard:

James L. Fisher, Jr.*
Project Group & Coordinating Committee Chairman (1974-1981)
Schrader Bellows Div.

Tobi Goldoftas
Coordinating Committee Chairman (1981-1982)
Penton/IPC

James C. White
Coordinating Committee Chairman (1983-1986)
Webster/Sta-Rite Industries Inc.

John Berninger
Coordinating Committee Vice Chairman
Parker Hannifin Corp.

Stanlee Meisinger
Coordinating Committee Secretary
Safe Way Hydraulics Inc.

Robert Mathias
Technical Auditor
Applied Power Inc.

Allen E. Tucker
Director of Technical Services
National Fluid Power Association

Ernest Cole*
Schrader Bellows Division

Ed Fletcher
John Deere PEC

William Hertel
Parker Hannifin Corp.

Z. Lansky
Parker Hannifin Corp.

H. Martin
Control Products Division

Bruce McCord
Aro Corp.

William Meisel
Denison Division/Abex Corp.

John Pippenger*
W. H. Nichols

Cliff Schaeffer*
J I Case

W. Smith
Illinois Institute of Technology

mkm

---

* Retired.

On 5 July 1985, ANSI/B93.2 was submitted to ANSI Committee B93 for Ballot. The Ballot resulted in two negative comments. One comment was withdrawn due to editorial corrections and the other will be discussed at the next meeting.

ANSI/B93.2, Fluid Power systems and products -Glossary, was approved by ANSI's Board of Standards Review on 16 July 1986.

The membership roster of Standards Committee B93 at the time of Ballot.

Jack McPherson
Chairman

Daniel B. Shore
Vice Chairman

Allen E. Tucker
Co-Secretary

Herb Kaufman
Co-Secretary

Compressed Air & Gas Institute
David E. Bonn
John Addington (alternate)

Fluid Controls Institute
Jude Pauli
E.C. Rutter (alternate)

Fluid Power Distributors Association
Thomas Neff

Fluid Power Society
Nick Beaver
J. Otto Byers
Robert L. Firth
Paul Gies
Harry R. Holsen
Dale L. Killen
Verne L. Middleton
David Prevallet
James C. White

Fluid Sealing Association
Alex Pilecki

Joint Industry Council
Robert Muhl

Material Handling Institute
Jack C. McPherson
Williard Chichester (alternate)

National Fluid Power Association
Richard N. Bailey
John Bowbin †
Walter Forster
Robert Kay
Z.J. Lansky
Paul Schacht

National Machine Tool Builders' Association
John B. Deam

Society of Automotive Engineers
William A. Hertel
John T. Parrett
Daniel B. Shore
W.L. Snyder
Robert W. White

US Department of Defense
S. Nguyen

Company Members
Logan Mathis
Lloyd L. Schmaltz
Don McGeachy
John Welker (alternate)

Individual Members
Dr. E. C. Fitch
Carroll Grigsby
John J. Pippenger
A.O. Roberts
Jack Walrad
Tom Wanke
Frank Yeaple

† Deceased.

- NOTES -

Fluid power systems and products - Glossary

0 INTRODUCTION

In fluid power systems, power is transmitted and controlled thru a fluid under pressure, within an enclosed circuit. (For the standard definition of fluid power, see the definition in this document 01.01.100.)

1 SCOPE AND FIELD OF APPLICATION

1.1 This Standard applies to fluid power systems and products excluding aerospace applications.

1.2 This Standard lists technical terms used in the Fluid Power industry.

1.3 This Standard provides:

- Definitions of technical terms used in the Fluid Power industry;
- Unified glossary for the Fluid Power industry, education programs and users of Fluid Power;
- Convenient reference for technicians.

1.4 This Standard reflects both domestic and international practice.

1.5 This Standard clarifies terms and definitions for beginners and encourages individuals to expand their vocabulary.

1.6 This Standard simplifies technical communications and reduces interpretation errors.

1.7 This Standard promotes:

- Common understanding of fluid power;
- Greater use of Fluid Power.

2 RULES

2.1 Terms

2.1.1 Provide pertinent terms having a unique meaning in Fluid Power technology.

2.1.2 Omit archaic, colloquial and proprietary terms.

2.1.3 Include common dictionary or engineering terms only when they are a generic root for a series of terms unique to the technology.

2.1.4 Refer synonymous terms to the preferred term.

2.1.5 List deprecated terms but define and clearly mark these terms "Deprecated." Indicate the preferred term.

2.1.6 Coin new terms only if specifically requested. This is the responsibility of appropriate groups who are to promote and underwrite the educational costs of launching new terms.

2.2 Definitions

2.2.1 Provide positive statements.

2.2.2 Provide short single sentences whenever possible.

2.2.3 Provide accurate definitions using simple definitive terms, whenever possible.

2.2.4 Provide clear definitions directed to the layman level, whenever possible.

2.2.5 Define what it is, in the case of physical adjectives.

2.2.6 Define its primary function in the case of general components and functional adjectives.

2.2.7 Include illustrations, if necessary, for clarity.

2.2.8 Include formulas, if necessary, for clarity.

2.2.9 Use supplementary sentences to provide useful related commentary.

2.3 Incorrect Definitions

2.3.1 Negative statements;

2.3.2 Comparative statements;

2.3.3 Statement of advantages or disadvantages;

2.3.4 Detailed engineering derivations.

GROUP 01 — PRIMARY TERMS

01.01.080 Fluid Logic

A branch of fluid power associated with digital signal sensing and information processing, using components with or without moving parts.

01.01.100 Fluid Power

Energy transmitted and controlled through use of a pressurized fluid.

01.01.200 Fluid Power System

A system that transmits and controls power through use of a pressurized fluid within an enclosed circuit.

01.01.300 Fluidics

Engineering science pertaining to the use of fluid dynamic phenomena to sense, control process information, and/or actuate.

01.01.400 Hydraulics

Engineering science pertaining to liquid pressure and flow.

01.01.500 Hydrodynamics

The engineering science which governs the movement of liquids and the forces opposing that movement.

01.01.600 Hydrokinetics

Engineering science pertaining to the energy of liquid flow and pressure.

01.01.700 Hydropneumatics

Pertaining to the combination of hydraulic and pneumatic fluid power.

01.01.800 Hydrostatics

Engineering science pertaining to the energy of liquids at rest.

01.01.801 Hydrostatic Transmission

Combination of one or more hydraulic pumps and motors forming a unit.

01.01.850 Moving Parts Logic

The technology of achieving logic control by means of fluid devices having moving parts.

01.01.900 Pneumatics

Engineering science pertaining to gaseous pressure and flow.

01.01.910 Time

An interval comprising a limited but continuous action.

01.01.911 Time, Actuated

The interval in which the component is under the influence of the actuating forces.

01.01.912 Time, Fall

The interval taken in a device for a quantity to change from a specified high level down to a specified lower level.

01.01.913 Time, Operation

The interval of an event measured from "start signal" to final "at rest".

01.01.914 Time, Released

The duration of time in which the component is not under the influence of the actuating forces.

01.01.915 Time, Relative Duty

(Expressed as a percentage)

$$\frac{t\ (actuated)}{t\ (actuated) + t\ (released)} \times 100$$

01.01.916 Time, Response

The elapsed time between the initiation of an action and the resulting reaction (measured under specified conditions).

01.01.918 Time, Rise

The interval in a device for a quantity to change from a specified low level up to a specified high level.

01.01.919 Time, Start Up

The interval needed to reach a steady state operating condition in the system from "start up".

01.01.930 Torque

Turning effort.

01.01.931 Torque, Derived

Torque corresponding to the derived hydraulic power.

01.01.932 Torque, Effective

Actual torque transmitted by the shaft under specified conditions.

01.01.933 Torque, Geometric

Torque corresponding to the geometric hydraulic power.

GROUP 02 - FLUID POWER LAWS AND RELATED TERMS

02.01.100 Bernoulli's Law

If no work is done on or by a flowing frictionless liquid its energy due to pressure and velocity remains constant at all points along the streamline.

02.01.200 Boyle's Law

The absolute pressure of a fixed mass of gas varies inversely as the volume, provided the temperature remains constant.

02.01.300 Charles' Law

The volume of a fixed mass of gas varies directly with absolute temperature, provided the pressure remains constant.

02.01.400 Continuity Equation

The mass rate of fluid flow into a fixed space is equal to the mass flow rate out. Hence, the mass flow rate of fluid past all cross sections of a conduit is equal.

02.01.500 Darcy's Formula

A formula used to determine the pressure drop due to flow friction through a conduit.

$$h_f = \frac{fLv^2}{2Dg}$$

h_f = Head loss, feet (metre)
f = Friction factor (see 02.01.600)
L = Length of conduit, feet (metre)
v = Mean velocity of flow, ft/sec. (metre/sec.)
D = Internal diameter of conduit, feet (metre)
g = Acceleration due to gravity, 32.2 ft./sec.^{2} (9.81 metre/sec.^{2})

02.01.600 Hagen Poiseuille Law

The friction factor of Darcy's Formula is a ratio of 64 to the Reynolds Numbers when flow is laminar.

$$f = \frac{64}{N_r}$$

f = Friction factor
N_r = Reynolds Number

02.01.700 Pascal's Law

A pressure applied to a confined fluid at rest is transmitted with equal intensity throughout the fluid.

02.01.800 Reynolds Number

A numerical ratio of the dynamic forces of mass flow to the shear stress due to viscosity. Flow usually changes from laminar to turbulent between Reynolds Number 2,000 and 4,000.

$$N_r = \frac{pvD}{u} = \frac{vD}{\nu}$$

N_r = Reynolds Number
p (Rho) = Fluid density, pounds(mass)/ft.^{3} (Kg/m^3)
v = Mean velocity of flow, ft./sec. (metre/sec.)
D = Internal diameter of conduit, feet (metre)
u (Mu) = Absolute viscosity, pounds/ft.sec. (Kg/cm^5)
ν (Nu) = Kinematic viscosity, ft.^{2}/sec. (mm^2/sec.)

02.01.900 Toricelli's Theorem

The liquid velocity at an outlet discharging into the free atmosphere is proportional to the square root of the head.

$$v = \sqrt{2gh}$$

v = Velocity (mean), ft./sec. (m/sec.)
g = Acceleration, 32.2 ft./sec.^{2} (9.81 m/sec.^{2})
h = Pressure head, ft(m)

GROUP 03 — FLOW TERMS

03.01.200 Cavitation

A localized gaseous condition within a liquid stream which occurs where the pressure is reduced to the vapor pressure.

03.01.210 Coanda Effect

The phenomenon, named after its discoverer, of the attachment of a free flowing turbulent jet to an adjacent, possibly curved, wall.

03.01.409 Flow

Movement of fluid generated by pressure differences.

03.01.410 Flow, Laminar (Streamline)

A flow situation in which fluid moves in parallel lamina or layers.

03.01.420 Flow, Metered

Flow at a controlled rate.

03.01.430 Flow, Steady State

A flow situation wherein conditions such as pressure, temperature, and velocity at any point in time do not change.

03.01.450 Flow, Turbulent

A flow situation in which the fluid particles move in a random fluctuating manner.

03.01.460 Flow, Upsteady

A flow situation wherein conditions such as pressure, temperature and velocity at points in the fluid change.

03.01.600 Shock Wave

A pressure wave front which moves at a sonic velocity.

03.01.800 Surge

A transient rise of pressure or flow.

GROUP 04 — MENSURATION TERMS

04.01.005 Air, Free

Air at ambient temperature, pressure, relative humidity, and density.

04.01.010 Air, Standard

Air at a temperature of $68°F$, a pressure of 14.70 pounds per square inch absolute, and a relative humidity of 36% (0.0750 pounds per cubic foot). In gas industries the temperature of "standard air" is usually given as $60°F$.

04.01.011 Amplification

The ratio between the output signal variations and the control signal variations (for analogue devices only).

04.01.012 Amplification, Flow

Ratio between the output flow and the input (control) flow.

04.01.013 Amplification, Power

The ratio between the output power variation and the corresponding input (control) power variation (for analogue devices only).

04.01.014 Amplification, Pressure

Ratio between the outlet pressure and the inlet (control) pressure.

04.01.020 Aniline Point

The lowest temperature at which a liquid is completely miscible with an equal volume of freshly distilled aniline (ASTM Designation D611-64).

04.01.025 Assurance Level

The minimum percentage of pressure containing envelopes of a verified design that will sustain 10 million applications of its Rated Fatigue Pressure.

04.01.028 Bistable

A binary circuit or device which has two stable states and which in each state requires an appropriate impulse to cause a transition to the other state.

04.01.030 Bulk Modulus

The measure of resistance to compressibility of a fluid. It is the reciprocal of the compressibility.

04.01.035 Capacity, Effective

Actual volume displaced under specified conditions.

04.01.036 Capacity, Geometric

Volume displaced, calculated geometrically without reference to tolerances, clearances or deformation.

04.01.040 Compressibility

The change in volume of a unit volume of a fluid when subjected to a unit change in pressure.

04.01.041 Confidence Level

The degree to which a manufacturer can be assured that the desired assurance level is attained.

04.01.043 Displacement, Volumetric

Volume absorbed or displaced per stroke of a cylinder (See 81.01.012) or per cycle of a pump or motor (43.01.005).

04.01.046 Efficiency

Ratio of output to the corresponding input.

04.01.047 Expectancy, Life

The predicted working period during which a component or system will maintain a specified level of performance under specified conditions. Sometimes expressed in statistical terms as a probability.

04.01.048 Fire Point

The temperature to which a fluid must be heated to ignite and burn for five seconds (minimum) in the presence of air when a small flame is applied under controlled conditions.

04.01.050 Flash Point

The temperature to which a liquid must be heated under specified conditions of the test method to give off sufficient vapor to form a mixture with air that can be ignited momentarily by a flame.

04.01.051 Flow Degradation Ratio

The ratio of stabilized flow rate after a contaminant injection to the initial measured pump flow (Qr).

04.01.052 Flow Factor

Characterizes the conductance of a pneumatic or hydraulic device, flowline or connection.

04.01.060 Flow Rate

The volume, mass or weight of a fluid passing through any conductor per unit of time.

04.01.061 Flow Rate, Relief

The rate at which fluid can flow through the unloading device for each specific increase in controlled pressure above the original setting, measured under specified conditions.

04.01.063 Flow, Supply Port

The flow of fluid through the supply ports of the device or system.

04.01.070 Fluid Friction

Friction due to the viscosity of fluids.

04.01.072 Frequency Response

The changes, under steady-state conditions, in the output variable which are caused by a sinusoidal input variable.

04.01.170 Hammer, Liquid

Pressure and depression waves created by relatively rapid flow changes and transmitted through the system.

04.01.100 Head

The height of a column or body of fluid above a given point expressed in linear units. Head is often used to indicate gage pressure. Pressure is equal to the height times the density of the fluid.

04.01.110 Head, Friction

The head required to overcome the friction at the interior surface of a conductor and between fluid particles in motion. It varies with flow, size, type and condition of conductors and fittings, and the fluid characteristics.

04.01.115 Head, Pressure

The pressure due to the height of a column or body of fluid. (It is usually expressed in inches [mm]).

04.01.120 Head, Static

The height of a column or body of fluid above a given point.

04.01.130 Head, Static Discharge

The static head from the centerline of the pump to the free discharge surface.

04.01.140 Head, Static Suction

The head from the surface of the supply source to the centerline of the pump.

04.01.150 Head, Total Suction

The static head from the surface of the supply source to the free discharge surface.

04.01.160 Head, Velocity

The equivalent head through which the liquid would have to fall to attain a given velocity. Mathematically it is equal to the square of the velocity (in feet) divided by 64.4 feet per second squared.

$$h = \frac{v^2}{2g}$$

h = Head, feet (metre)
g = Acceleration due to gravity, 32.2 ft./sec.^{2} (9.81 metre/sec.^{2})
v = Mean velocity of flow, ft./sec. (metre/sec.)

04.01.200 Hydraulic Power

Power computed from flow rate and pressure differential (drop).

Hydraulic Power = .000583 $q_v p$ (0,002234 $q_v p$) expressed as horsepower where:
q_v = Flow rate, gpm (L/min.)
p = Pressure, psi (bar)

Alternate formula = $\dfrac{q_v p}{1714}$ $\left(\dfrac{q_v p}{447.6}\right)$

04.01.210 Lift

The height of a column or body of fluid below a given point expressed in linear units. Lift is often used to indicate vacuum or pressure below atmosphere.

04.01.220 Lift, Static Suction

The lift from the centerline of the pump to the surface of the supply source. (See Head, Static Suction).

04.01.222 Linear Function

Describes a condition in which the relationship between two interdependent variables is constant.

04.01.223 Linear Region

The region of a given control characteristic over which the linearity remains within specified limits.

04.01.224 Linearity

The faithfulness with which an output signal of an electronic reproducing system reproduces an input signal.

04.01.230 Monostable

A binary circuit or device, which has one stable state and which requires an appropriate change of the input to cause a transition out of its stable state for a specified period of time. The specified period of time at which the circuit stays out of its stable state is independent of the duration of the appropriate change of the input signals.

04.01.250 Neutralization Number

A measure of the total acidity or basicity of an oil; this includes organic or inorganic acids or bases or a combination thereof (ASTM Designation D974-64).

04.01.260 Newt

A unit of kinematic viscosity in the English system. It is expressed in square inches per second (see 04.01.630 Stokes).

04.01.270 Poise

The standard unit of dynamic viscosity in the c.g.s. (centimeter-gram-second) system. It is the ratio of the shearing stress to the shear rate of fluid and is expressed in millipascal sec. (= 1 centipoise).

04.01.280 Pour Point

The lowest temperature at which a liquid will flow under specified conditions (ASTM Designation D97-66).

04.01.282 Power Consumption

The total power consumed by the device or system under specified conditions.

04.01.283 Power Supply, Fluid

Energy source which generates and maintains a flow of fluid under pressure.

04.01.290 Precipitation Number

The number of millilitres of precipitate formed when 10 mL of lubricating oil are mixed with 90 mL of ASTM precipitation naptha and centrifuged under prescribed conditions (ASTM Designation D91-61).

04.01.300 Pressure

Force per unit area, usually expressed in pounds per square inch (bar).

04.01.310 Pressure, Absolute

The pressure above zero absolute, i.e., the sum of atmospheric and gage pressure. In vacuum related work it is usually expressed in millimetres of mercury (mm Hg).

04.01.320 Pressure, Atmospheric

Pressure exerted by the atmosphere at any specific location. (Sea level pressure is approximately 14.7 pounds per square inch absolute, 1 bar = 14.5 psi).

04.01.330 Pressure, Back

The pressure encountered on the return side of a system.

04.01.340 Pressure, Breakloose (Breakout)

The minimum pressure which initiates movement.

04.01.360 Pressure, Burst

The pressure which causes failure of and consequential loss of fluid through the product envelope.

04.01.370 Pressure, Charge

The pressure at which replenishing fluid is forced into a fluid power system.

04.01.375 Pressure, Control Range

The permissible limits between which system pressure may be set.

04.01.380 Pressure, Cracking

The pressure at which a pressure operated valve begins to pass fluid.

04.01.385 Pressure, Cyclic Test

A pressure range applied in cyclic tests that are performed to verify a Rated Fatigue Pressure.

04.01.390 Pressure, Differential (Pressure Drop)

The difference in pressure between any two points of a system or a component.

04.01.400 Pressure, Gage

Pressure differential above or below ambient atmospheric pressure.

04.01.412 Pressure, Induced

Pressure generated by an externally applied force.

04.01.414 Pressure, Inlet

The pressure at the apparatus inlet port.

04.01.415 Pressure, Intensified

In a fluid power cylinder, the outlet pressure required to slow the piston rod extending under regulated pressure introduced at the cap end.

04.01.417 Pressure, Maximum Inlet

The maximum rated gage pressure applied to the inlet.

04.01.419 Pressure, Nominal

A pressure value assigned to a component or system for the purpose of convenient designation.

04.01.420 Pressure, Operating

The pressure at which a system is operated.

04.01.425 Pressure, Outlet

Pressure at the apparatus outlet port.

04.01.427 Pressure, Overrange Rating

The pressure to which a device can be subjected for extended time without change in operating characteristics, shift in set point, or damage to the device.

04.01.430 Pressure, Override

The difference between the cracking pressure of a valve and the pressure reached when the valve is passing its rated flow.

04.01.435 Pressure, Peak

The maximum pressure encountered in the operation of a component.

04.01.440 Pressure, Pilot

The pressure in the pilot circuit.

04.01.450 Pressure, Precharge

The pressure of compressed gas in an accumulator prior to the admission of a liquid.

04.01.460 Pressure, Proof

The non-destructive test pressure, in excess of the maximum rated operating pressure, which causes no permanent deformation, excessive external leakage, or other resulting malfuction.

04.01.470 Pressure, Rated

The qualified operating pressure which is recommended for a component or a system by the manufacturer.

04.01.472 Pressure, Recovery

The ratio of output pressure to the supply pressure.

04.01.473 Pressure, Rated Fatigue

A pressure that a pressure containing envelope is represented to sustain 10 million times without failure.

04.01.474 Pressure, Regulation of

Pertains to the control of pressure in a system.

04.01.476 Pressure, Rated Static

A pressure that a component pressure containing envelope is represented to sustain once, under test conditions without failure, after which the component must be discarded.

04.01.477 Pressure, Residual

The value of the output pressure in the "off" state of the device.

04.01.480 Pressure, Shock

The pressure existing in a wave moving at sonic velocity.

04.01.481 Pressure, Shockwave

A pressure pulse which moves at sonic speed in the liquid.

04.01.490 Pressure, Static

The pressure in a fluid at rest.

04.01.495 Pressure, Static Test

A pressure applied in a static test performed to verify a Rated Static Pressure.

04.01.500 Pressure, Suction

The absolute pressure of the fluid at the inlet of a pump.

04.01.505 Pressure, Supply

The pressure at the apparatus inlet port.

04.01.510 Pressure, Surge

The pressure resulting from surge conditions.

04.01.520 Pressure, System

The pressure which overcomes the total resistances in a system. It includes all losses as well as useful work.

04.01.530 Pressure, Vapor

The pressure, at a given fluid temperature, in which the liquid and gaseous phases are in equilibrium.

04.01.540 Pressure, Working

The pressure at which the apparatus is being operated in a given application.

04.01.541 Pressure, Range, Working

The tolerance (plus or minus) range of the working pressure.

04.01.610 Reyn

The standard unit of absolute viscosity in the English system. It is expressed in pound-seconds per square inch.

04.01.611 Rotation

The direction of rotation is always quoted as viewed looking at the shaft end. In dubious cases, provide a sketch.

04.01.612 Rotation, Anti-Clockwise

Rotation in the opposite sense to the clock.

04.01.613 Rotation, Clockwise

Forward rotation of the hands to the clock.

04.01.614 Rotational Frequency

Number of revolutions per unit of time.

04.01.620 Specific Gravity, Liquid

The ratio of the weight of a given volume of liquid to the weight of an equal volume of water.

04.01.630 Stokes

The standard unit kinematic viscosity in the c.g.s. (centimetre-gram-second) system. It is expressed in square centimetres per second; 1 centistokes equals .01 stokes.

04.01.640 Surface Tension

The surface force of a liquid in contact with a fluid by which it tends to assume a spherical form and to present the least possible surface. It is expressed in pounds per foot or dynes per centimetre.

04.01.641 Temperature, Ambient

The temperature of the environment in which the apparatus is working.

04.01.642 Temperature, Equipment

The temperature of the unit at a specified position and measured at a specified point.

04.01.643 Temperature, Fluid

The temperature of the pressure medium measured at a specified point.

04.01.644 Temperature, Inlet

Fluid temperature at the inlet port.

04.01.645 Temperature, Outlet

Fluid temperature at the outlet port.

04.01.646 Temperature Range

The permissible temperature range within which the apparatus or the fluid can operate satisfactorily.

04.01.647 Torr

A unit of pressure equal to 1/760 of an atmosphere and very nearly equal to 1mm Hg @ $0^{\circ}C$.

04.01.648 Unistable

A binary circuit or device, which has one stable state and in which the output changes state for the duration of the appropriate change of the input signal.

04.01.650 Vacuum

Pressure less than ambient atmospheric pressure.

04.01.657 Variability Factor

A multiplier applied to the Rated Fatigue Pressure for calculating the Cyclic Test Pressure to account for the variability in fatigue strength of metals. It is also applied to the Rated Static Pressure for calculating the Static Test Pressure.

04.01.700 Viscosity

A measure of the internal friction or the resistance of a fluid to flow.

04.01.710 Viscosity, Absolute

The ratio of the shearing stress to the shear rate of a fluid. It is usually expressed in centipoise.

04.01.720 Viscosity, Kinematic

The absolute viscosity divided by the density of the fluid. It is usually expressed in centistokes.

04.01.730 Viscosity, SAE Number

The Society of Automotive Engineers' arbitrary numbers for classifying fluids according to their viscosities. The numbers in no way indicate the viscosity index of fluids.

04.01.740 Viscosity, SUS

Saybolt Universal Second (SUS), which is the time in seconds for 60 millilitres of oil to flow through a standard orifice at a given temperature (ASTM Designation D88-56).

04.01.800 Viscosity Index

A measure of the viscosity-temperature characteristics of a fluid as referred to that of two arbitrary reference fluids (ASTM Designation D2270-64).

04.01.801 Viscosity Index Improver

A chemical compound added to a fluid to modify its temperature/viscosity relationship.

04.01.810 Vortex

Spiral motion of a fluid resulting in a radial pressure gradient. The trajectories are curves which encircle a single line (axis).

GROUP 05 — SYMBOLS

Section 01 — General
Section 02 — Symbol Types

SECTION 01 — GENERAL

05.01.100   Symbol, Fluid Power

A representation of the characteristics of a fluid power component by means of lines on a flat surface.

SECTION 02 — SYMBOL TYPES

05.02.100   Symbol, Combination

A symbol which combines graphical, cutaway and pictorial representations.

05.02.200   Symbol, Cutaway

A symbol showing principal internal parts, controls and actuating mechanisms, interconnecting lines, and functions of a component.

05.02.300   Symbol, Graphic (Schematic)

A simplified symbol which indicates essential characteristics applicable to all similar components.

05.02.400   Symbol, Pictorial

A symbol showing the actual shape of a component according to the manufacturer's description.

GROUP 10 — GENERAL FLUIDS

Section 01 — General
Section 02 — General Types
Section 03 — Fluid Stability
Section 04 — Fluid Characteristics

SECTION 01 — GENERAL

10.01.001   Additive

A chemical added to a fluid to impart new properties or to enhance those which already exist.

10.01.100   Fluid

A liquid, gas or combination thereof.

10.01.150   Inhibitor

Any substance which, when present in very small proportions, slows, prevents or modifies chemical reactions such as corrosion or oxidation.

SECTION 02 — GENERAL TYPES

10.02.010   Fluid, Aqueous

A fluid which contains water as a major constituent besides the organic material. The fire resistance properties are derived from the water content.

10.02.100   Fluid, Fatty Oil

A fluid composed of fats derived from animal, marine or vegetable origin. It may contain additives.

10.02.200   Fluid Fire Resistant (Non-Flammable)*
*Deprecated

A fluid difficult to ignite which shows little tendency to propagate flame.

10.02.300   Fluid, Hydraulic

A fluid suitable for use in a hydraulic system.

10.02.350   Fluid, Newtonian

Fluid having a viscosity that is always independent of the rate of shear.

10.02.400   Fluid, Pneumatic

A fluid suitable for use in a pneumatic system.

SECTION 03 — FLUID STABILITY

10.03.000   Fluid Stability

Resistance of a fluid to permanent changes in properties.

10.03.100   Fluid Stability, Chemical

Resistance of a fluid to chemical change.

10.03.150   Fluid Stability, Emulsion

The stability characteristics of an emulsion under defined storage conditions.

10.03.200   Fluid Stability, Hydrolytic

Resistance of a fluid to permanent changes in properties caused by chemical reaction with water.

10.03.300   Fluid Stability, Oxidation

Resistance of a fluid to permanent changes caused by chemical reaction with oxygen.

10.03.350   Fluid Stability, Shear

The ability of a fluid to maintain its viscosity under operating conditions.

10.03.400 Fluid Stability, Thermal

Resistance of a fluid to permanent changes caused soley by heat.

10.03.450 Incompatible Fluids

Fluid which when mixed in a system will have a deleterious effect on that system, its components, or its operation.

SECTION 04 — FLUID CHARACTERISTICS

10.04.010 Fluid, Aeration, Foam

A more or less stable extended air-liquid interface arising when bubbles persist at the surface of a fluid.

10.04.011 Fluid, Air Release

The ability of a fluid to release air bubbles dispersed therein.

10.04.012 Fluid, Anti-Corrosive

A fluid containing metal corrosion inhibitors.

10.04.013 Fluid, Ash Content

Percentage in weight of the residue after calcination of the fluid under defined conditions.

10.04.016 Fluid, Auto Ignition Temperature

The temperature at which a fluid will ignite when dripped by a pipette into a heated flask. A.I.T. is calculated at $5°C$ below the temperature at which ignition takes place.

10.04.100 Fluid Density

Quotient of the mass of a fluid by its volume at a specified temperature ($15°C$).

10.04.110 Fluid, Evaporation Deposits

Percentage of residue obtained after evaporation of the product in free air.

10.04.200 Fluid, Miscibility

Capacity of fluids to be mixed in any ratio without separation into phases.

10.04.270 Fluid, Rust Protection

Capacity of a fluid to prevent the formation of rust under specified conditions.

10.04.280 Fluid, Vapor Pressure

Pressure exerted at any temperature by a vapor from a fluid existing in equilibrium with its liquid phase.

10.04.290 Fluid, Water Content

Quantity of water contained in a mineral oil fluid other than as a contaminant.

GROUP 11 — EMULSIONS

Section 01 — General
Section 02 — Emulsion Types

SECTION 01 — GENERAL

11.01.050 Demulsibility, Water

Capacity of an emulsion of fluid and water to separate into two phases.

11.01.100 Emulsion

A homogeneous dispersion of two immiscible liquids.

11.01.400 Saponification, Value of

A measure of the free and combined acids in oils reacting with potassium hydroxide per gram of fluid.

SECTION 02 — EMULSION TYPES

11.02.525 Emulsion, Oil in Water

A dispersion of oil in a continuous phase of water.

11.02.550 Emulsion, Water in Oil

A dispersion of water in a continuous phase of oil.

GROUP 12 — PETROLEUM FLUIDS

Section 01 — General

SECTION 01 — GENERAL

12.01.100 Petroleum Fluid

A fluid composed of petroleum oil which may contain additives and/or inhibitors.

GROUP 13 — SYNTHETIC FLUIDS

Section 01 — General
Section 02 — Types of Synthetic Fluids

SECTION 01 — GENERAL

13.01.100 Synthetic Fluid

Fluid other than mineral oil which has been artificially compounded for use in a fluid power system.

SECTION 02 — TYPES OF SYNTHETIC FLUIDS

13.02.090   Synthetic Fluid, Chlorinated Hydrocarbon

An aromatic or paraffinic hydrocarbon fluid in which certain hydrogen atoms are replaced by chlorine. The fire resistance is derived from the chlorine present.

13.02.095   Synthetic Fluid, Di-Basic Ester

A fluid manufactured by the reaction of a di-basic acid with a monohydric alcohol.

13.02.100   Synthetic Fluid, Halogenated

A fluid composed of halogenated organic materials. It may contain additives.

13.02.200   Synthetic Fluid, Organic Ester

A fluid composed of esters which are compounds of carbon, hydrogen, and oxygen only (it may contain additives).

13.02.300   Synthetic Fluid, Phosphate Ester

A fluid composed of phosphate esters. It may contain additives.

13.02.400   Synthetic Fluid, Phosphate Ester Base

A fluid which contains a phosphate ester as one of the major components.

13.02.500   Synthetic Fluid, Polyglycol

A non-aqueous fluid composed of polyglycol derivatives. It may contain additives.

13.02.600   Synthetic Fluid, Silicate Ester

A fluid composed of organic silicates. It may contain additives.

13.02.700   Synthetic Fluid, Silicone

A fluid composed of silicones. It may contain additives.

GROUP 14 — WATER-GLYCOLS

    Section 01 — General

SECTION 01 — GENERAL

14.01.100   Water Glycol Fluid

A fluid whose major constituents are water and one or more glycols or polyglycols.

GROUP 15 — AIR

    Section 01 — General
    Section 02 — Types of Air

SECTION 01 — GENERAL

15.01.100   Air

A gas mixture consisting primarily of nitrogen, oxygen, argon, carbon dioxide, hydrogen, neon and helium.

15.01.102   Air, Contamination of

Contaminants in the air supplies to a system or device.

SECTION 02 — TYPES OF AIR

15.02.100   Air, Compressed (Pressure)

Air at any pressure greater than atmospheric pressure.

15.02.200   Air, Dried

Air with moisture content lower than the maximum allowable for a given application.

15.02.800   Air, Saturated

Air at 100 percent relative humidity, with a dew point equal to temperature.

Air, Free (See Mensuration Group)

Air, Standard (See Mensuration Group)

GROUP 16 — WATER

    Section 01 — General
    Section 02 — Types

SECTION 01 — GENERAL

16.01.100   Water

A fluid compound consisting of primarily two hydrogen atoms and one oxygen atom.

GROUP 21 — RESERVOIRS

    Section 01 — General
    Section 02 — Reservoir Types
    Section 03 — Reservoir Capacities

SECTION 01 — GENERAL

21.01.010   Capacitor

A device capable of storing a signal at a specific point in a pneumatic control or fluidic control circuit.

ANSI/B93.2

21.01.100 Reservoir

A container for storage of liquid in a fluid power system.

SECTION 02 — RESERVOIR TYPES

21.02.100 Reservoir, Atmospheric

A container for the storage of a fluid medium at atmospheric pressure.

21.02.200 Reservoir, Hydraulic

A reservoir for storing and conditioning a liquid in a hydraulic system.

21.02.300 Reservoir, Non-Integral

An independent or removable reservoir.

21.02.400 Reservoir, Pressure Sealed

A sealed reservoir for storage of fluids under pressure.

21.02.500 Reservoir, Sealed

A reservoir for storage of fluids isolated from atmospheric conditions.

21.02.600 Reservoir, Top Mounted

A reservoir with provisions for mounting the pump and components on top.

SECTION 03 — RESERVOIR CAPACITIES

21.03.100 Breathing Capacity

A measure of flow rate through an air breather.

21.03.400 Fluid Capacity

The liquid volume coincident with the "high" mark of the level indicator.

21.03.800 Reserve Capcity

The volume of air above the "high" mark of the level indicator.

GROUP 22 — TANKS

Section 01 — General
Section 02 — Tank Types

SECTION 01 — GENERAL

22.01.100 Tank

A container for the storage of fluid in a fluid power system.

SECTION 02 — TANK TYPES

22.02.100 Tank, Air-Oil

A tank in which pressurized air is used to force oil into the outlet port.

22.02.900 Tank, Vacuum

A tank for gas at less than atmospheric pressure.

GROUP 30 — FLUID CONDITIONERS — GENERAL

Section 01 — General
Section 02 — Classes of Contaminants
Section 03 — Types of Contamination
Section 04 — Causes of Contamination
Section 05 — Contaminant Dimensions
Section 06 — Contamination Analysis
Section 07 — Sampling Terms

SECTION 01 — GENERAL

30.01.050 Anti-Freeze

Any substance introduced into the working fluid which depresses the freezing point, applicable only to fluids which contain water.

30.01.200 Cleanliness Level

The antonym of contamination level.

30.01.201 Cleanliness, Roll-Off

The contamination level of a hydraulic system just prior to being released from the machine assembly area.

30.01.202 Conditioner, Air

An assembly comprising a filter, a pressure reducing valve with gage, and a lubricator, intended to deliver fluid in suitable condition.

30.01.300 Contamination Level

A quantitative term specifying the degree of contamination.

30.01.400 Decontamination

The process of removing unwanted material or substance; the reduction of contamination to an acceptable level.

30.01.500 Fluid Conditioning

Establishing and maintaining control of temperature and contamination level of a fluid.

SECTION 02 — CLASSES OF CONTAMINANTS

30.02.000 Contamination, Classes of

Having arbitrarily defined the grades, a certain number of classes can be established and a numerical grid fixed for each class and for each grade a numerical limit of particles not to be surpassed by a unit of volume. Having established the numerical distribution of contamination, it is said that a sample of fluid is contained in a class if for any grade, the number of particles does not surpass the maximum number given for the limit.

30.02.100 Contaminant (Contamination)

Any material or substance which is unwanted or adversely affects the fluid power system or components, or both.

30.02.200 Contaminant, Artificial

Contaminants of known composition and particle size distribution which are introduced into fluid systems or fluid systems components for test purposes. (The most commonly used artificial contaminants include standardized fine air cleaner test dust, standardized coarse air cleaner test dust, carbonyl iron, glass beads, cottonlinters, red iron oxide and black iron oxide).

30.02.400 Contaminant, Environmental

Contamination present in the immediate surroundings, inadvertantly introduced or ingested into a fluid power system or component.

30.02.500 Contaminant, Generated

Contamination created by the operation of a fluid system or component. (Generated contaminants are products of erosion, fretting, scoring, wear, corrosion, decomposition, oxidation and fluid-breakdown. Air bubbles may also be generated under some operating conditions.)

SECTION 03 — TYPES OF CONTAMINANTS

30.03.050 Dissolved Air

Air which is dispersed at a molecular level in hydraulic fluid to form a single phase.

30.03.100 Dissolved Water

Water which is dispersed at a molecular level in hydraulic fluid to form a single phase.

30.03.150 Entrained Air

A mechanical mixture of air bubbles having a tendency to separate from the liquid phase.

30.03.200 Free Air

Any compressible gas, air or vapor trapped within a hydraulic system that does not condense or dissolve to form a part of the system fluid.

30.03.250 Free Water

Water droplets or globules in the system fluid that tend to accumulate at the bottom or top of the system fluid depending on the fluids specific gravity.

30.03.300 Gel

A colloidal solution in which the intermingling of internal and external phases results in a system viscosity greater than the viscosity of the external phase alone, i.e., a jelly-like substance.

30.03.355 Initial

Initial residual contamination in a component, fluid, or system. Typical built-in contaminants are burrs, chips, flash, dirt, dust, fiber, sand, moisture, pipe dope, weld, spatter, paints and solvents, flushing solutions, incompatible fluids and operating fluid impurities.

30.03.390 Liquid

Contaminant in liquid form expressed in terms of weight, per weight of supplied air or gas.

30.03.400 Magnetic

Any substance that is excited when introduced into a magnetic field, i.e., attracted to a permanent magnet.

30.03.450 Micro-Biological

Pertaining to minute living organisms and vital processes.

30.03.500 Silt

Fine particulate matter, generally ranging from less than five to sub micrometres in size.

30.03.550 Slime

A soft viscous or glutinous deposit or coating.

30.03.600 Sludge

Particulate contaminant or a mixture of particulate and liquid contaminant separated from the fluid in an unconsolidated state.

30.03.648 Solid

Contamination in solid form expressed as percentage of solid particles per unit of mass.

30.03.649 Vapor

Contamination in vapor form expressed in terms of weight per weight at the specified operating temperature.

30.03.650 Varnish

Materials generated by the hydraulic fluid due to oxidation, thermal instability, hydrolytic instability, or other reactions. These materials are insoluble in the hydraulic fluid and are generally found as brownish deposits on the work surfaces.

SECTION 04 — CAUSES OF CONTAMINATION

30.04.050 Abrasion

A wearing, grinding, or rubbing away of material in mechanical elements. The products of abrasion will be introduced into the system as generated particulate contamination.

30.04.100 Corrosion

The chemical change in the mechanical elements caused by the interaction of fluid or contaminants, or both. More specifically related to chemical changes in metals. The products of change may be introduced into the system as generated particulate contamination.

30.04.150 Decomposition

Separation by chemical change into constituent parts, elements, or different compounds. More specifically related to fluid and seal chemical changes. The materials affected are primarily organic in nature. The products of change may be introduced into the system as contamination.

30.04.200 Erosion

The loss of material in mechanical elements caused by the impingement of fluid or fluid suspended particulate matter or both. The product of erosion will be introduced into the system as generated particulate contamination.

30.04.250 Fluid Breakdown

A change of chemical or mechanical properties of a fluid, or both. Some end products may be insoluble in the fluid.

30.04.300 Fretting

A type of wear resulting from minute reciprocal sliding motion which produces fine particulate contamination.

30.04.350 Fretting Corrosion

Oxidation of fretting wear debris.

30.04.400 Oxidation

The interaction of air and moisture on the surface of a mechanical element.

30.04.450 Scoring

Scratches in the direction of motion of mechanical parts caused by abrasive contaminants.

30.04.500 Silting

An accumulation of fine particles at a specific location in a fluid system.

SECTION 05 — CONTAMINANT DIMENSIONS

30.05.100 Effective Particle Diameter

The diameter of a circle having an area equivalent to the projected area of the particle.

30.05.200 Fiber

For the purpose of microscopic particle counting a fiber is a particle whose length is greater than 100 micrometres but at least ten times its width.

30.05.300 Longest Dimension

The greatest dimension of a particle equivalent to the diameter of a sphere enclosing the particle when tangent at a minimum of two points.

30.05.390 Mass Index

Mass (weight) of particles contained in a unit volume of fluid.

30.05.400 Micrometre (Micron*)
* Deprecated

Unit of measurement one millionth of a metre long, or approximately 0.00003937 inch expressed in English units.

30.05.500 Particle

A minute piece of matter with observable length, width, and thickness; usually measured in micrometres.

30.05.550 Second Longest Dimension

The greatest dimension perpendicular to the particles' longest dimension.

SECTION 06 — CONTAMINATION ANALYSIS

30.06.030 Agglomerate

A group of two or more particles combined, joined or clustered by any means.

30.06.060 Automatic Count

A particle count obtained by an electro-mechanical or electronic device as opposed to visual microscopic counting technique.

30.06.090 Background Contamination

The total of the extraneous particles which are introduced in the process of obtaining, storing, moving, transferring and analyzing the fluid sample.

30.06.120 Centrifuge Volume

The volume of contaminant (liquid or solid or both) separated from a volume of liquid exposed to centrifugal force.

30.06.150 Counting Calibration Factor

Ratio of the effective filtration area on the membrane to the area counted, SAE-ARP-598, paragraph 8.4.1.

30.06.180 Filterable Solids

The solids retained on a membrane for analysis by weight, count, or observation as it applies to the section on contamination measurement.

30.06.210 Globe and Circle Reticule

A reticule having lined circles and opaque discs of graduated diameters patterned to permit estimation (or measurement) of dimension or projected areas of particles.

ANSI/B93.2

30.06.240 Gravimetric Value

The weight of suspended solids per unit volume of fluid. A method employing membrane filters for this determination is outlined in Society of Automotive Engineers' Aerospace Recommended Practices number 785.

30.06.270 Linear Scale Reticule

A reticule having a straight scale marked to permit measurement or estimation of distance or length.

30.06.300 Microscopic Filar Eyepiece

A micrometer eyepiece containing a movable hairline connected to an external micrometer scale used to measure distance.

30.06.330 Microscope Gating

A particle counting and sizing technique in which the image of the particles are moved through the linear scale of a reticule for counting and sizing.

30.06.360 Microscope Reticule (Microscope Graticule)

A transparent disc inserted into the eyepiece of a microscope containing a scale or pattern used to measure or estimate the projected dimensions of particles when properly calibrated.

30.06.420 Microscopic

Particles whose diameter is below the threshold of normal vision, below forty micrometres for most individuals.

30.06.430 Multi-Pass Test

A test which requires the recirculation of unaltered effluent fluid through the filter element.

30.06.450 Non-Combustible Residue

Matter not changed to gaseous state when laboratory membranes are ashed at $1500°F$.

30.06.480 Non-Volatile Residue

The residue remaining on laboratory ware after the solvent or fluid has evaporated.

30.06.510 Optical Density

A method of expressing degree of contamination of a fluid by removal of contaminant by filtration and measuring change in optical transmission of the filter disc or fluid, or both.

30.06.540 Particle Count Blank

An allowance for the determinable background contamination.

30.06.570 Particle Size Distribution

The tabular or graphical listing of the number of particles according to particle size ranges.

30.06.600 Patch Test

Any method of evaluating fluid contamination wherein the sample is passed through a standardized laboratory filter, and the change in color, reflectivity, etc., of the laboratory filter is compared with previously established standards.

30.06.630 Precipitate

Particles separated from a fluid as a result of a chemical or physical change.

30.06.660 Raw Count

The actual number counted in each particle size range in a given sample.

30.06.690 Scoring Size, Particle

A particle whose dimensions are such that it is capable of entering a working clearance.

30.06.720 Total Statistical Count

The raw count multiplied by a counting calibration factor.

30.06.730 Visual Count

Any method allowing the evaluation of the contamination in a fluid by an optical counting procedure.

SECTION 07 — SAMPLING TERMS

30.07.150 Average Outgoing Quality Limit (A.O.Q.L.)

The maximum percentage of sample containers which may exceed the required cleanliness level as a process average.

30.07.250 Clean Fluid

Fluid which is compatible with particle counting method and the container used and does not contain more than one tenth the number of particles greater than 10 micrometres per millilitre that are allowed in the required cleanliness level.

30.07.300 Consecutive Acceptance Number (N)

The minimum number of initial qualifying inspections required to establish the acceptability of the cleaning process.

30.07.400 Fluid Sampling, Dynamic

The extraction of a sample fluid from a turbulent section of a flow stream.

30.07.450 Fluid Sampling, Static

The extraction of a sample fluid from a fluid at rest.

ANSI/B93.2

30.07.500   Inspection Ratio (R)

The ratio of number of randomly selected containers which are inspected to the number of containers processed.

30.07.800   Required Cleanliness Level (RCL)

The maximum number of particles greater than 10 micrometres per millilitre of sample container volume.

30.07.900   Sampler, Turbulent

A device for creating turbulence in the main stream while extracting a fluid sample.

GROUP 31 — AIR BLEEDERS & BREATHERS

     Section 01 — General

SECTION 01 — GENERAL

31.01.100   Air Bleeder

A device for removal of air.

31.01.200   Air Breather

A device permitting air movement between atmosphere and the component in which it is installed.

31.01.600   Vent

A passage to a reference pressure, usually the ambient pressure.

GROUP 32 — HEAT EXCHANGERS

     Section 01 — General
     Section 02 — Types of Heat Exchangers

SECTION 01 — GENERAL

32.01.050   Cooler

A heat exchanger which removes heat from a fluid.

32.01.070   Fan, Cooling

A device which mechanically creates a flow of air over a hot surface, usually used with a radiator in order to increase the rate of heat exchange.

32.01.100   Heat Exchanger

A device which transfers heat through a conducting wall from one fluid to another.

32.01.101   Heater

A device which transfers heat through a conducting wall from one fluid to another.

32.01.700   Temperature Controller

A device which maintains the fluid temperature within prescribed limits.

SECTION 02 — TYPES OF HEAT EXCHANGERS

32.02.220   Aftercooler

A device which cools a gas after it has been compressed.

32.02.240   Intercooler

A device which cools a gas between the compressive steps of a multiple stage compressor.

32.02.260   Precooler

A device which cools a gas before it is compressed.

32.02.270   Radiator

Device, usually of honeycomb or multi-tubular construction which transfers heat from a liquid to air, thereby acting as a liquid/air heat exchanger.

GROUP 33 — FILTER, PNEUMATIC & HYDRAULIC

     Section 01 — General
     Section 02 — Filter Types
     Section 04 — Filter Design and Installation
     Section 05 — Filter Components
     Section 06 — Filter Accessories
     Section 08 — Filter Element Medium
     Section 09 — Filter Element Parts & Related Items
     Section 10 — Filter Element Conditions
     Section 11 — Filter Performance

SECTION 01 — GENERAL

33.01.000   Filter

A device whose primary function is the removal of insoluble contaminants from a liquid or a gas.

SECTION 02 — FILTER TYPES

33.02.010   Filter, By-Pass (Reserve)

A filter which provides an alternate unfiltered flow path around the filter element when a preset differential pressure is reached.

33.02.015   Filter, Centrifugal

A filter in which separation of contaminants occurs when the fluid is accelerated in a circular path.

33.02.020   Filter, Disposable

A filter which is intended to be discarded and replaced after one service cycle.

33.02.030   Filter, Dual

A filter having two filter elements in parallel.

33.02.040 Filter, Duplex

An assembly of two filters with valving for selection of flow to either or both filters.

33.02.050 Filter, Fill Cap

A filter, usually hydraulic, which covers the fill opening to the reservoir and filters makeup fluid.

33.02.060 Filter, Filtered By-Pass

A filter, usually hydraulic, in which by-pass flow is filtered through a reserve filter element.

33.02.070 Filter, Full Flow

A filter which filters all influent flow.

33.02.075 Filter, With Full Flow By-Pass

A full flow filter which provides an alternative flow path around the filter element when a pre-set differential pressure is reached.

33.02.080 Filter, In-Line

A filter in which the inlet, outlet, and filter element axes are all in a straight line.

33.02.090 Filter, L-Type

A filter in which the inlet and outlet port axes are at right angles, and the filter element axis is parallel to either port axis.

33.02.100 Filter, Manifold

A filter containing multiple ports and integral related components which services more than one hydraulic circuit.

33.02.110 Filter, Modular

A filter which mounts to or within a manifold or subplate with flow passages at the interface.

33.02.120 Filter, Partial Flow

A filter which filters a portion of the influent flow.

33.02.140 Filter, Reservoir (Sump)

A filter, usually hydraulic, installed in a reservoir in series with a suction or return line.

33.02.150 Filter, Spin-On

A filter with an element sealed in its own pressure housing for independent mounting to the filter, spin-on.

33.02.160 Filter (Strainer)

A coarse hydraulic filter usually of woven wire construction. This may be in the form of a complete filter or just an element.

33.02.170 Filter, T-Type

A filter in which the inlet and outlet ports are located at one end of the filter with port axes in a straight line, and the filter element axis perpendicular to this line.

33.02.180 Filter, Two-Stage

A filter having two filter elements in series.

33.02.190 Filter, Wash, Hydraulic

A hydraulic filter in which a larger unfiltered portion of the fluid flowing parallel to the filter element axis is utilized to continuously clean the influent surface which filters the lesser flow.

33.02.200 Filter, Y-Type

A filter in which the inlet and outlet port axes are in a straight line, and the filter element axis is at an acute angle to this line.

SECTION 04 — FILTER DESIGN AND INSTALLATION

33.04.100 Effluent

The fluid leaving a component.

33.04.200 Element Removal Clearance

The minimum unobstructed distance required to remove the filter element.

33.04.300 Impingement

The direct high velocity impact of the fluid flow upon or against any internal portion of the filter.

33.04.400 Influent

The fluid entering a component.

33.04.500 Normal Flow

The intended direction of flow through filter.

33.04.800 Reverse Flow

Opposite to normal flow.

33.04.900 Starvation

Insufficient filter effluent to allow proper functioning of downstream components.

SECTION 05 — FILTER COMPONENTS

33.05.000 Filter Components

The parts that make up a filter.

33.05.010 Baffle

A device to prevent direct flow or impingement.

33.05.020 Base

The foundation or support for the filter which may also contain one or more ports.

33.05.030 Bowl (Shell)

A case that is closed at one end and mates with the filter head.

33.05.040 Cap

An end closure for the filter case.

33.05.060 Case (Shell)

A hollow part that provides a cavity for the filter element.

33.05.070 Cover

An end closure which provides access to the filter element.

33.05.080 Element (Cartridge)

The porous device which performs the actual process of filtration.

33.05.090 Head

An end closure for the filter case or bowl which contains one or more ports.

33.05.100 Housing

A ported enclosure which directs the flow through the filter element.

SECTION 06 — FILTER ACCESSORIES

33.06.000 Accessory, Filter

An auxiliary device incorporated into a filter to enhance its usefulness.

33.06.200 Indicator

A device which provides external visual evidence of sensed phenomena. (Filter Clogging)

33.06.210 Indicator, Clogging (By-Pass)

An indicator which signals alternate flow. (By-Passing)

33.06.220 Indicator, Differential Pressure

An indicator which signals a difference in pressure between two points which span the filter element.

33.06.230 Indicator, Pressure

An indicator which signals the presence or absence of pressure.

33.06.300 Magnetic Plug

A plug which attracts and holds ferromagnetic particles.

SECTION 08 — FILTER ELEMENT MEDIUM

33.08.010 Absorbent (Absorptive)

A filter medium that holds contaminant by mechanical means.

33.08.030 Adsorbent (Adsorptive)

A filter medium primarily intended to hold soluble and insoluble contaminants on its surface by molecular adhesion.

33.08.050 Combination (Composite)

A filter medium composed of two or more types, grades, or arrangements of filter media to provide properties which are not available in a single filter medium.

33.08.070 Deposited

A filter medium produced by chemical or electrolytic deposit.

33.08.080 Depth

A filter medium which primarily retains contaminant within tortuous passages.

33.08.090 Edge

A filter medium whose passages are formed by the adjacent surfaces of stacked discs, edgewound ribbons, or single-layer filaments.

33.08.100 Etched

A filter medium having passages produced by chemical or electrolytic removal.

33.08.110 Non-Woven

A filter medium composed of a mat of fibers.

33.08.120 Precoat

A filter medium in loose powder form (such as Fuller's or diatomaceous earth) introduced into the upstream fluid to condition a filter element.

33.08.130 Sintered

A metallic or non-metallic filter medium processed to cause diffusion bonds at all contacting points.

33.08.140 Surface

A filter medium which primarily retains contaminant on the influent face.

33.08.150 Wound

A filter medium comprised of layers of helical wraps of a continuous strand or filament in a predetermined pattern.

33.08.160 Woven

A filter medium made from strands of fiber, thread, or wire interlaced into a cloth on a loom.

SECTION 09 — FILTER ELEMENT PARTS & RELATED TERMS

33.09.010 Center Tube (Core)

The internal duct and filter media support.

33.09.040 Crest

The outer fold of a pleat.

33.09.050 End Cap

A ported or closed cover for the end of a filter element.

33.09.060 End Seal

The bond between the end cap and the filter medium. Also a sealing device which seals against the end cap by axial contact pressure.

33.09.070 External Support

A permeable structural enclosure which imparts rigidity to a filter element, and usually protects the filter medium.

33.09.080 Grooving

Shallow ridges in the filter medium perpendicular to the roots of the pleats.

33.09.100 Outer Wrapper

A permeable enclosure which protects the filter medium.

33.09.110 Pleats (Corrugations)

A series of folds in the filter medium usually of uniform height and spacing.

33.09.120 Root

The inner fold of a pleat.

33.09.130 Side Seal

The logitudinal seam of the filter medium in a filter element.

SECTION 10 — FILTER ELEMENT CONDITIONS

33.10.015 Element, Bi-Directional

A filter element designed for flow in both directions.

33.10.020 Element, Bridging

A condition of filter element loading in which contaminant spans the space between adjacent sections of a filter element thus blocking a portion of the useful filtration area.

33.10.030 Element, Burst

An outward structural failure of the filter element caused by excessive differential pressure.

33.10.040 Element, Clean

A new or properly cleaned filter element.

33.10.041 Element Cleanable

A filter element which when loaded can be restored by a suitable process to an acceptable percentage of its original dirt capacity.

33.10.045 Element, Clogged Filter

A filter element which has collected such a quantity of contaminant that it cannot maintain rated flow without excessive differential pressure increase.

33.10.046 Element, Clogging

Choking by deposits of solid or liquid particles.

33.10.050 Element, Collapsed

An inward structural failure of the filter element caused by excessive differential pressure.

33.10.060 Element, Contaminated

A filter element which releases into the effluent foreign particles resulting from handling, storage and fabrication.

33.10.070 Element, Dirty

A used filter element that is partially or completely loaded.

33.10.074 Element, Disposable (Throw-Away)

A filter element which is intended to be discarded and replaced after one service cycle.

33.10.076 Element Effective Filtration Area

Total area of the porous medium exposed to flow in a filter element.

33.10.078   Element, Extended Area

A filter element whose medium is pleated or otherwise formed to obtain more effective area within a given dimensional envelope.

33.10.080   Element, Fatigued

A structural failure of the filter medium due to flexing caused by cyclic differential pressure.

33.10.084   Element, Full System Differential Pressure

A filter element which will withstand a differential pressure at least equal to the maximum system operating pressure without structural or filter medium failure.

33.10.086   Element, Inside-Out Flow

A filter element designed for normal flow outward and perpendicular to the axis of the filter element.

33.10.090   Element, Loaded (Plugged) (See Clogged)

A filter element that has collected a sufficient quantity of insoluble contaminants such that is can no longer pass rated flow without excessive differential pressure.

33.10.092   Element, Magnetic

A filter element which in addition to its filter medium has a magnetic or magnets incorporated into its structure to attract and hold ferromagnetic particles.

33.10.094   Element, Modular (Plug-In)

A filter element which has no separate housings of its own, but whose housing is incorporated into the equipment which it services. It may also incorporate a suitable closure for the filter cavity.

33.10.096   Element, Outside-In Flow

A filter element designed for normal flow perpendicular and toward the axis of the filter element.

33.10.100   Element, Pinched Pleat

A pleat closed off by excessive differential pressure or crowding, thus reducing the effective area of the filter element.

33.10.103   Element, Plain

A filter element whose medium is not pleated or otherwise extended and has the geometric form of a cylinder, cone, disc, plate etc.

33.10.106   Element, Pleated (Corrugated)

A filter element whose medium consists of a series of uniform folds and has the geometric form of a cylinder, cone, disc, plate, etc.

33.10.109   Element, Primary

The first filter element in a series, or the main filter element of a filtered by-pass filter assembly.

33.10.112   Element, Renewable Filter

A filter element parts of which are replaced to restore the element to its "as new" flow/pressure differential characteristic.

33.10.115   Element, Reserve

A standby filter element.

33.10.120   Element, Ruptured

Any tear or split in the filter medium.

33.10.123   Element, Secondary

The second of two filter elements in series.

33.10.126   Element, Self Cleaning

A filter element designed to be cleaned without removing it from the filter assembly.

33.10.130   Element, Service Loaded

A filter element which is loaded from actual use.

33.10.133   Element, Two-Stage

A filter element assembly composed of two filter elements or media in series.

33.10.136   Element, Wash

A filter in which a larger unfiltered portion of the fluid flowing parallel to the filter element axis is utilized to continuously clean the influent surface which filters the lesser flow.

SECTION 11 — FILTER PERFORMANCE

33.11.000   Filter Performance

Those factors which describe the functions and attributes of a filter or filter element.

33.11.010 Absolute Filtration Rating (Largest Particle Passed)

The diameter of the largest hard spherical particle that will pass through a filter under specified test conditions. This is an indication of the largest opening in the filter element.

33.11.020 Air Inclusion

The volume of air introduced into a liquid system as a result of servicing a filter.

33.11.025 Apparent Capacity (α)

The actual weight (grams) of contaminant injected into the filter test system before the terminal pressure drop is reached.

33.11.030 Boil Point (Foam All Over) (Mass Bubble Point) (Open Bubble Point)

A differential gas pressure at which gas bubbles are profusely emitted from the entire surface of a wetted filter element under specified test conditions.

33.11.050 Burst Pressure

The pressure which causes rupture. Also, the inside-out differential pressure that causes outward structural or filter medium failure of a filter element.

33.11.060 Burst Pressure Rating

The maximum specified inside-out differential pressure which can be applied to a filter element without outward structural or filter medium failure.

33.11.070 Cleanability

The ability of a cleanable filter element to withstand repeated field cleanings and retain adequate direct capacity and service life.

33.11.080 Collapse Pressure

The outside-in differential pressure that causes structural failure of a filter element.

33.11.090 Collapse Pressure Rating

The maximum specified outside-in differential pressure which can be applied to a filter element without inward structural or filter medium failure.

33.11.120 Dirt Capacity (Dust Capacity) (Contaminant Capacity)

The weight of a specified artificial contaminant which must be added to the influent to produce a given differential pressure across a filter at specified conditions. Used as an indication of relative service life.

33.11.140 Effective Area

The total area of the porous medium exposed to flow in a filter element.

33.11.150 Efficiency

The ability, expressed as a percent, of a filter to remove specified artificial contaminant at a given contaminant concentration under specified test conditions.

33.11.151 Efficiency Curve

Filter curve with the efficiency as a function of the particle dimensions.

33.11.160 End Load

The axial force applied to the end of a filter element which may cause permanent deformation or seal failure.

33.11.170 End Load Rating

The maximum specified axial force which can be applied to a filter element without permanent deformation or seal failure.

33.11.176 Fabrication Integrity

The physical acceptability of a filter element relative to that designated by the filter manufacturer.

33.11.182 Filtration Ratio β_μ

The ratio of the number of particles greater than a given size (μ) in the influent fluid to the number of particles greater than the same size (μ) in the effluent fluid.

33.11.190 Flow Fatigue

The ability of a filter element to resist structural failure of the filter medium due to flexing caused by cyclic differential pressure.

33.11.240 Mean Filtration Rating

A measurement of the average size of the pores of the filter medium.

33.11.250 Migration

Contaminant released downstream.

33.11.251 Migration, Abrasion

Migration generated by parts that rub together and wear during vibration or shock induced by flow or other stimuli.

33.11.252 Migration, Built-In-Dirt

Migration composed of foreign materials introduced during handling, storage and fabrication.

33.11.253 Migration, Contaminant

Migration due to unloading.

33.11.254 Migration, Media

Migration composed of the materials making up the filter medium.

33.11.256 Net Pressure Drop

The difference between the terminal pressure and the pressure drop across a clean element.

33.11.260 Nominal Filtration Rating

An arbitrary micrometre value indicated by the filter manufacturer. Due to lack of reproducibility this rating is deprecated.

33.11.265 Normalized Capacity (θ)

The apparent capacity divided by the rated flow (litre/minute) of the element.

33.11.270 Open Area

The pore area of a filter medium often expressed as a percent of total area.

33.11.280 Open Area Ratio

The ratio of pore area to total area of a filter medium expressed as a percent of total area.

33.11.300 Permeability

The relationship of flow per unit area to differential pressure across a filter medium.

33.11.310 Pore Size Distribution

The ratio of the number of holes of a given size to the total number of holes per unit are expressed as a percent and as a function of hole size.

33.11.320 Porosity (Void Fraction)

The ratio of pore volume to total volume of a filter medium expressed as a percent.

33.11.330 Pressure Loss (Ψ)

The average pressure drop of the filter element throughout its useful life to terminal pressure drop.

33.11.350 Residual Dirt Capacity

The dirt capacity remaining in a service loaded filter element after use, but before cleaning, measured under the same conditions as the dirt capacity of a new filter element.

33.11.360 Sloughing Off

The release of contaminant from the upstream surface of a filter element to the upstream side of the filter enclosure.

33.11.365 Terminal Pressure Drop

The maximum pressure drop permitted across the filter element as designated by the manufacturer to limit useful performance.

33.11.370 Tortuosity

The ratio of the average effective flow path length to minimum theoretical flow path length (thickness) of a filter medium.

33.11.380 Total Area

The entire area of a porous medium, whether effective or not, in a filter element.

33.11.390 Unloading

The release of contaminant that was initially captured by the filter medium.

GROUP 34 — SEPARATORS

Section 01 — General
Section 02 — Separator Types

SECTION 01 — GENERAL

34.01.050 Deaerator

Equipment used to eliminate air or gas contained in the liquid in a hydraulic circuit.

34.01.100 Separator

A device whose primary function is to isolate contaminants by physical properties other than size. (Separators remove gas from liquid medium or remove liquid from gaseous medium).

SECTION 02 — SEPARATOR TYPES

34.02.100 Separator, Adsorbent

A separator that retains certain soluble and insoluble contaminants by molecular adhesion.

34.02.200 Separator, Centrifugal

A separator that removes non-miscible fluid and solid contaminants that have a different specific gravity than the fluid being purified by accelerating the fluid in a circular path and using the radial acceleration component to isolate these contaminants.

34.02.300 Separator, Coalescing

A separator that divides a mixture or emulsion of two non-miscible liquids using the interfacial tension between the two liquids and the difference in wetting of the two liquids on a particular porous medium.

34.02.400 Separator, Electrostatic

A separator that removes contaminant from dielectric fluids by applying an electrical charge to the contaminant which is then attracted to a collection device of different electrical charge.

34.02.500 Separator, Magnetic

A separator that uses a magnetic field to attract and hold ferromagnetic particles.

34.02.550 Separator, Oil Remover

A device which separates oil from a compressed gas.

34.02.600 Separator, Two Phase

A separator that is capable of dividing a liquid and gas mixture.

34.02.900 Separator, Vacuum

A separator that utilizes subatmospheric pressure to remove certain gases and liquids from another liquid because of their difference in vapor pressure.

34.02.910 Separator, Water Tap

A device fitted at a specific point in the installation to collect moisture and possibly other impurities from a pneumatic system.

GROUP 36 — LUBRICATORS

Section 01 — General

SECTION 01 — GENERAL

36.01.100 Lubricator

A device which adds controlled or metered amounts of lubricant into a fluid power system.

GROUP 37 — COMPRESSED AIR DRYERS

Section 01 — General
Section 02 — Types of Compressed Air Dryers
Section 04 — Operational Factors
Section 05 — Compressed Air Dryer Components

SECTION 01 — GENERAL

37.01.100 Compressed Air Dryer

A device for reducing the moisture vapor content of the working medium.

NOTE — "Dehumidifiers", "Dehydrators", and basic "Desiccators" are often erroneously considered as Compressed Air Dryers. Their primary function is to remove moisture from atmospheric pressure air or from solids and do not meet the strict definition of Compressed Air Dryers. Therefore, these terms are not included in the text.

SECTION 02 — TYPES OF COMPRESSED AIR DRYERS

37.02.000 Automatic

All functions are automatically controlled.

37.02.010 Deliquescent

Moisture is separated by using the absorptive properties of special hygroscopic compounds.

37.02.013 Demand Cycle

A compressed air dryer which switches chambers in accordance with the drying load on the unit, rather than a fixed cycle.

37.02.016 Desiccant

A compressed air dryer which lowers the dew point of compressed air by passing air through a drying agent.

37.02.400 Fixed Cycle

A compressed air dryer which operates on a time-controlled cycle.

37.02.500 Manual

A compressed air dryer in which all functions are manually controlled.

37.02.600 Refrigerated

Moisture is separated by lowering the air temperature by means of a refrigeration compressor and heat exchanger.

37.02.620 Refrigerated, Cycling

A refrigerated compressed air dryer which automatically cycles on and off according to load.

37.02.660 Refrigerated, Non-Cycling

A refrigerated compressed air dryer that runs continuously.

37.02.700 Regenerative

The capacity of the dryer to separate moisture can be restored without replacing the drying compound.

37.02.710 Regenerative, Closed System

A regenerative compressed air dryer which does not exhaust purge air to the atmosphere.

37.02.720 Regenerative, Dual Tower

A regenerative compressed air dryer which includes two desiccant chambers which are alternately cycled from drying period to reactivation.

37.02.725 Regenerative, Heat

Heat is applied to the saturated drying compound to drive off collected moisture and thereby regenerate it.

37.02.730 Regenerative, Externally (Convection) Heated

A regenerative compressed air dryer which utilizes a heater (s) located externally to the desiccant bed.

37.02.740 Regenerative, Internally Heated

A regenerative compressed air dryer which utilizes a heater (s) located within the desiccant bed.

37.02.745 Regenerative, Heatless

Air previously dried under pressure is expanded to atmospheric pressure and allowed to flow through the saturated compound to drive off collected moisture and thereby regenerate it.

37.02.750 Regenerative, Open System

A regenerative compressed air dryer which exhausts purge air to the atmosphere.

37.02.800 Semi-Automatic

A compressed air dryer in which some functions are automatically controlled while others are manually controlled.

SECTION 04 — OPERATIONAL FACTORS

37.04.125 Ambient Temperature Range

The range of temperature of the air surrounding the dryer in which the equipment will perform as recommended.

37.04.190 Capacity, Compressed Air Dryer

The amount of dry air delivered at recommended operating conditions.

37.04.200 Condensation

The process of changing a vapor into a liquid condensate by the extraction of heat.

37.02.250 Contact Time

The time required for a molecule in a stream of air to pass completely through a desiccant bed based on superficial bed velocity.

37.04.280 Cycle, Adsorbent Dryer

The time required for an adsorbent desiccant bed to pass through one drying period and one regeneration period.

37.04.300 Dew Point

The temperature at which vapors in a gas condense. For practical purposes, it must be referred to a stated pressure.

37.04.310 Dew Point, Atmosphere

The dew point in the air at atmospheric pressure.

37.04.340 Dew Point, Pressure

The dew point in the air at the actual operating pressure.

37.04.350 Dew Point Depression

The difference between inlet and outlet dew points of a compressed air dryer referred to the same operating conditions.

37.04.400 Evaporator Freeze-Up

Blocking of the air passages through the evaporator due to freezing of the condensate.

37.04.450 Fluidization Velocity

The rate of air flow upward through a desiccant bed, which if exceeded, will physically disturb the desiccant.

37.04.560 Period, Cooling

That portion of the regeneration cycle during which the desiccant is cooled.

37.04.570 Period, Drying (Sorption)

That portion of the cycle during which the desiccant bed is on-stream in drying service.

37.04.580 Period, Heating

The portion of the regeneration cycle during which the desiccant is heated.

37.04.590   Period, Regeneration (Desorption)

That portion of the cycle during which the desiccant bed is removed from drying service and its efficiency is restored.

37.04.600   Purge Flow

A flow of air to regenerate a desiccant.

37.04.620   Purge Flow, Co-Current

Purge flow is in the same direction through the desiccant as drying flow.

37.04.630   Purge Flow, Countercurrent

Purge flow direction through the desiccant is opposite to the direction of drying flow.

37.04.650   Regeneration (Reactivation)

The process of restoring the capacity of adsorbent desiccant.

37.04.660   Regeneration, Atmospheric

Reactivation of the desiccant bed at atmospheric pressure.

37.04.670   Regeneration, Heat

Reactivation of desiccant by increasing its temperature.

37.04.680   Regeneration, Heatless

Reactivation of desiccant without heat. It is usually done with dry air purge flow.

37.04.690   Regeneration, Pressure

Reactivation of the desiccant bed at or near operating pressure.

37.04.700   Repressurization

Return of the regenerated desiccant chamber from atmospheric pressure to operating pressure prior to chamber switchover.

37.04.750   Superficial Bed Velocity

The rate of air flow through the cross-sectional area of the desiccant bed, without regard for the area occupied by the desiccant.

SECTION 05 — COMPRESSED AIR DRYER COMPONENTS

37.05.050   Afterfilter

A filter which follows the compressed air dryer and usually for the protection of downstream equipment from desiccant dust.

37.05.100   Automatic Drain

A device which automatically discharges condensate from the moisture separator.

37.05.200   Condensing Unit

A specific refrigerant machine combination for a given refrigerant consisting of one or more power driven compressors, air or water cooled condensers, etc.

37.05.250   Desiccant

Material that tends to remove moisture from compressed air.

37.05.255   Desiccant, Absorbent (Deliquescent)

A desiccant that dissolves into the moisture it removes from the compressed air and is slowly consumed in the process.

37.05.260   Desiccant, Adsorbent

A solid desiccant which is capable of removing moisture from compressed air by adherence of moisture to its surface.

37.05.280   Desiccant, Tabular Support

Inert material which supports a desiccant and diffuses air.

37.05.300   Evaporator

The heat exchanger where the refrigerant absorbs heat.

37.05.350   Evaporator Back-Pressure Valve

A valve which maintains the refrigerant pressure, and consequently its temperature, in the evaporator at a predetermined level.

37.05.400   Expansion Device

A device which controls expansion of high pressure liquid refrigerant to a lower pressure. It may be a capillary tube, thermostatic expansion valve, or automatic expansion valve.

37.05.440   High Side Components

Parts of a refrigeration system under condenser pressures or higher.

37.05.450   Hot Gas By-Pass Valve

A modulating valve which by-passes hot refrigerant gas from the high pressure to the low pressure side of the system in order to reduce refrigeration capacity commensurate with reduction in load and to control evaporation temperature.

37.05.490   Low Side Components

Parts of a refrigeration system at or below evaporation pressure.

37.05.500   Moisture Separator

A device which removes liquids from an air system.

37.05.550   Precooler Reheater

A heat exchanger which lowers the temperature of the inlet air and raises the temperature of the exciting air.

37.05.600   Prefilter

A filter which precedes the compressed air dryer and usually for the protection of desiccant or heat transfer surfaces.

37.05.650   Refrigerant

A substance which produces a cooling effect by its absorption of heat while expanding or vaporizing.

37.05.700   Refrigerant Gauge

A pressure gauge calibrated to indicate the corresponding temperature of saturated refrigerant.

37.05.750   Refrigerant Receiver

A liquid refrigerant storage tank in a refrigeration system.

37.05.800   Refrigeration Compressor

The part of a refrigeration system which takes refrigerant at low pressure and compresses it to a smaller volume at a higher pressure.

37.05.830   Refrigeration Compressor, Hermetic

A refrigeration compressor which is sealed in a housing with its driving motor.

37.05.840   Refrigeration Compressor, Open Type

A refrigeration compressor which is driven by a physically-separated power source through mechanical power transmission equipment.

GROUP 41 — INTENSIFIERS

Section 01 — General
Section 02 — Intensifier

SECTION 01 — GENERAL

41.01.099   Intensification, Ratio of

The ratio of the secondary pressure to the primary pressure or of the primary flow rate to the secondary flow rate.

41.01.100   Intensifier

A device which converts low pressure fluid power into higher pressure fluid power.

41.01.700   Intensifier, Primary Fluid

Descriptive of fluid at low pressure applied to the inlet port of the pressure intensifier, and of anything concerned with the fluid flow, pressure, circuit.

SECTION 02 — INTENSIFIER TYPES

41.02.100   Intensifier, Continuous

An intensifier in which continuous application of primary fluid to the inlet port can produce a continual flow of secondary fluid.

41.02.120   Intensifier, Double Acting

A unit which intensifies the secondary fluid pressure whatever the direction of flow of the primary fluid.

41.02.125   Intensifier, Dual Fluid

An intensifier in which different types of fluid are used in the primary and secondary circuits.

41.02.130   Intensifier, Secondary Fluid

Descriptive of fluid at high pressure applied to the outlet port of the pressure intensifier, and of anything concerned with the fluid: flow, pressure, circuit.

41.02.250　　　　　Intensifier, Single Acting

A unit which only intensifies the fluid pressure in one direction of flow of the primary fluid.

41.02.255　　　　　Intensifier, Single Fluid

An intensifier in which fluid of similar type is used in both primary and secondary circuits.

41.02.260　　　　　Intensifier, Single Shot

An intensifier in which the continuous application of primary fluid at the inlet port can only give a limited volume of secondary fluid.

GROUP 42 — POWER UNITS

Section 01 — General
Section 02 — Types of Power Units

SECTION 01 — GENERAL

42.01.100　　　　　Power Unit

A combination of pump, pump drive, reservoir, controls and conditioning components which may be required for its application.

SECTION 02 — TYPES OF POWER UNITS

42.02.400　　　　　Power Unit, Hydraulic

An assembled group of components which facilitate fluid storage, conditioning and delivery under conditions of controlled pressure and flow to the discharge port of the pump, including maximum pressure controls and sensing devices when applicable. (Circuitry components, although sometimes mounted on the reservoir, are not considered part of the power unit.)

GROUP 43 — HYDRAULIC PUMPS

Section 01 — General
Section 02 — Hydraulic Pump Types
Section 03 — Hydraulic Pump Mounting

SECTION 01 — GENERAL

43.01.005　　　　　Capacity, Derived

Volume displaced at defined minimum working pressure calculated from two measurements at different speeds.

43.01.015　　　　　Efficiency, Mechanical

Ratio of derived torque to absorbed torque.

43.01.016　　　　　Efficiency, Overall

Ratio of the effective power to absorbed power.

43.01.017　　　　　Efficiency, Volumetric

Ratio of the effective output flow to the derived output flow.

43.01.018　　　　　Extension, Pump Shaft

The part of the drive shaft which extends outside the unit and which includes the means whereby the drive is affected; e.g., key, taper, spline, etc.

43.01.020　　　　　Flow, Output

Flow rate discharged at the outlet port.

43.01.021　　　　　Flow, Output, Derived

Product of the derived capacity by the number of revolutions or cycles per unit of time.

43.01.022　　　　　Flow, Output, Effective

Actual output flow at the pump outlet measured at the pressure and temperature at that point.

43.01.023　　　　　Flow, Output, Geometric

Product of the geometric capacity by the number of revolutions or cycles per unit of time.

43.01.100　　　　　Hydraulic Pump

A device which converts mechanical force and motion into hydraulic fluid power.

43.01.150　　　　　Losses, Hydrodynamic

Losses due to motion of the fluid.

43.01.151　　　　　Losses, Mechanical

Losses due to mechanical friction.

43.01.152　　　　　Losses, Pump

Portion of the absorbed power not transformed into fluid power.

43.01.153　　　　　Losses, Volumetric

   a) Losses due to imperfect filling of the pumping chambers.
   b) Internal Leakage.
   c) External drainage.
   d) Volumetric compressibility losses.

43.01.170　　　　　Power

Increase of hydraulic energy per unit time between inlet and outlet ports of the pump.

43.01.171　　　　　Power, Derived

Hydraulic power calculated from the derived output.

ANSI/B93.2

43.01.172   Power, Effective

Hydraulic power calculated from the effective output.

43.01.173   Power, Geometric

Hydraulic power calculated from the geometric output.

43.01.174   Power, Input

Power applied to the driving shaft of the pump.

43.01.175   Power, Installed

Rated power of the driving motor.

43.01.176   Power, Required

Power which is necessary to drive the pump shaft under specified conditions.

43.01.180   Pressure, Deadhead

Output pressure without flow (stagnation pressure).

43.01.300   Torque

Torque transmitted by the driving shaft of the pump.

SECTION 02 — HYDRAULIC PUMP TYPES

43.02.100   Centrifugal

A hydraulic pump which produces fluid velocity and converts it to pressure head.

43.02.130   Centrifugal, Diffuser (Concentric)

A centrifugal hydraulic pump in which fluid enters at the center of the impeller, is accelerated radially, and leaves through vanes arranged to provide a gradually enlarging flow passage.

43.02.145   Centrifugal, Peripheral

A centrifugal hydraulic pump in which fluid enters, follows, and leaves the periphery of the impeller.

43.02.175   Centrifugal, Volute (Spiral)

A centrifugal hydraulic pump in which fluid enters at the center of the impeller, is accelerated radially, and leaves through a gradually enlarging flow passage.

43.02.200   Fixed Displacement

A hydraulc pump in which the volume displaced per cycle cannot be varied.

43.02.250   Gear

Pump in which two or more gears act in engagement as pumping members.

43.02.251   Gear, External

Pump with two or more external gears.

43.02.252   Gear, Fixed Clearance

Pump in which the side clearance of the gears is fixed.

43.02.253   Gear, Internal

Pump with an internal gear in engagement with one or more external gears.

43.02.254   Gear, with Pressure Loading

Pump in which the side clearance of the gears is controlled as a function of the delivery pressure.

43.02.300   Hand

A hand operated hydraulic pump.

43.02.301   Hand, Double Acting

A hand pump in which there are two alternate discharge strokes per cycle.

43.02.302   Hand, Single Acting

A hand pump in which the suction takes place during one part of the cycle and delivery during the remaining part of the cycle.

43.02.350   Multiple Stage

Two or more hydraulic pumps in series.

43.02.396   Piston

Pump in which the fluid volume is displaced by one or more reciprocating pistons.

43.02.397   Piston, Angled

Axial piston pump in which the drive shaft is at an angle to the common axis.

43.02.398   Piston, Axial

Pump having several pistons with mutually parallel axes which are arranged around and parallel to a common axis.

43.02.399   Piston, Inline

Pump having several pistons with mutually parallel axes arranged on a common plane.

43.02.400   Piston, Radial

Pump having several pistons arranged to operate radially.

43.02.410   Pump-Motor

Unit which functions either as a pump or as a rotary motor.

ANSI/B93.2

43.02.450   Reciprocating Duplex

A hydraulic pump having two reciprocating pistons.

43.02.500   Reciprocating Single Piston

A hydraulic pump having a single reciprocating piston.

43.02.550   Screw

A hydraulic pump having one or more screws rotating in a housing.

43.02.600   Vane

A hydraulic pump having multiple radial vanes within a supporting rotor.

43.02.601   Vane, Balanced

Pump in which the transverse forces on the rotor are balanced.

43.02.602   Vane, Unbalanced

Pump in which the transverse forces on the rotor are not balanced.

43.02.650   Variable Displacement

A hydraulic pump in which the volume displaced per cycle can be varied.

SECTION 03 — PUMP MOUNTING

43.03.100   Flange

Mounted by a flange with the supporting face at right angles to the driving shaft.

43.03.120   Foot

Mounting with the supporting face parallel to the driving shaft.

43.03.300   Pilot (Spigot)

Unit located by a spigot (pilot) however mounted.

GROUP 44 — COMPRESSORS

   Section 01 — General
   Section 02 — Compressor Types

SECTION 01 — GENERAL

44.01.100   Compressor

A device which converts mechanical force and motion into pneumatic fluid power.

44.01.101   Installation

An assembly of motor, compressor, receiver, regulator, etc.

44.01.800   Unloading Device

A device which allows the compressor to run with no load when a predetermined pressure is reached.

SECTION 02 — COMPRESSOR TYPES

44.02.300   Compressor, Multiple Stage

A compressor having two or more compressive steps in which the discharge from each supplies the next in series.

GROUP 45 — VACUUM PUMPS

   Section 01 — General

SECTION 01 — GENERAL

45.01.100   Vacuum Pump

A device which uses mechanical force and motion to evacuate gas from a connected chamber to create subatmospheric pressure.

GROUP 50 — GENERAL CONDUCTOR TERMS

   Section 01 — General
   Section 02 — General Components
   Section 03 — Functional Lines
   Section 04 — Port Types
   Section 05 — Port Functions
   Section 06 — Port Threads
   Section 07 — Physical Characteristics

SECTION 01 — GENERAL

50.01.200   Channel

A fluid passage, the length of which is large with respect to its cross-sectional area.

50.01.201   Channel, Control

Channel through which the control or input signal enters the device.

50.01.202   Channel, Output

The channel through which the output signal leaves the device.

50.01.600   Passage

A machined or cored fluid-conducting path which lies within or passes through a component.

50.01.610   Flow Passage, Controlled

A flow passage whose ability to pass fluid can be changed by the influence of a signal.

50.01.625   Flow Path

A series of conductors and passages which convey fluid.

ANSI/B93.2

50.01.700   Port

A terminus of a passage in a component to which conductors can be connected.

50.01.701   Port, Area of

Minimum area of fluid passage through a port.

SECTION 02 — GENERAL COMPONENTS

50.02.100   Conductor

A component whose primary function is to contain and direct fluid.

50.02.200   Conduit

Any confining element employed to transfer fluid.

50.02.300   Line

A tube, pipe, or hose for conducting fluid.

50.02.400   Nipple

A short length of pipe or tube.

SECTION 03 — FUNCTIONAL LINES

50.03.050   Line, Bleed

Line through which air is purged from pipes (conductors) containing liquid.

50.03.100   Line, Drain

A line returning leakage fluid independently to the reservoir or vented manifold.

50.03.200   Line, Exhaust

A line returning power or control fluid back to the reservoir or atmosphere.

50.03.300   Line, Joining

Lines which connect in a circuit.

50.03.350   Line, Make-Up

A pipeline (conductor) to supply working fluid to a circuit to make up losses as required.

50.03.400   Lines, Passing

Lines which cross but do not connect in a circuit.

50.03.500   Line, Pilot

A line which conducts control fluid.

50.03.510   Line, Pump Inlet

A pipe (conductor) connected to the inlet port of a pump and carrying the supply of working fluid to the pump.

50.03.560   Line, Return

A pipe (conductor) to return the working fluid to the reservoir.

50.03.600   Line, Suction

A supply line at sub-atmospheric pressure to a pump, compressor, or other component.

50.03.700   Line, Working

A line which conducts fluid power.

SECTION 04 — PORT TYPES

50.04.200   AND 10050

A United States Air Force-Navy Aeronautical Design Standard in which a straight thread port is used to attach tube fittings to various components. It employs an "O" ring seal compressed in a special cavity.

50.04.400   Pipe

A port which conforms to pipe thread standards.

50.04.600   Plain "O" Ring

A flat-faced port which uses bolts for attaching the conductor coupling and which includes an O-ring in a recessed groove against the flat face of the port.

50.04.800   SAE J 514

A straight thread port used to attach tube and hose fittings. It employs an "O" ring compressed in a wedge-shaped cavity. A standard of the Society of Automotive Engineers J514 and ANSI/B116.1.

50.04.850   Take-Off Point

Auxiliary connection on units or pipes (conductors) for fluid supply or measurement.

SECTION 05 — PORT FUNCTIONS

50.05.030   Port, Bias

The port at which a biasing signal is applied.

50.05.050   Port, Bleed

A port which provides a passage for the purging of gas from a system or components.

50.05.100   Port, Control

A port which provides passage for a control signal.

50.05.150 Port, Cylinder

A port which provides a passage to or from an actuator.

50.05.180 Port, Differential Pressure

A port(s) which provides a passage to the upstream and downstream sides of a component.

50.05.200 Port, Discharge

A port which provides a passage for fluid power to the system.

50.05.250 Port, Drain

A port for removal of fluid from a component, open to atmosphere, or connected to an unrestricted line.

50.05.251 Port, Airline Drain

Port which enables liquid to be drained from pneumatic circuits.

50.05.300 Port, Exhaust

A port which provides a passage to the atmosphere.

50.05.350 Port, Fill

A port which provides a passage for filling purposes.

50.05.360 Port, Flanged

Port arranged to accept flanged connections.

50.05.400 Port, Inlet

A port which provides a passage for the influent.

50.05.416 Port, Manifold

Connection made through a mounting face.

50.05.430 Port, Outlet (Output)

A port which provides a passage for the effluent.

50.05.450 Port, Pressure

A port which provides a passage from the source of fluid.

50.05.600 Port, Suction

A port which provides a passage for atmospheric charging of a pump or compressor.

50.05.620 Port, Supply

The port at which power is provided to an active device.

50.05.650 Port, Tank (Reservoir) (Return)

A port which provides a passage to the fluid source.

50.05.700 Port, Vent

A port which provides a passage to a reference pressure, usually the ambient pressure.

SECTION 06 — PORT THREADS

50.06.000 Port, Threaded

Port arranged to accept screw thread connections.

50.06.400 Pipe Thread

Screw threads for joining pipe.

50.06.420 Pipe Thread, Dryseal

Tapered pipe threads in which sealing is a function of root and crest interference.

50.06.440 Pipe Thread, Tapered

Pipe threads in which the pitch diameter follows a helical cone to provide interference in tightening.

SECTION 07 — PHYSICAL CHARACTERISTICS

50.07.200 Back Connected

Where connections are made to normally unexposed surfaces of components.

50.07.400 Front Connected

Where connections are made to normally exposed surfaces of components.

50.07.600 Port-to-Port Dimension

The distance between two ports measured from face to face or between center lines.

GROUP 51 — FITTINGS

Section 01 — General
Section 02 — Types of Fittings

SECTION 01 — GENERAL

51.01.100 Fitting

A connector or closure for fluid power lines and passages.

SECTION 02 — TYPES OF FITTINGS

51.02.050 Fitting, Bushing

A short externally threaded connector with a smaller size internal thread.

51.02.100 Fitting, Cap

A cover for fluid passage.

51.02.150 Fitting, Closure

A cap or a plug.

51.02.200 Fitting, Compression

A fitting which seals and grips by manual adjustable deformation.

51.02.250 Fitting, Connector

A fitting for joining a conductor to a component port or to one or more other conductors.

51.02.300 Fitting, Coupling

A straight connector for fluid lines.

51.02.350 Fitting, Cross

A fitting with four ports arranged in pairs, each pair on one axis, and the axes at right angles.

51.02.400 Fitting, Elbow

A fitting that makes an angle between mating lines. The angle is always 90 degrees unless another angle is specified.

51.02.440 Fitting, Female Thread

Connection with internal thread.

51.02.450 Fitting, Flange

A fitting which utilizes a radially extending collar for sealing and connection.

51.02.500 Fitting, Flared

A fitting which seals and grips by a pre-formed flare at the end of the tube.

51.02.550 Fitting, Flared AN

A United States Air Force-Navy 37° flared tube fitting Design Standard.

51.02.600 Fitting, Flareless

A fitting which seals and grips by means other than a flare.

51.02.620 Fitting, Male Thread

Connection with external thread.

51.02.650 Fitting, Reusable Hose

A hose fitting that can be removed from a hose and reused.

51.02.700 Fitting, Welded

A fitting attached by welding.

51.02.750 Fitting, Plug

A closure which fits into a fluid passage.

51.02.760 Fitting, Plug, Dryseal Pipe

A plug made with a thread which conforms to Dryseal Pipe Thread Standards.

51.02.770 Fitting, Plug, Short Pipe Thread

A plug which conforms in all respects to standard pipe threads except that the full thread has been shortened one full thread from the small end.

51.02.780 Fitting, Plug, Standard Pipe Thread

A plug with American (National) tapered pipe threads.

51.02.790 Fitting, Plug, Straight Thread

A plug with straight thread conforming to Unified Thread Standards.

51.02.795 Fitting, Pneumatic

Leakproof devices to connect pipelines (conductors) to one another, or the equipment.

51.02.800 Fitting, Reducer

A fitting having a smaller line size at one end than the other.

51.02.840 Fitting, Tailpiece

A fitting inserted into a flexible tube and secured.

51.02.850 Fitting, Tee

A fitting with three ports, a pair on one axis with one side outlet at right angles to this axis.

51.02.860 Fitting, Threaded Union

A straight connector or adaptor with external or internal threads.

51.02.900 Fitting, Union

A fitting which permits lines to be joined or separated without requiring the lines to be rotated.

51.02.950 Fitting, Wye (Y)

A fitting with three ports, a pair on one axis with one side outlet at any angle other than right angles to this axis. The side outlet is usually 45°, unless another angle is specified.

GROUP 52 — HOSE

Section 01 — General
Section 02 — Types of Hose

SECTION 01 — GENERAL

52.01.100 Hose

A flexible line or conductor whose nomimal size is its inside diameter.

SECTION 02 — TYPES OF HOSE

52.02.800 Hose, Wire Braided

Hose consisting of a flexible material reinforced with woven wire braid. (Other types of hose construction are available.)

GROUP 53 — PIPE

Section 01 — General
Section 02 — Types of Pipe

SECTION 01 — GENERAL

53.01.100 Pipe

A conductor whose outside diameter is standardized for threading. Pipe is available in Standard, Extra Standard, Double Extra Strong or Schedule wall thickness.

53.01.500 Pipe (Conductor) Clamp

Device to hold and support pipe lines (conductors).

GROUP 54 — TUBE

Section 01 — General
Section 02 — Types of Tubing

SECTION 01 — GENERAL

54.01.100 Tube

A conductor whose size is its outside diameter. Tube is available in varied wall thickness and materials.

GROUP 55 — QUICK DISCONNECTS

Section 01 — General
Section 02 — Types of Quick Disconnects

SECTION 01 — GENERAL

55.01.010 Air Inclusion

The ambient atmosphere forced or trapped into the system during connection of the quick disconnect coupling halves.

55.01.020 Break-Away

Automatic separation of a mounted quick disconnect coupling when a force is applied axially to the unmounted coupling half.

55.01.030 Connect Under Pressure

Ability to connect coupling halves with internal line pressure applied to either both sides or one side.

55.01.080 Spillage

The fluid removed from the system during disconnection of a coupling assembly.

55.01.100 Quick Disconnect Coupling

A component which can quickly join or separate a fluid line without the use of tools or special devices.

SECTION 02 — TYPES OF QUICK DISCONNECTS

55.02.200 Quick Disconnect, Break-away

A quick disconnect which provides automatic separation of the coupling halves, when a predetermined axial force is applied.

55.02.220 Quick Disconnect, Claw Type

A connection which is joined by the rotation of one part with respect to the other.

55.02.400 Quick Disconnect, One Valve

A quick disconnect with a shut-off valve in one half only.

55.02.600 Quick Disconnect, Un-valved*
Deprecated*

A quick disconnect with no shut-off valves.

55.02.800 Quick Disconnect, Valved

A quick disconnect with a shut-off valve in each half.

GROUP 56 — SWIVELS, ROTATING JOINTS & JOINTS

Section 01 — General
Section 02 — Types of Joints

SECTION 01 — GENERAL

56.01.100 Joint

A line positioning connector.

SECTION 02 — TYPES OF JOINTS

56.02.300 Joint, Rotary

A joint connecting lines which have relative operational rotation.

56.02.598 Joint, Spherical

Pipe junction which allows relative movement in any direction about a point.

56.02.600 Joint, Swivel

A joint which permits variable operational positioning of lines.

56.02.602   Joint, Telescopic

A junction consisting of two tubes sliding longitudinally one within the other, to convey the working medium to the equipment.

GROUP 57 — MANIFOLDS

    Section 01 — General
    Section 02 — Types of Manifolds

SECTION 01 — GENERAL

57.01.100   Manifold

A conductor which provides multiple connection ports.

57.01.101   Manifold Block, Valve

A base which forms the sub-plate for two or more subplate mounted valves, incorporating the various ports for connection of the external pipelines. It can also embody flow paths for interconnecting the various valves mounted thereon.

SECTION 02 — TYPES OF MANIFOLDS

57.02.800   Manifold, Vented

A manifold which is open to the atmosphere and returns fluid to the reservoir.

GROUP 58 — SUBPLATES

    Section 01 — General
    Section 02 — Types of Subplates

58.01.100   Subplate (Sub-Base)

Mounting to which a simple valve is fitted and which includes external ports for fluid connections.

58.01.101   Subplate (Sub-Base), Ganged Valve

Similar sub-bases of which two or more can be clamped together by tie bolts or other means. It can be arranged for the mating faces of the sub-bases to have matching ports. thus providing for a common supply and/or exhaust system. The sub-bases incorporate the various ports for connection of the external pipelines.

58.01.102   Subplate (Sub-Base), Multiple Valve

Mounting with appropriate ports matching those of two or more similar valves which are fitted to it and which include external ports for pipe connections.

GROUP 61 — ACCUMULATORS

    Section 01 — General
    Section 02 — Accumulator Types

SECTION 01 — GENERAL

61.01.100   Accumulator

A container in which fluid is stored under pressure as a source of fluid power.

SECTION 02 — ACCUMULATOR TYPES

61.02.400   Accumulator, Hydropneumatic

An accumulator in which compressed gas applies force to the stored liquid.

61.02.402   Accumulator, Gas Loaded, Transfer Type

A gas loaded accumulator for use with additional gas capacity contained in one or more supplementary gas bottles connected to the gas side of the transfer accumulator by a common pipe line.

61.02.420   Accumulator, Hydropneumatic Bladder

A hydropneumatic accumulator in which the liquid and gas are separated by an elastic bag or bladder.

61.02.440   Accumulator, Hydropneumatic, Diaphragm

A hydropneumatic accumulator in which the liquid and gas are separated by a flexible diaphragm.

61.02.460   Accumulator, Hydropneumatic, Non-Separator

A hydropneumatic accumulator in which the compressed gas operates directly on liquid within the pressure chamber.

61.02.480   Accumulator, Hydropneumatic, Piston

A hydropneumatic accumulator in which the liquid and gas are separated by a floating piston.

61.02.800   Accumulator, Mechanical

An accumulator incorporating a mechanical device which applies force to the stored liquid.

61.02.840   Accumulator, Mechanical, Spring

A mechanical accumulator in which springs apply force to the stored fluid.

61.02.880   Accumulator, Mechanical, Weighted

A mechanical accumulator in which the gravitational force acting upon weights applies force to the stored fluid.

GROUP 62 — PRESSURE VESSELS

Section 01 — General
Section 02 — Pressure Vessel Types

SECTION 01 — GENERAL

62.01.100   Pressure Vessel

A container which holds fluid under pressure.

GROUP 63 — CUSHIONS & SNUBBERS

Section 01 — General
Section 02 — Types of Cushions & Snubbers

SECTION 01 — GENERAL

63.01.100   Cushion

A device which provides controlled resistance to motion.

63.01.150   Cushion, Pressure, Damping

Pressure generated by the damping device to decelerate the total moving mass.

SECTION 02 — TYPES OF CUSHIONS & SNUBBERS

63.02.210   Cushion, Cylinder

A cushion built into a cylinder to restrict flow at the outlet port thereby arresting the motion of the piston rod.

63.02.220   Cushion, Die

A cushion installed with a die on a press to provide controlled resistance against the work. The return motion of the cushion is sometimes used to eject the work.

63.02.230   Cushion, Hydraulic

A cushion in which resistance is developed hydraulically.

63.02.240   Cushion, Hydropneumatic

A cushion in which resistance is developed hydraulically and pneumatically.

63.02.250   Cushion, Pneumatic

A cushion in which resistance is developed pneumatically.

GROUP 70 — GENERAL CONTROLS

Section 01 — General
Section 02 — Types of General Controls
Section 03 — Restrictors
Section 04 — Logic Characteristics

SECTION 01 — GENERAL

70.01.100   Control

A device used to regulate the function of a component or system.

70.01.101   Controller

A device which senses a change of fluid state and automatically makes adjustments to maintain the state of the fluid between predetermined limits, e.g., pressures, temperatures, etc.

70.01.860   Fluid Memory, Off Return

Fluid memory which receives a momentary signal and produces a change of state which continues to exist after the initiating signal has disappeared, providing an input is present at the supply port of the device. Upon loss of the supply pressure, the device reverts to its initial state.

70.01.870   Fluid Memory, Retentive

Fluid memory which receives a momentary signal and produces a change of state which continues to exist after the initiating signal has disappeared regardless of the presence or absence of supply pressure to the device. The device returns to its original state only upon receipt of a second reset control signal.

70.01.900   Fluid Signal

Fluid pressure or flow which can be detected or sensed.

70.01.908   Fluid Signal, Maintained

A fluid signal which exists indefinitely until caused to disappear by a secondary control action.

70.01.910   Fluid Signal, Momentary

A fluid signal which exists briefly and then disappears.

70.01.914   Fluid Signal, Timed

A fluid signal which exists for a definite period of time and then disappears.

SECTION 02 — TYPES OF GENERAL CONTROLS

70.02.105   Control, Automatic

A control which actuates equipment in a predetermined manner.

70.02.106   Control, Auxiliary

A device, usually manual, fitted to a valve to provide an alternative method of control.

70.02.110   Control, Combination

A combination of more than one basic control.

70.02.115   Control, Cylinder

A control in which a fluid cylinder is the actuating device.

70.02.117   Control, Detent

A device which retains the moving part in position by means of artificially created resistance. Movement to a different position is achieved either by release of the detent, or by the application of sufficient force to overcome it.

70.02.120   Control Electric

A control actuated electrically.

70.02.121   Control, Emergency

A device, usually manual, fitted to a valve or circuit providing an alternative method of control in the case of failure of the normal method of control.

70.02.122   Control, Feedback

The means whereby the state of the controlled element is signaled.

70.02.123   Control, Feedback, Mechanical

Feedback using a mechanical transmission.

70.02.124   Control, Feedback, Hydraulic

Feedback using a hydraulic circuit.

70.02.125   Control, Feedback, Electric

Feedback using an electrical signal.

70.02.126   Control, Feedback, Pneumatic

Feedback using a pneumatic circuit.

70.02.127   Control, Force Motor

A type of electro-mechanical transducer having linear motion.

70.02.130   Control, Hydraulic

A control actuated by a liquid.

70.02.133   Control, Impulse Generator

A device so arranged that, if a continuous pneumatic signal is applied to the input port, a single pulse is produced at the output port.

70.02.134   Control, Latch

The moving parts are retained in a fixed position by means of a locking device which must be externally released.

70.02.135   Control, Lever

A pivoted arm which is hand operated by pushing or pulling.

70.02.136   Control, Linkage

Means of mechanical connection.

70.02.137   Control, Liquid-Level

A device which controls the liquid level by a float switch or other means.

70.02.139   Control, Manual

A control device which is manually operated.

70.02.140   Control, Mechanical

A control actuated by linkages, gears, screws, cams or other mechanical elements.

70.02.141   Control, One-Way Trip

A mechanism which will allow movement in one direction only of the actuating force.

70.02.142   Control, Over Center

The moving parts cannot be stopped in an intermediate position (dead center position).

70.02.143   Control, Override

An alternative method of control fitted to a valve, which takes precedence over the normal method of control.

70.02.144   Control, Pedal

A control device foot operated in one direction only.

70.02.145   Control, Plunger

A rod acting in the direct line of the application of force.

70.02.146   Control, Pneumatic

A control actuated by air or other gas pressure.

70.02.147   Control, Pneumatic Programmer

A calibrated device so arranged that, if a continuous signal is applied to the input port, one or more output signals will be produced. The duration of an interval between the outputs can be predetermined.

70.02.148   Control, Pneumatic Programmer Cyclic

Apparatus comprising a number of valves controlled by programming device with a repetitive action. The program may be either fixed or variable.

70.02.149   Control, Pneumatic Time Delay

A device so arranged that, if continuous pneumatic signal is applied to or removed from the input port, a signal will be produced at the output port after a predetermined time has elapsed. The time delay may be fixed or variable.

70.02.150   Control, Pressure

A control method operated by a change of fluid pressure in a pilot line.

70.02.151   Control, Pressure Compensated

A control in which a pressure signal operates a compensating device.

70.02.152 Control, Pressure, Direct

Control method in which the position of the moving parts is controlled directly by alteration of the control pressure.

70.02.153 Control, Pressure, Indirect

Control method in which the position of the moving parts is controlled by a change of the control pressure to a pivot device.

70.02.154 Control, Pressure Pulse Generator

A device so arranged that, if a continous pneumatic signal is applied to the input port, repetitive pulses are produced at the output port.

70.02.155 Control, Pump

A control applied to a positive displacement variable delivery pump to adjust the volumetric output or direction of flow.

70.02.156 Control, Push-Pull Button

A control device which is palm or finger operated by pushing or by pulling.

70.02.157 Control, Roller

A roller is attached to the operating mechanism to permit operation by means of a cam or slide acting at right angles to the mechanism.

70.02.158 Control, Roller Lever

A lever with roller attached transmitting movement.

70.02.159 Control, Roller Plunger

A plunger with roller attached permitting the transmission of movement at right angles to the mechanism.

70.02.160 Control, Roller Rocker

A lever pivoted between rollers attached at each end, transmitting movement in both directions of the operating mechanism.

70.02.161 Control, Rotating Shaft

A rotary mechanical control component.

70.02.162 Control, Servo

A control actuated by a feedback system which compares the output with the reference signal and makes corrections to reduce the difference.

70.02.163 Control, Solenoid, Single Acting (One - Way)

An electro-magnetic mechanism which has two positions, being operated against a bias to one extreme position by energizing the coil.

70.02.164 Control, Solenoid, Double Acting (Two - Way)

An electro-magnetic mechanism which can take up two or three positions and is operated to either extreme position by energizing the appropriate coil.

70.02.165 Control, Spring Return (Offset)

The moving parts of the unit are returned to the initial position by spring force after the actuating forces are removed.

70.02.166 Control, Stepping Motor

An electric motor designed to provide displacement or speed variation in successive steps.

70.02.167 Control, Torque Motor

A type of electro-mechanical transducer having rotary motion.

70.02.168 Control, Tracer

A control operated by a system which follows the contours of master pattern.

70.02.169 Control, Treadle

A control device foot operated in two directions.

70.02.700 Logic Devices

The general category of components which perform logic functions; for example, AND, NAND, OR, and NOR. They can permit or inhibit signal transmission with certain combinations of control signals.

70.02.710 Logical State

Signal levels in logic devices are characterized by two stable states, the logical 1 (one) state and the logical 0 (zero) state. The designation of the two states is chosen arbitrarily. Commonly the logical 1 state represents an "on" signal, and the 0 state represents an "off" signal.

70.02.880 Sensing Device

A component which measures the state of a system variable such as fluid level, viscosity, temperature, pressure, or flow rate.

70.02.900 Sensor

A device which detects a condition in a system and produces a signal.

ANSI/B93.2

SECTION 03 — RESTRICTORS

70.03.050   Diode, Fluid

A device with a passage for fluid flow with high, to infinitely high resistance in one direction and low resistance in the opposite direction.

70.03.100   Restrictor

A device which reduces the cross-sectional flow area.

70.03.130   Restrictor, Choke

A restrictor, the length of which is relatively large with respect to its cross-sectional area.

70.03.160   Restrictor, Orifice

A restrictor, the length of which is relatively small with respect to its cross-sectional area. The orifice may be fixed or variable. Variable types of non-compensated, pressure compensated, or pressure and temperature compensated.

SECTION 04 — LOGIC CHARACTERISTICS

70.04.100   AND Device

A control device which has its output in the logical 1 state if and only if all the control signals assume the logical 1 state.

70.04.400   Flip Flop

A digital component or circuit with two stable states and sufficient hysteresis so that it has "memory". Its state is changed with a control pulse; a continuous control signal is not necessary for it to remain in a given state.

70.04.700   NAND Device

A control device which has its output in the logical 0 state if and only if all the control signals assume the logical 1 (one) state.

70.04.720   NOR Device

A control device which has its output in the logical 1 state if and only if all the control signals assume the logical 0 state.

70.04.740   NOT Device

A control device which has its output in the logical 1 state if and only if the control signal assumes the logical 0 state. The NOT device is a single input NOR device.

70.04.760   OR Device

A control device which has its output in the logical 0 state if and only if all the control signals assume the logical 0 state.

GROUP 71 — VALVES

    Section 01 — General
    Section 02 — Functional Types
    Section 03 — Basic Designs
    Section 04 — Positions
    Section 05 — Flow Conditions
    Section 06 — Actuators
    Section 07 — Mountings
    Section 08 — Features

SECTION 01 — GENERAL

71.01.100   Valve

A device which controls fluid flow direction, pressure, or flow rate.

71.01.500   Valve, Monoblock

Unit comprising a number of similar valves in a common housing.

SECTION 02 — FUNCTIONAL TYPES

71.02.050   Valve, Air

A valve for controlling air.

71.02.055   Valve, Ball, Seat Action

A valve design which utilizes a solid ball to obstruct the flow path.

71.02.056   Valve, Ball, Shear Action

A valve design which utilizes a ported ball that rotates on an axis normal to the flow path.

71.02.060   Valve, Butterfly

A straight-through shut-off in which the valve element consists of a flat disc rotating about a diametrical axis perpendicular to the flow of fluid.

71.02.070   Valve, Cartridge

A valve with working parts contained in a cylindrical body. The cylindrical body must be inserted into a housing for use. Ports through the body cooperate with ports in the containing housing.

71.02.080   Valve, Diaphragm Type

A valve in which the element is moved by forces acting on a diaphragm.

71.02.100   Valve, Directional Control

A valve whose primary function is to direct or prevent flow through selected passages.

71.02.110   Valve, Directional Control, Check

A directional control valve which permits flow of fluid in only one direction.

71.02.111   Valve, Directional Control, Check, Spring Loaded

A valve in which flow may occur in one direction only when fluid pressure overcomes spring pressure.

71.02.112   Valve, Directional Control, Check, Pilot Operated

A check valve in which the opening or closing is controlled by a pilot signal.

71.02.113   Valve, Directional Control, Check, Cushioned

A check valve in which the movement of the check device is damped, for use in systems with pulsating pressures.

71.02.130   Valve, Directional Control, Four Way

A directional control valve whose primary function is to pressurize and exhaust two ports.

71.02.140   Valve, Directional Control, Selector (Diversion)

A directional control valve whose primary function is to selectively interconnect two or more ports.

71.02.160   Valve, Directional Control, Straightway

A two port directional control valve.

71.02.170   Valve, Directional Control, Three Way

A directional control valve whose primary function is to pressurize and exhaust a port.

71.02.172   Valve, Directly Operated

A valve in which the controlling forces acting on the element directly influence the movement of the control elements.

71.02.174   Valve, Disc (Globe)

A shut-off valve in which the flow at one point is at right angles to the normal direction of flow. The valve member is a flat disc which is lifted or seated to open or close the flow path.

71.02.175   Valve, Disc (Swing)

A shut-off valve design which utilizes a hinged disc to obstruct the flow path.

71.02.190   Valve, Flow Combining, Pressure Compensated

Pressure compensated valve which combines two input flow rates maintaining a pre-selected output.

71.02.200   Valve, Flow Control (Flow Metering)

A valve whose primary function is to control flow rate.

71.02.201   Valve, Flow Control, Bypass

A pressure compensated flow control valve which regulates the working flow diverting surplus fluid to reservoir or to a second service.

71.02.202   Valve, Flow Control Adjustable Restrictor

Valve in which the inlet and outlet ports are interconnected through a restricted passageway whose cross-sectional area can be varied within limits.

71.02.203   Valve, Flow Control Fixed Restrictor

A valve in which the inlet and outlet ports are interconnected through a restricted passageway whose cross-sectional area cannot be altered.

71.02.210   Valve, Flow Control, Deceleration

A flow control valve which gradually reduces flow rate to provide deceleration.

71.02.216   Valve, Flow Control, One-Way Restrictor

A valve which allows free flow in one direction and restricted flow in the other direction. Restricted flow path may be fixed or variable.

71.02.220   Valve, Flow Control, Pressure Compensated

A flow control valve which controls the rate of flow independent of system pressure.

71.02.230   Valve, Flow Control, Pressure-Temperature Compensated

A pressure compensated flow control valve which controls the rate of flow independent of fluid temperature.

71.02.250   Valve, Flow Dividing

A valve which divides the flow from a single source into two or more branches.

71.02.280   Valve, Flow Dividing, Pressure Compensated

A flow dividing valve which divides the flow at constant ratio regardless of the difference in the resistances of the branches.

71.02.300 Valve, Gate

A straight-through shut-off valve in which the valve element moves perpendicularly to the axis of the flow to control opening and closing.

71.02.301 Valve, Gate, Spreader

A gate valve which utilizes two companion discs which are positively seated by common spreaders to obstruct the flow path.

71.02.302 Valve, Gate, Wedge

A gate valve which utilizes a solid wedge shaped gate to obstruct the flow path.

71.02.320 Valve, Globe

(See: Disc Valve - 71.02.174)

71.02.350 Valve, Hydraulic

A valve for controlling liquid.

71.02.380 Valve, Needle

A flow control valve in which the adjustable control element is a tapered needle. Its usual purpose is the accurate control of the rate of volume of flow.

71.02.400 Valve, Pilot

A valve applied to operate another valve or control.

71.02.401 Valve, Pilot Operated (Indirect)

A valve in which a relatively small flow through an integral vent line relief (pilot) controls the movement of the main element.

71.02.410 Valve, Pinch

A straight-through shut-off valve in which the valve element consists of a flexible sleeve which is distorted to control the flow of the fluid.

71.02.420 Valve, Piston

A valve in which the element is moved by forces acting on a piston.

71.02.430 Valve, Plug

A shut-off valve in which ports are connected or sealed off by a rotating plug containing flow paths.

71.02.431 Valve, Plug, Cylinder

A valve in which the surface of contact between the plug and the valve body is cylindrical and requires a method of sealing.

71.02.432 Valve, Plug, Shear Action

A valve design which utilizes a ported plug that rotates on an axis normal to the flow path.

71.02.433 Valve, Plug, Spherical

A valve in which the surface of contact between the plug and the valve body is spherical and requires a method of sealing.

71.02.434 Valve, Plug, Tapered

A valve in which the surface of contact between the plug and the valve body is conical and provides the sealing surface.

71.02.450 Valve, Pneumatic

A valve for controlling compressed air.

71.02.460 Valve, Poppet

A valve in which the flow paths are opened or closed as the valve element (poppet) is lifted or seated.

71.02.470 Valve, Power Control

A valve which controls fluid power operating working devices.

71.02.500 Valve, Prefill

A valve which permits full flow from a tank to a cylinder during the advance portion of a cycle, permits the operating pressure to be applied to the cylinder during the working portion of the cycle, and permits free flow from the cylinder to the tank during the return portion of the cycle.

71.02.550 Valve, Pressure Control

A valve whose primary function is to control pressure.

71.02.560 Valve, Pressure Control, Counterbalance

A pressure control valve which maintains back pressure to prevent a load from falling.

71.02.570 Valve, Pressure Control, Decompression

A pressure control valve that controls the rate at which the contained energy of the compressed fluid is released.

71.02.580 Valve, Pressure Control, Load Dividing

A pressure control valve used to proportion pressure between two pumps in series.

71.02.590   Valve, Pressure Control, Pressure Reducing

A pressure control valve whose primary function is to limit outlet pressure.

71.02.591   Valve, Pressure Control, Reducing & Relieving

A valve which limits maximum pressure by exhausting fluid when the required pressure is reached.

71.02.600   Valve, Pressure Control, Relief

A pressure control valve whose primary function is to limit system pressure.

71.02.620   Valve, Pressure Control Relief, Safety

A relief valve whose primary function is to provide pressure limitation after malfunction.

71.02.630   Valve, Pressure Control, Unloading

A pressure control valve whose primary function is to permit a pump or compressor to operate at maximum load.

71.02.635   Valve, Pressure Proportioning

A pressure reducing valve in which the outlet pressure is maintained at a fixed ratio to the inlet pressure.

71.02.640   Valve, Pressure Sensing

A device similar to an electrical pressure switch, in which a signal to be sensed enters a control point, and actuates a mechanism which, at the proper pressure level, causes one or more flow passages to change condition. Removal of the signal allows the pressure sensing valve to reset.

71.02.650   Valve, Priority

A valve which directs flow to one operating circuit at a fixed rate and directs excess flow to another operating circuit.

71.02.655   Valve, Quick Exhaust

Valve in which, when air pressure falls at the inlet, the outlet is automatically opened to exhaust.

71.02.660   Valve, Relay

A logic device which receives control signals and changes flow conditions in one or more controlled flow passages.

71.02.662   Valve, Relay, Free Floating

A relay valve wherein the internal element moves freely without restraint and normally utilizes bias pressure at one control point.

71.02.664   Valve, Relay, One Shot

A relay valve wherein controlled flow passages immediately change conditions when a control point is pressurized by a maintained signal. After a period of time, the controlled flow passages return to their original conditions, even though the control point is pressurized. When the control signal is removed it resets for another operation.

71.02.666   Valve, Relay, Time Delay After Exhausting a Control Point

A relay valve with one control point which receives a maintained signal and causes immediate actuation of controlled flow passages. When the control signal is removed, a time delay occurs before controlled flow passages are reset.

71.02.668   Valve, Relay, Time Delay After Pressurizing a Control Point

A relay valve with one control point which receives a maintained signal and causes a time delay before the controlled flow passages are actuated. The device resets immediately upon exhausting the control point.

71.02.670   Valve, Relay, Time Delay

A relay valve which creates a time interval between the pressurizing of a control point and a change in the controlled flow passages.

71.02.672   Valve, Relay, Time Delay, Detented

A time delay relay valve having "A" and "B" control points arranged to accept and act on momentary signals. A momentary signal into the "A" control point starts actuation and a momentary signal into the "B" control points starts reset.

71.02.674   Valve, Relay, Time Delay, Detented, Delayed Action

A detented time delay relay valve in which a momentary signal into the "A" control point creates a time delay before controlled flow passages are actuated. The device resets immediately upon receipt of a signal in the "B" control point.

71.02.676   Valve, Relay, Time Delay, Detented, Delayed Reset

A detented time delay relay valve in which a momentary signal into the "A" control point produces immediate actuation of the controlled flow passages. A momentary signal to the "B" control points starts the reset action with a time interval before controlled flow passages reset.

71.02.680   Valve, Rotary Selector

A valve which utilizes rotary actuation to connect the inlet to any one of a number of outlets.

71.02.690   Valve, Separator Drain

A device whereby solid or liquid impurities which have collected in the installation can be removed. May be actuated automatically or manually.

71.02.700 Valve, Pressure Control, Sequence

When the inlet pressure exceeds the preset value, the valve opens to permit flow through the outlet port, (The effective setting, is not affected by the pressure on the outlet port.)

71.02.750 Valve, Shutoff

A valve which operates fully open or fully closed.

71.02.751 Valve, Shut Off, Automatic

A valve which closes automatically when the pressure drop across the valve, caused by increased flow, exceeds a predetermined amount.

71.02.752 Valve, Shut Off Sliding

A shut off valve whose flow paths are connected together or sealed off by means of a moveable sliding member. The movement may be axial, radial or both.

71.02.753 Valve, Shut Off, Sliding, Flat

A shut off valve in which the flow paths are connected together or sealed off by means of a movable flat faced valve member sliding on a flat seat.

71.02.754 Valve, Shut Off, Sliding, Spool

A shut off valve in which the flow paths are connected or sealed off by a cylindrical spool which slides within the matching bore of the valve body.

71.02.800 Valve, Shuttle

A connective valve which selects one of two or more circuits because flow or pressure changes between the circuits.

71.02.801 Valve, Shuttle, High Pressure

A valve in which the inlet at higher pressure is connected to the outlet, the other inlet is closed. The position is maintained under reverse flow.

71.02.802 Valve, Shuttle, Low Pressure

A valve in which inlet at lower pressure is connected to the outlet, the other inlet is closed. The position is maintained under reverse flow.

71.02.810 Valve, Slide

A valve in which the flow paths are connected or isolated by means of a flat movable sliding member. The movement may be axial, rotary, or both.

71.02.811 Valve, Slide, Linear

A valve in which the flow paths are connected or isolated by means of a flat faced valve member which slides on a flat seat.

71.02.812 Valve, Slide, Rotary

A sliding plate shear action valve design in which the motion of the plate is rotary.

71.02.820 Valve, Spool

A shear action valve design which utilizes a spool that slides through the flow path.

71.02.850 Valve, Surge Damping

A valve which reduces shock by limiting the rate of acceleration of fluid flow.

71.02.900 Valve, Time Delay

A valve in which the change of flow occurs only after a desired time interval has elapsed.

SECTION 03 — BASIC DESIGNS

71.03.200 Flapper Action

A valve design in which output control pressure is regulated by a pivoted flapper in relation to one or two orifices.

71.03.300 Jet Action

A valve design in which flow effect is controlled by the relative position of a nozzle and a receiver.

71.03.500 Seating Action

A valve design in which flow is stopped by a seated obstruction in the flow path.

71.03.700 Shear Action

A valve design in which flow is modulated by an element which slides across the flow path.

SECTION 04 — POSITIONS

71.04.000 Valve Position

The point at which flow directing elements provide a specific flow condition in a valve.

71.04.002 Valve Position, Actuated

One of the final positions of the valving element when under the influence of the actuating forces.

71.04.100 Valve Position, Center

The selective mid-position in a directional control valve.

71.04.200 Valve Position, Detent

A predetermined position maintained by a holding device acting on the flow-directing elements of a directional control valve.

71.04.220   Valve Position, Initial

The position of the valving element after main pressure is admitted and before the intended operating cycle begins under the influence of the actuating forces.

71.04.222   Valve Position, Intermediate (Transit)

Any position between the initial and the actuated position.

71.04.300   Valve Position, Normal

The valve position when signal or actuating force is not being applied.

71.04.400   Valve Position, Offset

An off-center position in a directional control valve.

71.04.500   Valve Position, Return

The initial valve position.

71.04.600   Valve, Four Position

A directional control valve having four positions to give four selections of flow conditions.

71.04.700   Valve, Three Position

A directional control valve having three positions to give three selections of flow conditions.

71.04.701   Valve, Three Position, Closed Center

All ports are closed in the initial position.

71.04.702   Valve, Three Position Open Center

All ports are connected in the initial position.

71.04.800   Valve, Two Position

A directional control valve having two positions to give two selections of flow conditions.

SECTION 05 — FLOW CONDITIONS

71.05.000   Valve Flow Condition

A flow pattern in a directional control valve.

71.05.100   Valve Flow Condition, Closed

All ports are closed.

71.05.200   Valve Flow Condition, Float

Working ports are connected to exhaust or return.

71.05.300   Valve Flow Condition, Hold

Working ports are blocked to hold a powered device in a fixed position.

71.05.400   Valve Flow Condition, Open

All ports are open.

71.05.500   Valve Flow Condition, Regenerative

Working ports are connected to supply.

71.05.600   Valve Flow Condition, Tandem

Working ports are blocked and supply is connected to the return port.

SECTION 06 — VALVE ACTUATORS

71.06.000   Valve Actuator

The valve part(s) through which force is applied to move or position flow-directing elements.

71.06.200   Valve Actuator, Manual

A valve actuator consisting of a hand lever, palm button, foot treadle, or other manual energizing devices.

71.06.400   Valve Actuator, Mechanical

A valve actuator consisting of a cam, lever, roller, screw, spring, stem, or other mechanical energizing devices.

71.06.600   Valve Actuator, Pilot

A valve actuator which utilizes pilot fluid.

71.06.610   Valve Actuator, Pilot, Barrier

A pilot valve actuator wherein the working fluid is isolated from the actuator.

71.06.620   Valve Actuator, Pilot, Differential Area

A pilot valve actuator wherein pilot fluid acts on unequal areas.

71.06.630   Valve Actuator, Pilot, Differential Pressure

A pilot actuator wherein pilot fluid acts at unequal pressure.

71.06.640   Valve Actuator, Pilot, External

A pilot valve actuator wherein fluid is received from an external source.

ANSI/B93.2

NFPA/T2.1.1 R1

71.06.650   Valve Actuator, Pilot, Internal

A pilot valve actuator wherein pilot fluid is received from within the valve.

71.06.660   Valve Actuator, Pilot, Solenoid, Controlled

A pilot valve actuator wherein pilot fluid is controlled by the action of one or more solenoids.

71.06.800   Valve Actuator, Solenoid

A valve actuator which utilizes one or more solenoids.

SECTION 07 — VALVE MOUNTINGS

71.07.000   Valve Mounting

The mounting characteristics of a valve.

71.07.200   Valve Mounting, Base

The valve is mounted on a plate which has top and side ports.

71.07.201   Valve Mounting, Base, Gang

Unit consisting of an assembly of a number of similar valves banked together, often with common supply and/or exhaust systems.

71.07.400   Valve Mounting, Line

The valve is mounted directly to system lines.

71.07.600   Valve Mounting, Manifold

The valve is mounted to a plate which provides multiple connection ports for two or more valves.

71.07.800   Valve Mounting, Sub-Plate

The valve is mounted to a plate which provides straight-through top and bottom ports.

SECTION 08 — FEATURES

71.08.200   Flow Path (Gallery)

Passage through which fluid flows within a device.

GROUP 72 — FLUIDIC DEVICES

Section 01 — General
Section 02 — Types of Fluidic Devices
Section 03 — Amplification
Section 04 — Gain
Section 05 — Interface Devices
Section 06 — Pneumatic Power for Fluidics
Section 07 — Fluidic Response Factors

SECTION 01 — GENERAL

72.01.100   Analog

Of or pertaining to the general class of fluidic devices or circuits whose output varies as a continuous function of its input.

72.01.105   Aspect Ratio, Nozzle

The ratio of nozzle depth to nozzle width.

72.01.250   Capacitance, Fluid

Ratio of mass flow to rate of change of pressure drop.

72.01.254   Characteristic, Switching

Curve expressing output quantity as a function of control quantity.

72.01.255   Collector (Receiver)

Nozzle located downstream of a free flowing jet, normally used to catch the energy of the flowing medium of the jet.

72.01.257   Conductance, Fluid

Ratio between steady state mass flow and pressure drop (reciprocal value of fluid resistance).

72.01.300   Digital

Of or pertaining to the general class of fluidic devices or circuits whose output varies in discrete steps (i.e., pulses or "on-off" characteristics).

72.01.400   Fan In Ratio

The number of operating controls in a single fluidic device which individually and in combination will produce the same output.

72.01.500   Fan Out Ratio

The number of like devices to which operating controls are supplied by the output of the fluidic device.

72.01.600   Fluidic

Of or pertaining to devices, systems, assemblies, etc., utilizing fluidic components.

72.01.650   Impedance, Capacitive, Fluid

Imaginary ratio of pressure drop and transient mass flow in which pressure drop leads flow.

72.01.651   Impedance, Fluid

Complex ratio between pressure drop and transient mass flow.

72.01.652   Impedance, Fluidic Input

The impedance measured at an input port.

72.01.653   Impedance, Fluidic Output

The impedance measured at an output port.

72.01.654   Impedance, Inductive, Fluid

Imaginary ratio of pressure drop and transient mass flow in which pressure drop leads flow by phase.

72.01.659   Inductance, Fluid

Ratio of pressure drop and rate of change of mass flow.

72.01.700   Interface

A point or component where a transition is made between medium, power levels, modes of operation, etc.

72.01.710   Jet

Emission of a fluid from an orifice.

72.01.711   Jet, Attached

Jet which is attached to a wall by Coanda effect.

72.01.712   Jet, Confined

A jet influenced by its physical surrounding.

72.01.713   Jet, Free

A jet not influenced by its surroundings.

72.01.714   Jet, Main Power

A laminar or turbulent flow of fluid emitted from the supply channel or nozzle of a fluidic device.

72.01.800   Load Line

Curve expressing output pressure as a function of output flow. (The derivative of this curve is the expression of the output impedance.)

72.01.801   Logic Threshold

The minimum number of signals required at the inputs of a multi input device to change the output condition.

72.01.900   Ratio, Series

The number of identical devices mounted in series, which can be controlled by the output of a device.

72.01.920   Ratio, Signal to Noise

The ratio of the signal strength to that of the noise strength.

72.01.930   Splitter

That part of a fluid amplifier which separates alternative outputs.

72.01.950   Volume, Fluidic Control

The volume of the input chamber, including the pilot line.

SECTION 02 — TYPES OF FLUIDIC DEVICES

72.02.050   Fluidic Device, Active

A device which requires a power supply independent of the value of input signals.

72.02.100   Fluidic Amplifier

A device which enables one or more fluid dynamic signals to control a source of power and thus is capable of delivering at its output an enlarged reproduction of the essential characteristics of the signal.

72.02.110   Fluidic Amplifier, Closed

A fluidic amplifier which has no vent port.

72.02.111   Fluidic Amplifier, Flow

A device which amplifies flow by use of a small valve which acts as a pilot for a larger one.

72.02.115   Fluidic Amplifier, Digital

An amplifier whose output varies in descrete steps related to the control signal.

72.02.120   Fluidic Amplifier, Impact Modulator

A fluidic amplifier in which the impact plane position of two opposed streams is controlled to alter the output.

72.02.122   Fluidic Amplifier, Momentum

An amplifier which functions on the interaction of momentum of the power and control jets.

72.02.125   Fluidic Amplifier, Open

A fluidic amplifier which has a vent port.

72.02.130   Fluidic Amplifier, Stream Deflection

A fluidic amplifier which utilizes one or more control streams to deflect a power stream, altering the output. It is usually analog.

72.02.135   Fluidic Amplifier, Turbulence

A fluidic amplifier in which the power jet is at a pressure such that it is in the transition region of laminar stability and can be caused to become turbulent by a secondary jet or by sound.

72.02.140   Fluidic Amplifier, Vortex

An amplifier which senses the pressure drop across a vortex, modulating the main flow.

72.02.145   Fluidic Amplifier, Wall Attachment

A fluidic amplifier in which the control of the attachment of a stream to a wall(s) alters the output. It is usually digital.

72.02.700   Fluidic Device, Passive

The general class of fluidic devices that operates on signal power alone.

72.02.710   Fluidic Device, Interface

A device which converts information between different types or levels of energy.

SECTION 03 — AMPLIFICATION (GAIN)

72.03.200   Gain, Flow

The ratio of the change of output flow to the change of control flow at a given point.

72.03.400   Gain, Power

The ratio of the change of output power to the change of control power at a given point.

72.03.600   Gain, Pressure

The ratio of output pressure change to control pressure change at a given point.

SECTION 06 — PNEUMATIC POWER FOR FLUIDICS

72.06.050   Damping Parameter

A measure of the time required as a function of the maximum pressure excursion of the power supply output to attain essentially steady state operation after an abrupt disturbance. Specifically, it is the transient recovery time divided by the maximum excursion.

72.06.100   Drift

The percentage above and below the operating pressure at a constant flow rate over a specified length of time.

72.06.120   Flow, Minimum Control

Flow through the control port at minimum control pressure.

72.06.121   Flow/Pressure Characteristic

The change of the specified controlled pressure due to change in the flow rate of the fluid, measured at specified pressure conditions.

72.06.140   Interaction Region, Jet

Chamber in which the power jet is affected by one or more control jets.

72.06.150   Maximum Excursion

The maximum pressure deviation from the operating pressure after an abrupt disturbance.

72.06.199   Noise, Acoustic

Spurious signals generated by external acoustic disturbances.

72.06.200   Noise, Fluidic

Random fluctuations of the signal level which may cause undesirable spurious signals in a circuit.

72.06.250   Operating Band

The range of pressures above and below the operating pressure within which it is desired to keep the supply output.

72.06.300   Power Capacity

The total volume of gas available at the operating pressure (applies to compressed gas storage supply source).

72.06.325   Power, Output

The power recoverable at the output port.

72.06.350   Power Supply

That component or group of components which supplies and processes the fluid for operating fluidic systems.

72.06.400   Rated Flow

The maximum flow that the power supply system is capable of maintaining at a specific operating pressure.

72.06.410   Recovery, Flow Rate

Ratio of no-load flow at the output to the supply flow.

72.06.420   Recovery, Power

A maximum ratio of power recovered at the output port to the supply power.

72.06.450 Ripple

A periodic variation of the pressure above and below the operating pressure. It is defined as a percentage of the operating pressure in terms of the maximum peak-to-peak value obtained at the point of rating.

72.06.500 Start-Up Time

The period of time needed to reach a steady state condition within the operating band starting from a long term off condition.

72.06.550 Steady State Pressure Regulation

A band indicating maximum and minimum pressure or a single curve with maximum deviation indicated in percent of operating pressure, all as a function of flow.

72.06.600 Transient Recovery Time

The period of time required for an abrupt change in the power supply output pressure to dampen out to within the operating band.

SECTION 07 — FLUIDIC RESPONSE FACTORS

72.07.200 Decay

A falling pressure.

72.07.300 Decay Rate

The ratio of pressure decay to time.

72.07.600 Output, Active

Output power which, in all possible states of the device, is derived from supply power.

72.07.601 Output, Passive

Output the power of which in one or more states of the devices is derived solely from the input signals.

72.07.700 Response Time

Interval between the initiation of an operation and its completion.

72.07.800 Rise Rate

The ratio of pressure rise to time.

GROUP 73 — SERVOVALVES

Section 01 — General
Section 02 — Types of Servovalves
Section 03 — Construction Features
Section 04 — Electrical Characteristics
Section 05 — Steady State Characteristics
Section 06 — Dynamic Characteristics

SECTION 01 — GENERAL

73.01.100 Servovalve

A valve which modulates output as a function of an input command.

SECTION 02 — TYPES OF SERVOVALVES

73.02.200 Servovalve, Electrohydraulic

A servovalve which is capable of continuously controlling hydraulic output as a function of an electrical input.

73.02.300 Servovalve, Electrohydraulic, Flow Control

An electrohydraulic servovalve whose primary function is control of output flow.

73.02.400 Servovalve, Four-Way

A multi-orifice flow control valve with supply, return and two control ports arranged so that the valve action in one direction opens supply to control port 1 and opens control port 2 to return. Reversed valve action opens supply to control port 2 and opens control port 1 to return.

73.02.450 Servovalve, Mechanical Hydraulic

A hydraulic servovalve in which the input command is mechanical.

73.02.460 Servovalve, Pressure Control

A hydraulic servovalve whose primary function is the control of output pressure.

73.02.500 Servovalve, Three-Way

A multi-orifice flow control valve with supply, return and one control port arranged so that valve action in one direction opens supply to control port and reversed valve action opens the control port to return.

73.02.600 Servovalve, Two-Way

A single orifice flow control valve with supply and one control port arranged so that action is in one direction only, from supply to control port.

SECTION 03 — CONSTRUCTION FEATURES

73.03.100 Force Motor

A type of electromechanical transducer having linear motion used in the input stages of servovalves.

73.03.200 Hydraulic Amplifier

A fluid device which enables one or more inputs to control a source of fluid power and thus is capable of delivering at its output an enlarged reproduction of the essential characteristics of the input. Hydraulic amplifiers may utilize sliding spools, nozzle-flappers, jet pipes, etc.

73.03.300 Output Stage

The final stage of hydraulic amplification used in a servovalve.

73.03.400 Stage

A hydraulic amplifier used in a servovalve. Servovalves may be single stage, two stage, three stage, etc.

73.03.500 Torque Motor

A type of electromechanical transducer having rotary motion used in the input stages of servovalves.

SECTION 04 — ELECTRICAL CHARACTERISTICS

73.04.100 Dither

A low amplitude, relatively high frequency periodic electrical signal, sometimes superimposed on the servovalve input to improve system resolution. Dither is expressed by the dither frequency (Hz) and the peak-to-peak dither current amplitude.

SECTION 05 — STEADY STATE CHARACTERISTICS

73.05.050 Servovalve Control Flow

The flow through the servovalve control ports. Conventional test equipment normally measures no-load flow.

73.05.060 Servovalve Control Flow, Loaded

The flow through the servovalve control ports when there is load pressure drop.

73.05.070 Servovalve Control Flow, No-Load

The flow through the servovalve control ports when there is zero load pressure drop.

73.05.100 Servovalve Flow Curve

The graphical representation of control flow versus input current of a servovalve. This is usually a continuous plot of a complete cycle between plus and minus rated current valves (See Fig. 73.1).

73.05.130 Servovalve Flow Curve, Normal

The locus of the midpoints of the complete cycle flow curve, which is the zero hysteresis flow curve, of a servovalve. Usually valve hysteresis is sufficiently low, such that one side of the flow curve can be used for the normal flow curve. (See Fig. 73.1).

73.05.150 Servovalve Flow Gain

The slope of the control flow versus input current curve in any specific operating region, of a servovalve. Three operating regions are usually significant with flow-control servovalves: (1) the null region, (2) the region of normal flow control, and (3) the region where flow saturation effects may occur. Where this term is used without qualification, it is assumed to mean normal flow gain. (See Fig. 73.2)

73.05.160 Servovalve Flow Gain, No-Load

The normal flow gain of a servovalve with zero load pressure drop.

73.05.170 Servovalve Flow Gain, Normal

The slope of a straight line drawn from the zero flow point of the servovalve normal flow curve, throughout the range of rated flow current of one polarity, and drawn to minimize deviations of the normal flow curve from the straight line. Flow gain may vary with the polarity of the input, with the magnitude of load differential pressure and with changes in operating conditions. (See Fig. 73.2)

73.05.200 Servovalve Flow Limit

The condition wherein control flow no longer increases with increasing input current. Flow limitation may be deliberately introduced within the servovalve.

73.05.250 Servovalve Flow Saturation Region

The region where flow gain decreases with increasing input current, in a servovalve. (See Fig. 73.2)

73.05.300 Servovalve Hysteresis

The difference in the servovalve input currents required to produce the same output during a single cycle of valve input current when cycled at a rate below that at which dynamic effects are important. Hysteresis is normally specified as the maximum difference occuring in the flow curve throughout plus or minus rated current, and is expressed as percent of rated current. (See Fig. 73.1)

73.05.350 Servovalve Internal Leakage

The total internal servovalve flow from pressure to return with zero control flow. It is usually measured with control ports blocked. Leakage flow will vary with input pressure and input current.

73.05.400 Servovalve Lap

In a sliding spool servovalve the relative axial position relationship between the fixed and moveable flow metering edges with the spool at null. Lap is measured as the total separation at zero flow of straight line extensions of the nearly straight portions of the normal flow curve, drawn separately for each polarity, expressed as percent of rated current.

73.05.405 Servovalve Null Leakage

Total internal leakage from the valve in the null position

73.05.410 Servovalve, Overlap

The lap condition which results in a decreased slope of the normal flow curve in the null region. (See Fig. 73.3)

73.05.420 Servovalve, Underlap

The lap condition which results in an increased slope of the normal flow curve in the null region. (See Fig. 73.4)

73.05.430 Servovalve Zero Lap

The lap condition in which there is no separation of the straight line extensions of the nomal flow curve. (See Fig. 73.5).

73.05.450 Servovalve Flow Linearity

The degree to which the normal flow curve conforms to the normal flow gain line with other operational variables held constant in a servovalve. Linearity is measured as the maximum deviation of the normal flow curve from the normal flow gain line, expressed as percent of rated current.

73.05.500 Servovalve Null

The condition in which the servovalve supplies zero control flow at zero load pressure drop.

73.05.510 Servovalve Null Bias

The input current required to bring the servovalve to null, excluding the effects of valve hysteresis, expressed as percent of rated current. (See Fig. 73.1).

73.05.520 Servovalve Null Pressure

The pressure existing at both control ports in a servovalve at null.

73.05.530 Servovalve Null Region

The region in a servovalve about null wherein effects of lap in the output stage predominate.

73.05.540 Servovalve Null Shift

A change in null bias in a servovalve, expressed as percent of rated current. Null shift may occur with changes in supply pressure minus the return pressure minus the load pressure drop.

73.05.550 Servovalve Pressure Drop

The sum of the differential pressures across the control orifices of the output stage in a servovalve. Pressure drop will equal the supply pressure minus the return pressure minus the load pressure drop.

73.05.570 Servovalve Load Pressure Drop

The differential pressure between the control ports of a servovalve.

73.05.600 Servovalve Pressure Gain

The change in load pressure drop per unit input current with zero control flow (control ports blocked). Pressure gain is specified as the average slope of the curve of load pressure drop versus input current in the region between ± 40% of maximum load pressure drop (See Fig. 73.7).

73.05.650 Servovalve Rated Flow

The specified control flow corresponding to rated current and specified servovalve pressure drop.

73.05.660 Servovalve Resolution

The increment of input signal required to produce a change in valve output at a specified signal level, expressed as a percentage of rated signal. Resolution is normally specified as the minimum signal required to cause either an increase or a decrease of valve output. If these signals differ, the larger of the two should be quoted.

73.05.700 Servovalve Symmetry

The degree of equality between the servovalve normal flow gain of one polarity and that of the reversed polarity. Symmetry is measured as the difference in normal flow gain of each polarity, expressed as percent of the greater.

73.05.750 Servovalve Threshold

The increment of input current required to produce a change in servovalve output, expressed as percent of rated current. Threshold is normally specified as the current increment required to revert from a condition of increasing output to a condition of decreasing output.

SECTION 06 — DYNAMIC CHARACTERISTICS

73.06.200 Servovalve Amplitude Ratio

The ratio of the servovalve control flow amplitude to a sinusoidal input current-amplitude at a particular frequency divided by the same ratio at the same input-current amplitude at a specified low frequency. Amplitude ratio may be expressed in decibels where db = 20 $\log_{10}$ AR (See Fig. 73.6).

73.06.500 Servovalve Frequency Response

The complex ratio of servovalve control flow to input current as the current is varied sinusoidally over a range of frequencies. Frequency response is normally measured with constant input current amplitude and zero load pressure drop, expressed as amplitude ratio, and phase lag. Servovalve frequency response may vary with the input-current amplitude, temperature, supply pressure, and other operating conditions (See Fig. 73.6).

73.06.800 Servovalve Phase Lag

The instantaneous time by which the servovalve sinusoidal flow follows the sinusoidal input current, measured at a specified frequency and expressed in degrees (See Fig. 73.6).

SERVOVALVES
FLOW CURVE

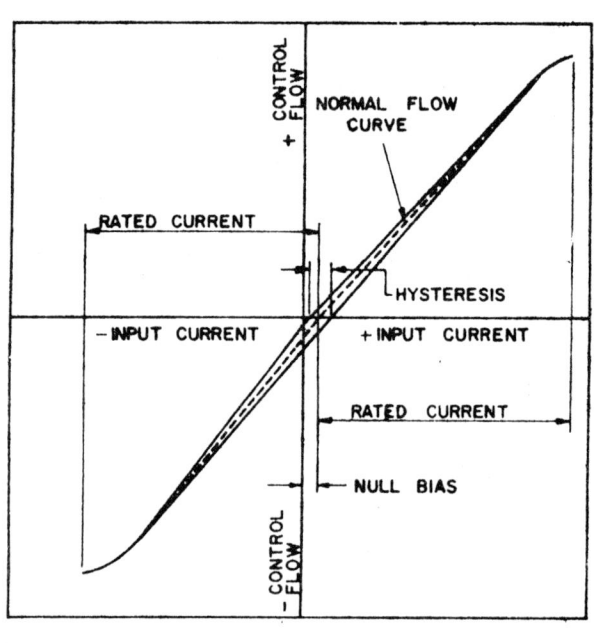

Fig. 73.1

SERVOVALVES
FLOW GAIN

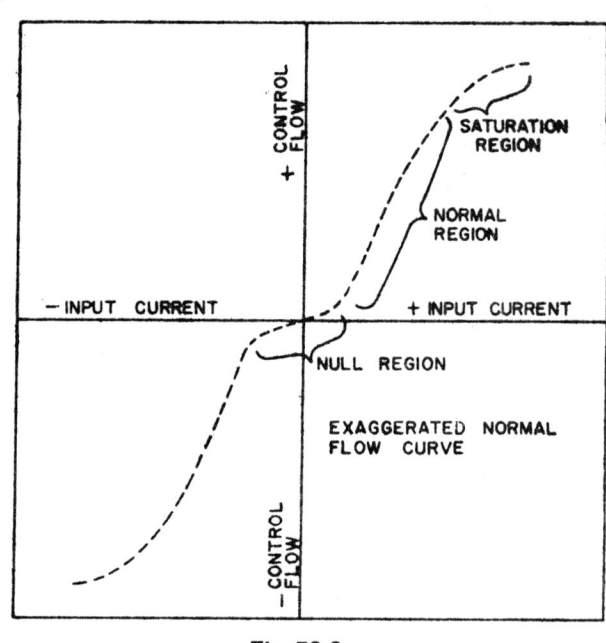

Fig. 73.2

SERVOVALVES
OVERLAP CONDITION

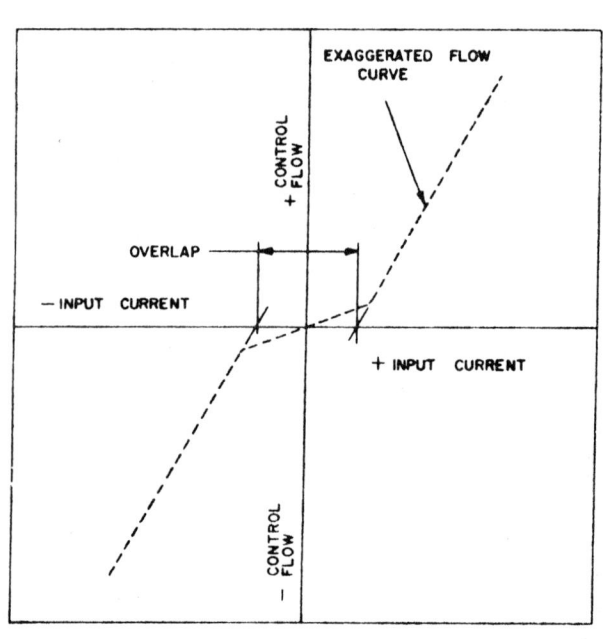

Fig. 73.3

SERVOVALVES
UNDERLAP CONDITION

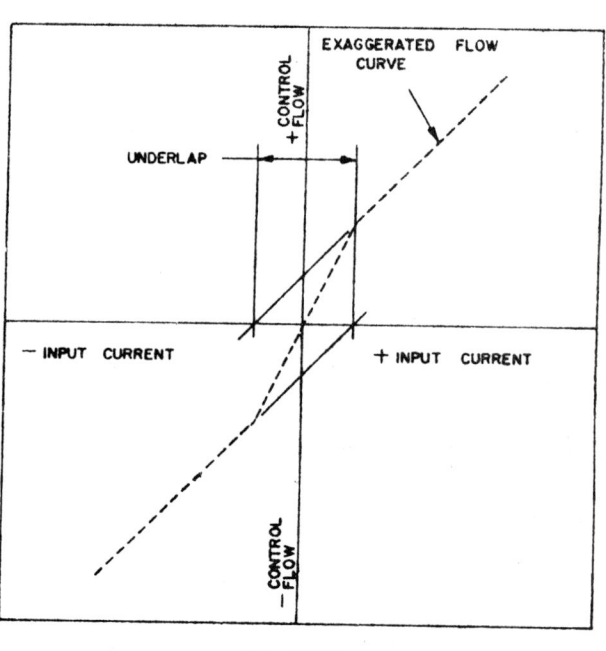

Fig. 73.4

SERVOVALVES
ZERO LAP CONDITION

SERVOVALVES
EXAMPLE OF SINUSOIDAL FREQUENCY RESPONSE CHARACTERISTICS

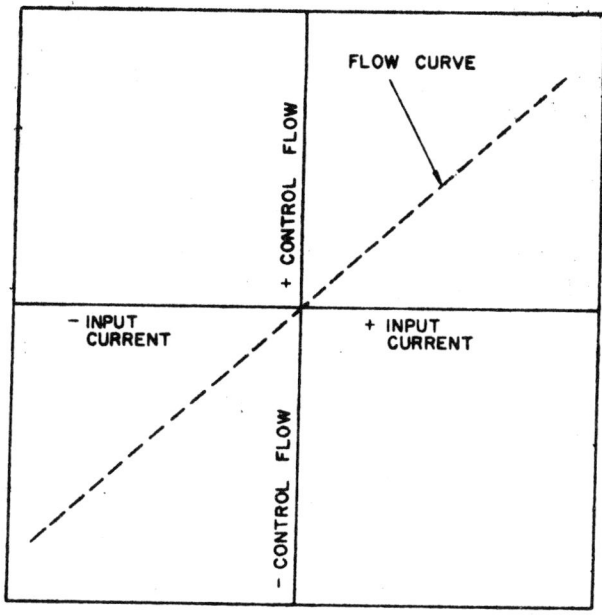

Fig. 73.5

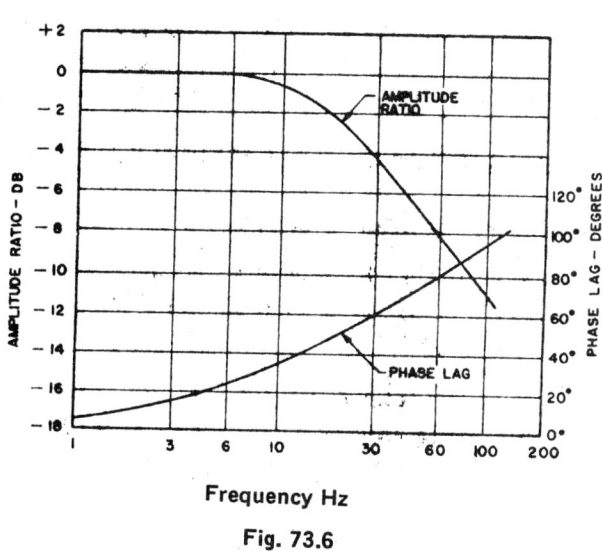

Fig. 73.6

PRESSURE GAIN

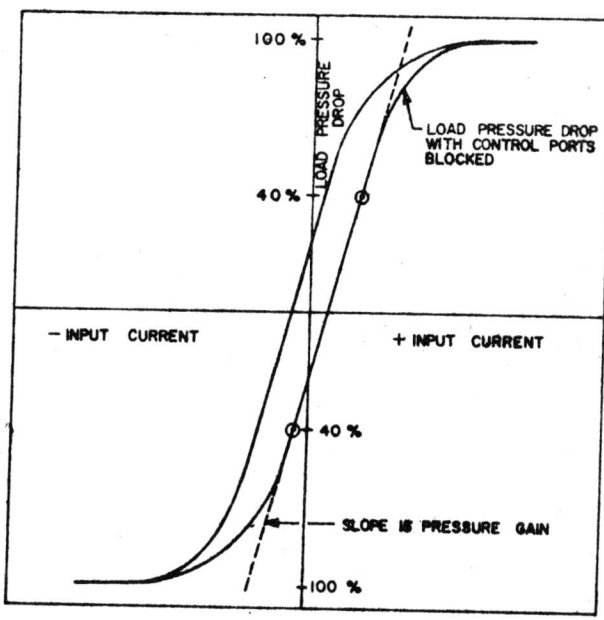

Fig. 73.7

GROUP 74 — REGULATORS

Section 01 — General
Section 02 — Regulator Types
Section 03 — Operational Factors

SECTION 01 — REGULATOR TYPES

74.02.100 Regulator, Air Line Pressure

A regulator which transforms a fluctuating air pressure supply to provide a constant lower pressure output.

74.02.120 Regulator, Constant Bleed Air Line Pressure

An Air Line Pressure Regulator that depends upon a bleed to atmosphere for proper operation.

74.02.160 Regulator, Pressure Relieving Air Line

An Air Line Pressure Regulator which automatically vents over-pressures applied to the regulated (secondary) pressure.

SECTION 03 — OPERATIONAL FACTORS

74.03.100 Droop

The deviation between no flow secondary pressure and secondary pressure at a given flow.

74.03.200 Flow Characteristic Curve

The change in regulated (secondary) pressure occuring as a result of a change in the rate of air flow over the operating range of the regulator.

74.03.300 Maximum Inlet Pressure

The maximum rated gage pressure applied to the inlet port of the regulator.

74.03.400 Reduced Pressure Range

The adjustment range of the regulator.

74.03.500 Regulation Characteristic Curve

The change in regulated (secondary) pressure occuring as a result of a change in the supply (primary) pressure to a regulator.

74.03.501 Pressure, Regulation Steady State Supply

The region between the permitted limits of the supply pressure.

74.03.600 Relief Characteristic Curve

The change in the relief flow rate in a relieving type air line regulator which occurs as a result of an increase in regulated (secondary) pressure over set pressure.

GROUP 75 — SWITCHES

Section 01 — General
Section 02 — Switch Type

SECTION 02 — SWITCH TYPES

75.02.400 Switch, Float

An electric switch which is responsive to liquid level.

75.02.500 Switch, Flow

An electric switch operated by fluid flow.

75.02.600 Switch, Liquid Level

A device incorporating an electrical switch in which actuation of the contacts is effected at a predetermined level of the liquid.

75.02.700 Switch, Pressure

An electric switch operated by fluid pressure.

75.02.750 Switch, Pressure Differential

An electric switch operated by a difference in pressure.

GROUP 76 — MUFFLERS & SILENCERS

Section 01 — General
Section 02 — Types of Mufflers & Silencers

SECTION 01 — GENERAL

76.01.100 Muffler

A device for reducing gas flow noise. Noise is decreased by back pressure control of gas expansion.

76.01.600 Silencer

A device for reducing gas flow noise. Noise is decreased by tuned resonant control of gas expansion.

GROUP 81 — CYLINDERS

Section 01 — General
Section 02 — Cylinder Types
Section 05 — Cylinder Components
Section 15 — Cylinder Mountings

SECTION 01 — GENERAL

81.01.000 Actuator, Pneumatic/Hydraulic

A device in which power is transferred from one pressurized medium (pneumatic) to another (hydraulic) without intensification.

81.01.001 Cylinder

A device which converts fluid power into linear mechanical force and motion. It usually consists of a moveable element such as a piston and piston rod, plunger or ram, operating within a cylindrical bore.

81.01.004   Cylinder, Area, Head End, Effective

The annulus area between the bore and the piston rod diameter.

81.01.005   Cylinder, Area, Piston, Effective

Area upon which fluid pressure acts to provide a mechanical force.

81.01.006   Cylinder, Area, Piston Rod

Cross-sectional area of the piston rod.

81.01.008   Cylinder, Attachment, Piston Rod

Method by which the piston rod transmits force; e.g., threaded, plain, eye, clevis.

81.01.010   Cylinder, Bore

The internal diameter of the cylinder body.

81.01.012   Cylinder, Capacity

The volume of a theoretically incompressible fluid that would be displaced by the piston during a complete stroke. (For double acting cylinders it must be given for both directions of stroke.)

81.01.013   Cylinder, Capacity, Extending

Volume required for one full extension of a cylinder.

81.01.014   Cylinder, Capacity, Retracting

Volume (annular) absorbed by one full retraction of the cylinder.

81.01.200   Cylinder, Force, Piston Rod

The force transmitted by the piston rod.

81.01.201   Cylinder, Force, Actual

The force actually transmitted by the piston rod.

81.01.202   Cylinder, Force, Nominal

The force exerted by the piston rod allowing for all frictional losses when the cylinder is operating under standard conditions. For single acting cylinders with a return spring, the value must be given at both beginning and end of the stroke.

81.01.203   Cylinder, Force, Theoretical

The pressure multiplied by the effective piston area, ignoring friction. For double acting cylinders the value must be given for both directions of stroke.

81.01.600   Cylinder, Input Power

Decrease in fluid energy per unit time between the inlet and outlet of the cylinder. In the case of a single acting cylinder this is the total energy at the inlet.

81.01.601   Cylinder, Output Power

Mechanical power transmitted by the piston rod.

81.01.610   Cylinder, Pressure

Static pressure at a stated point during actual operation.

81.01.700   Cylinder Stroke, Extend (Out)

The outward movement of the piston rod.

81.01.701   Cylinder Stroke, Full

The distance travelled by the piston in moving from one extreme position to another.

81.01.702   Cylinder Stroke, Retract (In)

The inward movement of the piston rod.

81.01.703   Cylinder Stroke, Working

The distance travelled by the piston in moving between two defined positions during actual operation.

81.01.800   Cylinder, Time, Extend Stroke

The time from commencement of movement to the completion of the outward stroke measured under "no load" or specified load conditions.

81.01.802   Cylinder, Time, Retract Stroke

The time from commencement of movement to the completion of the inward stroke, measure under "no load" or specified conditions.

SECTION 02 — CYLINDER TYPES

81.02.050   Cylinder, Adjustable Stroke

A cylinder equipped with adjustable stops at one or both ends to limit piston travel.

81.02.075   Cylinder, Bolted

A cylinder with head and cap end closures that are secured by bolt fastening methods.

ANSI/B93.2

81.02.100 Cylinder, Cushioned

A cylinder with a piston-assembly deceleration device at one or both ends of the stroke.

81.02.101 Cylinder, Dashpot

A hydraulic damping device which acts as a variable speed regulator for a pneumatic cylinder.

81.02.102 Cylinder, Diaphragm Type

A cylinder in which the mechanical force is produced by fluid pressure acting on a diaphragm.

81.02.104 Cylinder, Differential

A double acting cylinder in which the ratio of the area of the bore to the annular area between the bore and the piston rod is significant in circuit function.

81.02.150 Cylinder, Double Acting

A cylinder in which fluid force can be applied to the moveable element in either direction.

81.02.200 Cylinder, Double Rod

A cylinder with a single piston and a piston rod extending from each end.

81.02.250 Cylinder, Dual Stroke

A cylinder combination which provides two working strokes.

81.02.252 Cylinder, Duplex

A unit comprised of two cylinders with independent control, mechanically connnected on a common axis to provide three or four positions depending on the method of application.

81.02.254 Cylinder, Multiposition

An arrangement of at least two pistons on the same axis, moving within a common cylinder body divided into several independently controlled chambers, to permit the selection of a variety of positions.

81.02.290 Cylinder, Non-Rotating

A cylinder in which relative rotation of the cylinder housing and the piston and the piston rod, plunger or ram, is not recommended.

81.02.300 Cylinder, Piston Type

A cylinder in which the piston has a greater cross-sectional area than the piston rod.

81.02.350 Cylinder, Plunger (Ram)

A cylinder in which the piston has the same cross-sectional area as the piston rod.

81.02.450 Cylinder, Adjustable Stroke

A cylinder in which the stop can be temporarily changed to permit the length of the stroke to be varied.

81.02.460 Cylinder, Rotating

A cylinder in which the piston and piston rod, plunger or ram, is permitted to rotate with reference to the cylinder housing.

81.02.500 Cylinder, Single Acting

A cylinder in which the fluid force can be applied to the moveable element in only one direction.

81.02.510 Cylinder, Single Acting, Gravity Return

A single acting cylinder returned by gravity.

81.02.550 Cylinder, Single Rod

A cylinder with a piston rod extending from one end.

81.02.600 Cylinder, Spring Return

A cylinder in which a spring returns the piston assembly.

81.02.650 Cylinder, Tandem

Arrangement of at least two pistons on the same rod moving in separate chambers on the same cylinder body allowing the compounding of force on the piston rod.

81.02.700 Cylinder, Telescoping

Cylinder with two or more stages or extensions, achieved by hollow piston rods sliding one within the other (may be single or double acting).

81.02.750 Cylinder, Tie Rod

A cylinder with head and cap end closures that are secured by tie rods.

SECTION 05 – CYLINDER COMPONENTS

(Wherever the term "piston rod" appears it is intended to also mean "plunger" or "ram")

81.05.030 Angle

A mounting device which is angular in cross section. It usually made from a $90°$ angle.

81.05.180 Cap (Back End) (Blind End) (Blind Head) (Rear End) (Rear Head)

A cylinder end closure which completely covers the bore area.

81.05.210 Clevis (Hinge) (Pendulum)

A "U" shaped mounting device which contains a common pin hole at right angle or normal to the axis of symmetry through each extension. A clevis usually connects with an eye.

81.05.240 End

Either of two envelope surfaces at right angle or normal to the piston rod centerline.

81.05.270 Eye (Hinge) (Pendulum)

A mounting device consisting of a single extension which contains a mounting pin hole at right angle or normal to the axis of symmetry. An eye usually connects with a clevis.

81.05.280 Efficiency, Overall

Ratio between mechanical power output and power input.

81.05.281 Efficiency, Speed

Ratio between the effective piston speed and the theoretical speed.

81.05.282 Efficiency, Thrust

Ratio between the effective force and the theoretical force.

81.05.300 Flange

A mounting device consisting of a plate or collar extending past the basic cylinder profile to provide an attachment area for bolt fastening methods. A flange is usually at a right angle (normal) to the cylinder centerline.

81.05.450 Head (Front End) (Front Face) (Front Head) (Rod Head)

The cylinder end closure which covers the differential area between the bore area and the piston rod area.

81.05.510 Lug (Foot)

A mounting device consisting of a block extending past the basic cylinder profile. The block usually has a tapped or through mounting hole at right angles to the cylinder axis.

81.05.550 Piston Rod

The element transmitting mechanical force and motion from the piston.

81.05.570 Rabbet

A mounting device which utilizes matching male and female forms (usually coaxial circular) between the cylinder and its mating element.

81.05.720 Side (Base)

An envelope surface which is parallel to the piston rod centerline.

81.05.750 Tie Rod

An axial external cylinder element which traverses the length of the cylinder. It is prestressed at assembly to hold the ends of the cylinder against the tubing. Tie rod extensions can be a mounting device.

81.05.780 Trunnion

A mounting device consisting of a pair of opposite projecting cylindrical pivots. The cylindrical pivot pins are at right angle or normal to the piston rod centerline to permit the cylinder to swing in a plane.

SECTION 15 — CYLINDER MOUNTINGS

(Wherever the term "piston rod" appears it is intended to also mean "plunger" or "ram")

81.15.000 Cylinder Mounting

A device by which a cylinder is fastened to its mating element.

81.15.100 Cylinder Mounting, Centerline

A mounting which permits connection on a plane in line with the piston rod centerline.

81.15.150 Cylinder Mounting, Centerline, Lug

A centerline mounting consisting of two opposite lugs at each end of the cylinder.

81.15.200 Cylinder Mounting, End

A mounting which permits connection at either or both ends of a cylinder.

81.15.204 Cylinder Mounting, End, Both

A mounting at both ends of the cylinder.

81.15.208 Cylinder Mounting, End, Both, Tie Rods Extended

A cylinder mounted at both ends by means of extended tie rods.

81.15.212 Cylinder Mounting, End, Cap

A mounting which permits connection at the cap end.

81.15.216 Cylinder Mounting, End, Cap, Circular

A direct circular cap mounting.

81.15.220 Cylinder Mounting, End, Cap, Circular Flange

A cap mounting consisting of a supplementary circular flange plate.

81.15.224 Cylinder Mounting, End, Cap Detachable Clevis

A cap mounting consisting of a clevis which can be removed or rotated.

81.15.228 Cylinder Mounting, End, Cap, Detachable Eye

A cap mounting consisting of an eye which can be removed or rotated.

81.15.232 Cylinder Mounting, Clevis

A U-shaped mounting device which accepts a lug and through which a pin or bolt passes, to make a pivot mounting.

81.15.236 Cylinder Mounting, End, Cap, Fixed Eye

A cap mounting consisting of an eye integral with the cap to maintain a fixed eye-port relationship.

81.15.239 Cylinder Mounting, End, Cap Flange

A mounting consisting of a suitable shaped plate or collar usually extending beyond the cylinder profile, secured to or forming part of the cylinder. (It is usually provided with suitable holes for fixing).

81.15.240 Cylinder Mounting, End, Cap, Rectangular Flange

A cap mounting consisting of a supplementary rectangular flange plate.

81.15.242 Cylinder Mounting, End, Cap, Spherical

Arrangement swivelling around a point allowing angular movement of the cylinder in any plane including its axis.

81.15.244 Cylinder Mounting, End, Cap, Square

A direct square cap mounting.

81.15.248 Cylinder Mounting, End, Cap, Square Flange

A cap mounting consisting of a supplementary square flange plate.

81.15.250 Cylinder Mounting, End, Cap, Threaded

Mounting by means of threaded projections or recesses coaxial with cylinder axis to permit mounting.

81.15.252 Cylinder Mounting, End, Cap, Tie Rods Extended

A cap mounting consisting of extended tie rods.

81.15.256 Cylinder Mounting, End, Cap, Trunnion

A cap mounting consisting of trunnion pins near or at the cap end of the cylinder.

81.15.260 Cylinder, Mounting, End, Head

A mounting which provides cylinder connection at the head end.

81.15.264 Cylinder Mounting, End, Head, Circular

A direct circular head mounting.

81.15.268 Cylinder Mounting, End, Head, Circular Flange

A head mounting consisting of a supplementary circular flange plate.

81.15.272 Cylinder Mounting, End, Head, Female Rabbet

A head mounting consisting of a female pilot recess.

81.15.276 Cylinder Mounting, End, Head, Male Rabbet

A head mounting consisting of a male pilot extension.

81.15.280 Cylinder Mounting, End, Head, Rectangular Flange

A head mounting consisting of a supplementary rectangular flange plate.

81.15.284 Cylinder Mounting, End, Head, Square

A direct square head mounting.

81.15.288	Cylinder Mounting, End, Head, Square Flange

A head mounting consisting of a supplementary square flange plate.

81.15.290	Cylinder Mounting, End, Head, Threaded, Neck or Nose

A threaded projection coaxial with the cylinder axis at the rod end, to permit mounting.

81.15.292	Cylinder Mounting, End, Head, Tie Rods Extended

A head mounting consisting of extended tie rods.

81.15.296	Cylinder Mounting, End, Head, Trunnion

A head mounting consisting of trunnion pins near or at the head end of the cylinder.

81.15.300	Cylinder Mounting, Fixed

A mounting which provides rigid connection between the cylinder and the mating element wherein the piston rod reciprocates in a fixed line.

81.15.400	Cylinder Mounting, Intermediate

A mounting which provides cylinder connection at an intermediate external position along the piston rod centerline between ends.

81.15.410	Cylinder Mounting, Side, Trunnion

Mounting device consisting of a pair of male or female pivots on opposite sides of the cylinder whose axis intersects the cylinder axis at right angles.

81.15.425	Cylinder, Mounting, Intermediate, Fixed Trunnions

An intermediate mounting consisting of trunnion pins which cannot be repositioned.

81.15.460	Cylinder Mounting, Intermediate, Side Trunnion

A mounting device consisting of a pair of male or female pivots on opposite sides of the cylinder whose axis intersects the cylinder axis at right angles.

81.15.500	Cylinder Mounting, Pivot

A mounting which permits a cylinder to change its alignment in a plane.

81.15.600	Cylinder Mounting, Side

A mounting which provides cylinder connection at one of its sides.

81.15.615	Cylinder Mounting, Side, End Angles

A side mounting consisting of an angle at each end of the cylinder with the free legs facing a common side.

81.15.630	Cylinder Mounting, Side, End Lugs

A side mounting consisting of one or more lugs at each end of the cylinder and facing a common side.

81.15.645	Cylinder Mounting, Side, End Plates

A side mounting consisting of an extended plate at each cylinder end and facing a common side.

81.15.660	Cylinder Mounting, Side, Lugs

A side mounting consisting of two opposite lugs at each cylinder end facing a common side.

81.15.675	Cylinder Mounting, Side, Tapped

A side mounting consisting of one or more tapped holes at each cylinder end facing a common side.

81.15.690	Cylinder Mounting, Side, Through Holes

A side mounting consisting of holes drilled across both ends to a common side.

81.15.700	Cylinder Mounting, Universal

A mounting which permits a cylinder to change its alignment in all directions.

GROUP 82 — HYDRAULIC MOTORS

Section 01 — General
Section 02 — Types of Hydraulic Motor

SECTION 01 — GENERAL

82.01.000	Hydraulic Motor

A device which converts hydraulic fluid power into mechanical force and motion. It usually provides rotary mechanical motion.

82.01.020	Hydraulic Motor, Capacity, Derived

Volume absorbed at defined minimum working pressure obtained from measurements at two different speeds.

82.01.030	Hydraulic Motor, Efficiency, Hydromechanical

Ratio of the effective torque to the derived torque.

82.01.031 Hydraulic Motor, Efficiency, Overall

Ratio of the output power to the effective hydraulic power.

82.01.032 Hydraulic Motor, Efficiency, Volumetric

Ratio of the derived input flow to the effective input flow.

82.01.040 Hydraulic Motor, Flow, Input

Flow rate crossing the transverse plane of the inlet port.

82.01.041 Hydraulic Motor, Flow, Input, Derived

The product of the derived capacity and the number of revolutions.

82.01.042 Hydraulic Motor, Flow, Input, Effective

Actual flow at inlet measured at the pressure and temperature at that point.

82.01.043 Hydraulic Motor, Flow, Input, Geometric

The product of the geometric capacity and the number of revolutions per unit time.

82.01.100 Hydraulic Motor, Power

Decrease in hydraulic energy per unit time between the inlet and outlet of the motor.

82.01.101 Hydraulic Motor, Power, Derived

Hydraulic power calculated from the derived capacity.

82.01.102 Hydraulic Motor, Power, Effective

Hydraulic power calculated from the effective capacity.

82.01.103 Hydraulic Motor, Power, Geometric

Hydraulic power calculated from the geometric capacity.

82.01.104 Hydraulic Motor, Output Power

Mechanical power transmitted by the shaft of the motor.

82.01.110 Hydraulic Motor, Losses, Motor

The portion of the effective hydraulic (input) power not transformed into output power. Volumetric losses - Hydrodynamic losses - Mechanical losses. (See 3.0.2.2)

82.01.125 Hydraulic Motor, Rated Speed

The maximum continuous motor shaft speed specified by the manufacturer.

82.01.150 Hydraulic Motor, Speed Degradation Ratio

The ratio of the stabilized motor speed after a contaminant injection to the rated motor speed.

82.01.200 Slip

Difference between shaft speeds at specified input flow and under different loading conditions.

SECTION 02 — TYPES OF HYDRAULIC MOTORS

82.02.010 Hydraulic Motor, Axial Piston

A motor having several pistons with mutually parallel axes which are arranged around and parallel to a common axis.

82.02.050 Hydraulic Motor, Displacement

A motor in which the quantity of fluid absorbed is related to shaft speed.

82.02.100 Hydraulic Motor, Fixed Displacement

A hydraulic motor in which the displacement per unit of output motion cannot be varied.

82.02.150 Hydraulic Motor, Gear

A motor in which two or more gears act in arrangement as working members.

82.02.151 Hydraulic Motor, Gear, External

A motor having two or more external gears.

82.02.152 Hydraulic Motor, Gear, Internal

A motor with an internal gear in engagement with one or more external gears.

82.02.176 Hydraulic Stepping Motor

A hydraulic motor which follows the commands of a stepped input signal to achieve positional accuracy.

82.02.200 Hydraulic Motor, Linear

A fluid power cylinder with built in control by which the piston rod is automatically reciprocated.

82.02.210 Hydraulic Motor, Multiple

Two or more motors having a common shaft.

82.02.220 Hydraulic Motor, Over-Center

A motor in which the rotation of the output may be changed without changing the direction of the input flow.

82.02.231 Hydraulic Motor, Radial Piston

A motor having several pistons arranged to operate radially.

82.02.250   Hydraulic Motor, Reciprocating

A motor in which the working members reciprocate.

82.02.260   Hydraulic Motor, Reversible

A motor in which the direction of rotation of the output may be reversed by changing the direction of the input flow.

82.02.300   Hydraulic Motor, Rotary

A hydraulic motor capable of continuous rotary motion.

82.02.301   Hydraulic Motor, Rotary, Limited

A hydraulic rotary motor having limited motion.

82.02.310   Hydraulic Motor, Torque

The torque transmitted by the shaft of the motor.

82.02.311   Hydraulic Motor, Torque, Starting

The minimum torque available at the motor shaft when starting from rest for a given pressure differential under specified conditions.

82.02.320   Hydraulic Motor, Vane

A motor in which the fluid under pressure acting on a set of radial vanes causes rotation of an internal member.

82.02.321   Hydraulic Motor, Vane, Balanced

A motor in which the transverse forces on the rotor are balanced.

82.02.322   Hydraulic Motor, Vane, Unbalanced

A motor in which the transverse forces acting on the rotor are not balanced.

82.02.600   Hydraulic Motor, Variable Displacement

A hydraulic motor in which the displacement per unit of output motion can be varied.

82.02.620   Hydraulic Motor, Assembly

A combination of hydraulic motor, pressure relief valves and control valve.

GROUP 84 — AIR MOTORS

    Section 01 — General
    Section 02 — Types of Air Motors

SECTION 01 — GENERAL

84.01.100   Air Motor

A device which converts pneumatic fluid power into mechanical force and motion. It usually provides rotary mechanical motion.

SECTION 02 — TYPES OF AIR MOTORS

84.02.100   Air Motor, Piston

An air motor which usually has pistons driving a shaft. Pressure acts successively on each piston through a valve gear governed by the rotation of the monitor.

84.02.101   Air Motor, Vane

An air motor which uses a stator and a rotor with grooves parallel to the axis of rotation mounted eccentrically within the stator. Air pressure acts on the vanes which slide in the grooves.

84.02.102   Air Motor, Gerotor

An air motor which consists of one or more gerotor power elements of which the inner member rotates concentrically with the axis of the output shaft. Pressure and flow are valved so as to permit the outer member to orbit eccentrically and rotate about the inner member and in turn transmit the torque to the motor shaft causing it to turn. Because the outer member has more lobes than the inner member, torque is multiplied and speed reduced.

84.02.500   Air Motor, Assembly

A combination of pneumatic motor, pressure relief valve and control valve.

GROUP 85 — ACTUATOR, ROTARY

    Section 01 — General

SECTION 01 — GENERAL

85.01.000   Cylinder, Rotary Actuator

A cylinder which translates piston receprocation into oscillation of an output shaft.

GROUP 91 — INSTRUMENTATION

    Section 01 — General
    Section 02 — Instrument (Gage)
    Section 03 — Meters
    Section 04 — Recorder
    Section 05 — Sensor

SECTION 01 — GENERAL

91.01.280   Dipstick

A removable graduated rod for indicating the level of the contents of a reservoir.

91.01.300   Gage Damper (Snubber)

A device employing a fixed or variable restrictor inserted in the pipeline to a pressure gage, to prevent damage to the gage mechanism caused by rapid fluctuations of fluid pressure.

91.01.301   Gage Protector

A device inserted in the pipeline to a pressure gage and arranged to isolate the pressure gage from the fluid pressure if this exceeds a predetermined limit. The device can usually be adjusted to suit the range of the pressure gage.

91.01.400   Indicator

A device which provides external visual evidence of sensed phenomena.

91.01.700   Repeatability

Quantitative expression of the random error associated with a single tester in a given laboratory obtaining successive results with the same apparatus under constant operating conditions on identical test material.

SECTION 02 — INSTRUMENT (GAGE)

91.02.050   Gage, (Instrument)

An instrument or device for measuring, indicating, or comparing a physical characteristic.

91.02.051   Annunciator (Indicator)

A device which shows the presence or absence of a phenomenon such as pressure or flow, but which does not measure it.

91.02.052   Annunciator, Flow

A device employing a ball, vane or other means inside a transparent cover. Motion of the ball or vane indicates that fluid is flowing through the pipeline.

91.02.053   Annunciator, Pressure

A device which changes fluid pressure into mechanical motion and usually by means of a plunger indicates the presence or absence of pressure. These devices usually do not indicate the exact pressure but show whether the fluid pressure is above or below a predetermined level.

91.02.100   Gage, Bellows

A gage in which the sensing element is a convoluted closed cylinder. A pressure differential between outside and inside causes the cylinder to expand or contract axially.

91.02.200   Gage, Bourdon Tube

A pressure gage in which the sensing element is a curved tube that tends to straighten out when subjected to internal fluid pressure.

91.02.300   Gage, Diaphragm

A gage in which the sensing element is relatively thin and its inner portion is free to deflect with respect to its periphery.

91.02.400   Gage, Fluid Level

A gage which indicates the fluid level at all times.

91.02.500   Gage, Manometer

A differential pressure gage in which pressure is indicated by the height of a liquid column of known density. Pressure is equal to the difference in vertical height between two connected columns multiplied by the density of the manometer liquid. Some forms of manometers are "U" tube, inclined tube, well, and bell types.

91.02.600   Gage, Piston

A pressure gage in which the sensing element is a piston operating against a spring.

91.02.650   Gage, Pitot Tube

A velocity-sensing tubular probe with one end facing fluid flow and the other end connected to a gage.

91.02.700   Gage, Pressure

A gage which indicates the pressure in the system to which it is connected.

91.02.730   Gage, Reservoir

A device which senses either the liquid height, weight or pressure in a reservoir and displays it, usually by mechanical pointer on a scale graduated in tank contents. It can also consist of a sight glass with a scale calibrated in cubic content instead of depth of fluid.

91.02.800   Gage, Vacuum

A pressure gage for pressures less than atmospheric.

91.02.850   Instrument, Absolute Pressure Measuring

An instrument which shows the absolute pressure of the fluid in relation to a theoretically perfect vacuum.

91.02.852   Instrument, Differential Pressure Measuring

An instrument which measures the difference between two pressures.

91.02.853   Instrument, Flow Measuring

A device which measures the flow rate of a fluid.

91.02.860   Instrument, Liquid Level Measuring

A device which indicates the level of liquid.

SECTION 03 — METERS

91.03.300 Manometer

A device which indicates fluid pressure by liquid levels usually employing a U-tube filled with mercury or water. The level of the liquid above or below a reference point indicates the fluid pressure relative to ambient pressure. Unless otherwise specified, it indicates the fluid pressure relative to ambient atmospheric pressure.

91.03.400 Meter, Flow

A device which indicates either flow rate, total flow, or a combination of both.

91.03.401 Meter, Flow, Integrating

An instrument which indicates the total quantity of the fluid that has flowed past the measuring point.

SECTION 04 — RECORDER

91.04.100 Recorder, Flow

An instrument which provides a permanent record of fluid flow (usually on paper, film or tape).

91.04.200 Recorder, Pressure

An instrument which provides a permanent record of pressure (usually on paper, film, or tape).

SECTION 05 — SENSOR

91.05.000 Sensor

A device which detects and transmits changes in external conditions.

91.05.100 Transducer, Flow

A device which converts fluid flow to an electrical signal.

91.05.600 Transducer, Pressure

A device which converts fluid pressure to an electrical signal.

GROUP 94 — SEALING DEVICES

Section 01 — General
Section 02 — Types of Sealing Devices
Section 06 — Auxiliary Devices
Section 15 — Physical Criteria

SECTION 01 — GENERAL

94.01.020 Compatibility, Seal

Ability of an elastomer to resist the action of a fluid on its dimensional and mechanical properties.

94.01.030 Durometer Hardness

An arbitrary indication of hardness determined by an indentor.

94.01.080 Pack

To install packing.

94.01.100 Sealing Device (Seal)

A device which prevents or controls the escape of a fluid or entry of a foreign material.

SECTION 02 — TYPES OF SEALING DEVICES

94.02.200 Packing

A sealing device consisting of bulk deformable material of one or more mating deformable elements, reshaped by manually adjustable compression to obtain and maintain effectiveness. It usually uses axial compression to obtain radial sealing.

94.02.220 Packing, Coil

Packing in coil form.

94.02.240 Packing, U

A packing in which the deformable element has a U shaped cross-section.

94.02.260 Packing, V

A packing in which the deformable element has a V shaped cross-section.

94.02.280 Packing, W

A packing in which the deformable element has a W shaped cross-section.

94.02.410 Ring, O

A ring which has a round cross-section.

94.02.420 Ring, Piston

A piston sealing ring. It is usually one of a series and is often split to facilitate expansion or contraction.

94.02.430 Ring, Scraper

A ring which removes material by a scraping action.

94.02.440 Ring, "U"

A ring which has a "U" shaped cross-section.

94.02.450 Ring, "V"

A ring which has a "V" shaped cross-section.

94.02.460 Ring, "W"

A ring which has a "W" shaped cross-section.

94.02.470 Ring, Wiper

A ring which removes material by a wiping action.

Seal (See Sealing Device)

94.02.603 Seal, Axial (End) (Face) (Shoulder)

A sealing device which seals by axial contact pressure.

94.02.604 Seal, Butyl

A material composed of copolymer of isobutylene and isoprene exhibiting good chemical and ozone resistance and low permeability to gas. (Resistant to a number of phosphate ester fluids but not to petroleum based fluids.)

94.02.605 Seal, Chevron

A radial seal comprising several mating elements with cross-sections of V form.

94.02.606 Seal, Composite

A sealing device having elements of different materials.

94.02.607 Seal, Cork

Normally comprises cork granules bonded together with minor amounts of natural or synthetic rubber or resins. Fluid resistance will depend mainly on the bonding medium.

94.02.608 Seal, Cup

A sealing device with a radial base integral with an axial cylindrical projection at its outer diameter.

94.02.609 Seal, Diaphragm (Flat Diaphragm)

A relatively thin, flat or molded sealing device fastened and sealed at its periphery with its inner portion free to move.

94.02.610 Seal, Dished Diaphragm

A diaphragm in which the central area is depressed in a free state. (It permits longer travel than a flat comparable diaphragm.)

94.02.612 Seal, Dynamic

A sealing device used between parts that have relative motion.

94.02.613 Seal, Elastomer

A material having rubber-like properties; i.e., having the capacity for large deformation and rapid and substantially complete recovery on release from the deforming force.

94.02.617 Seal, Ethylene Propylene

Copolymer of ethylene and propylene. Resistant to phosphate ester fluid but not to mineral oils.

94.02.621 Seal, Flange (Hat)

A sealing device with a radial base integral with an axial projection at its inner diameter.

94.02.622 Seal, Fluorinated (Viton®) [1]

A fluorinated rubber having outstanding chemical and fluid resistance. Resistance to high temperatures is excellent. Low temperature characteristics are poor.

94.02.630 Seal, Lip

A sealing device which has a flexible sealing projection.

94.02.631 Seal, Leather

Leather for hydraulic duty which is usually chrome tanned or impregnated with waxes, rubbers or resins to reduce fluid permeability.

94.02.633 Seal, Lubricant

A sealing device which uses lubricant as a sealing barrier.

94.02.636 Seal, Mechanical

A sealing device in which sealing action is aided by mechanical force.

94.02.638 Seal, Nitrile

A material composed of copolymers of butadiene and acrylonitrile (resistance to petroleum base fluids varies according to the acrylonitrile content of the polymer).

94.02.639 Seal, Oil

A sealing device which retains oil.

94.02.642 Seal, "O" Ring

A sealing ring which has a round cross-section.

94.02.645 Seal, Piston

A sealing device installed on a piston to maintain a sealing fit with a cylinder bore.

94.02.646 Seal, Polyacrylic

A copolymer of acrylate of ethyl. Good resistance to mineral oils. Resistance to heat is better than that of nitrile rubbers.

94.02.647 Seal, Polyamid (Nylon)

Polyamid thermoplastic materials characterized by their high strength and resistance to abrasion.

---

[1] This is a Trademark of du Pont.

94.02.648 Seal, Polychloroprene

A material composed of polychloroprene. (Has fair to good resistance to pretroleum base fluids and good resistance to ozone and weathering).

94.02.649 Seal, Polytetrafluorethylene (PTFE)

A thermoplastic polymer which is virtually immune to chemical attack and which may be used over a very wide temperature range. Co-efficient of friction is very low but flexibility is limited and recovery characteristics only moderate.

94.02.650 Seal, Polyurethane

Material comprising mainly isocyanate having good resistance to petroleum base fluids and to abrasion, liable to degradation in the presence of water at moderate temperatures.

94.02.651 Seal, Pressure Actuated

A sealing device in which sealing action is aided by fluid pressure.

94.02.654 Seal, Radial

A sealing device which seals by radial contact pressure.

94.02.657 Seal, Ram (Plunger)

A sealing device which seals the periphery of a ram.

94.02.660 Seal, Rod (Shaft) (Stem)

A sealing device which seals the periphery of a piston rod.

94.02.663 Seal, Rotary (Shaft)

A sealing device used between parts that have relative rotary motion.

94.02.665 Seal, Silicone

Polysiloxanes having inorganic molecular chains with attached organic groupings. They are outstanding amongst rubbers in their retention of rubber-like properties over a very wide temperature range.

94.02.672 Seal, Sliding

A sealing device used between parts that has relative reciprocating motion.

94.02.674 Seal, Ring (Square)

A seal which has a square-shaped cross-section.

94.02.675 Seal, Static (Gasket)

A sealing device used between parts that have no relative motion.

94.02.676 Seal, U Ring

A seal which has a U shaped cross-section.

94.02.677 Seal, V Ring

A seal which has a V shaped cross-section.

94.02.678 Seal, W Ring

A seal which has a W shaped cross-section.

94.02.681 Seal, Water

A sealing device which uses water as a sealing barrier.

94.02.684 Seal, Wiper

A sealing device which operates by a wiping action.

94.02.685 Seal, X Ring

A seal which has an X shaped cross-section.

SECTION 06 — AUXILIARY DEVICES

94.06.100 Adapter (Support Ring)

A seal support shaped to conform with the contour of the seal and the mating element.

94.06.125 Adapter, Female

An adapter with a concave seal support. Referred to as top adapter with "V" type seals.

94.06.150 Adapter, Male

An adapter with a convex seal support. Referred to as bottom adapter with "V" type seals.

94.06.175 Adapter, Pedestal

An adapter usually used to support a "U" type seal.

94.06.180 Back-Up Ring (Anti-Extrusion Ring) (Junket Ring) (Bull Ring)

A ring which bridges a clearance to minimize seal extrusion.

94.06.190 Exclusion Device

A ring employed to remove fluid, mud, etc., from a reciprocating member offering protection of seals or packings.

94.06.200 Filler Ring

A ring which fills the recess of a "V" or "U" types seal.

94.06.210 Gland

The cavity of a stuffing box.

| 94.06.250 | Gland Follower |

The closure for a stuffing box.

| 94.06.360 | Lantern Ring (Seal Cage) |

A ring in line with a port in a gland to introduce a lubricant or a coolant to the packing and stuffing box.

| 94.06.400 | Shell |

A structural form to which a sealing element is assembled or bonded.

| 94.06.510 | Spring, Expander |

A spring which produces outward radial force.

| 94.06.520 | Spring, Finger (Lug) |

A spring with flexible fingers which produce force.

| 94.06.530 | Spring, Garter |

A compression or tension ring formed from helical wire spring with connected ends to produce force.

| 94.06.560 | Spring, Spreader |

A spring which produces sealing force against both lips of "U" or "V" seals.

| 94.06.570 | Spring, Wave (Marcel) (Wave Washer) |

A compression spring of wave configuration which produces force.

| 94.06.600 | Stuffing Box |

A cavity and closure for a sealing device.

SECTION 15 — PHYSICAL CRITERIA

| 94.15.300 | Excluder Bore |

The major or outside diameter of a groove carved in a gland or in one member of a concentric joint to accommodate an exclusion device.

| 94.15.320 | Excluder Cavity |

The confining annulus of a gland which accomodates an exclusion device.

| 94.15.560 | Radial Cavity |

The distance, in a rod sealed cylinder, between the piston rod and the outside diameter of the gland; or in the case of a piston sealed cylinder, the distance between the inside diameter of the piston groove and cylinder wall.

GROUP 96 — SYSTEMS & RELATED TERMS

Section 01 — General
Section 02 — Systems
Section 03 — Diagrams
Section 04 — Circuits

SECTION 01 — GENERAL

| 96.01.090 | Commissioning |

The act of operating, testing and adjusting a system or unit for the first time to ensure that it functions according to the specified performance. Functional tests include the extremes of the required specification.

| 96.01.100 | Compartment |

A space within the base, frame or column of the equipment.

| 96.01.150 | Connector, Electrical |

A two-piece instrument (plug and socket) which, when joined, provides electrical continuity.

| 96.01.200 | Cycle |

One complete set of events or conditions which repeat in an identical manner.

| 96.01.210 | Cycle, Automatic |

A cycle of operation which once started is repeated indefinitely until stopped.

| 96.01.220 | Cycle, Manual |

A cycle which is manually started and controlled through all phases.

| 96.01.230 | Cycle, Semi-automatic |

A cycle which is started upon a given signal, proceeds through a predetermined sequence, and stops with all elements in their initial position.

| 96.01.231 | Cycle, Speed of |

Number of cycles completed per unit of time under stated conditions.

| 96.01.232 | Cycle, Speed of, Maximum |

Maximum number of cycles per unit time under specified conditions.

| 96.01.233 | Cycle, Working |

A cycle during which work is performed.

| 96.01.300 | Enclosure |

A housing for components.

96.01.350 Maintenance, System

The process of inspecting or checking a unit or system to ensure it is functioning correctly, or to determine why it is not operating correctly. Repairing or renewing defective or damaged components, adjusting settings and controls to maintain the required or specified performance.

96.01.351 Manual, Commissioning

A document detailing the quantity and type of fluid, electrical or other services and procedures to be followed before starting equipment for the first time. It will also detail the sequence of operations and observations to be made to ensure correct function of the equipment when first operated.

96.01.352 Manual, Installation

A document detailing all materials and services required, mounting facilities relative disposition of units and means of connecting equipment in preparation to commissioning and starting up of a new installation or piece of equipment.

96.01.353 Manual, Maintenance

A document detailing the disciplines and procedures to be followed to maintain an item of equipment.

96.01.354 Manual, Operating

A document detailing the sequence of operations, adjustments and observations to be made ensuring correct use and operation of equipment, following satisfactory installation and commissioning.

96.01.600 Panel

A plate or surface for mounting components.

96.01.610 Panel, Control

A grouping of components mounted on a panel or integrally built into an assembled unit having a single mounting surface.

96.01.620 Panel, Mounting

A panel on which a number of components may be mounted.

96.01.630 Parts List

A document listing and identifying components and assemblies which make up the unit, module or system.

96.01.631 Parts List, Spares

A document detailing the quantity and type of components, sub-assemblies and units recommended by the equipment manufacturer to be held in store for preventive maintenance and general repair to keep the equipment in good working condition.

96.01.700 Phase

A distinct functional operation during a cycle. Some typical sequential phases are: neutral, rapid advance, feed or pressure stroke, dwell and rapid return.

96.01.710 Phase, Dwell

The phase of a cycle where a specified motion is stopped for a pre-determined length of time.

96.01.720 Phase, Feed

The phase of a cycle where work is performed on the workpiece.

96.01.730 Phase, Neutral

The phase of a cycle from which the work sequence begins.

96.01.740 Phase, Rapid Advance

The phase of a cycle where tools or workpiece approach at high speed to the feed position.

96.01.750 Phase, Rapid Return

The phase of a cycle where tools or workpiece return at high speed to the cycle starting position.

96.01.770 Phase, Working

A phase, during which the work is accomplished.

96.01.810 Stop, Positive Position

A structural member which accurately stops motion.

96.01.820 Stop, Positive Safety

A structural member which confines maximum travel to safe limits.

96.01.850 Systems, Fluid Power

An arrangement of interconnected components which transmits and controls power by the use of pressurized fluid within an enclosed circuit.

96.01.910 Reversal, Position

A reversal of direction of movement initiated by a signal given at some predetermined point of movement.

96.01.920 Reversal, Pressure

A reversal of direction of movement initiated by a signal responsive to rise in pressure.

SECTION 02 — SYSTEMS

96.02.050 Condition, Acceptable

A condition which permits a tolerable standard of performance and life.

96.02.051 Condition, Actual

The condition observed during operation.

96.02.052 Condition, Continuous Working

The continuous working condition is indicated by the values of the various factors which permit the unit to operate continuously. They are indicated: qc, Pc, etc. Often equals rated conditions.

96.02.053 Condition, Cycle Stabilized

The condition in which the relevant parameters vary in a repetitive manner, similar conditions repeating at regular intervals.

96.02.054 Condition, Discontinuous (Unstable)

The condition in which the relevant parameters do not attain stabilization.

96.02.055 Condition, Instantaneous

The condition which exists at a specified point in time.

96.02.056 Condition, Intermittent

The condition in which periods of use are separated by periods of rest (either stopped or idling).

96.02.057 Condition, Limiting

The limiting condition is indicated by the minimum or maximum values of various factors which permit the unit to operate in extreme cases. The other effective factors and the duration of load being precisely defined. Limiting conditions are indicated: q min, 1 max, etc.

96.02.058 Condition, Operating

The operating condition is indicated by the numerical values of the various factors relating to any given specific application of a unit. These factors may vary during the course of operations.

96.02.059 Condition, Rated

The steady state condition for which a component or system is recommended as a result of specified testing. The "rated characteristics" are in general shown in catalogs and are indicated qn, Pn, etc.

96.02.060 Condition, Specified

The condition required to be met in service.

96.02.061 Condition, Steady State

The condition in which relevant variable parameters do not change appreciably after a period for stabilization.

96.02.067 Cooling System

Means whereby unwanted heat is removed from the working fluid or from components.

96.02.068 Cooling System Air

A cooling system in which air is the transfer medium.

96.02.069 Cooling System Refrigeration

A cooling system which uses refrigeration techniques; i.e., uses a liquid which, when caused to assume the vapour state, absorbs heat at a high rate during the change of state.

96.02.070 Cooling System Water

A cooling system in which water is the transfer medium by which heat is removed. Ultimately the water is air cooled. Note: Other coolant liquids can be used but these are unusual.

96.02.150 Installation

The arrangement of a fluid power system in relation to its associated machinery and site.

96.02.151 Installation, Fixed

Installation in a permanent immoveable location.

96.02.152 Installation, Integral

Installation in which major components of the fluid power system form part of the structure of the machine.

96.02.153 Installation, Mobile

Installation on a machine which moves while operating.

96.02.154 Installation, Pipe

Arrangement of the fluid pipe lines (conductors) in the fluid power system and its associated machine.

96.02.155 Installation, Portable

Installation in equipment which may be moved between working locations but which does not operate while being moved.

96.02.600 Prime Mover

The device which serves as the source of mechanical power for the fluid power system; i.e., that which drives the pump or compressor (electric motor, internal combustion engine).

96.02.650 Specification, Performance

Document detailing the functional requirements of a piece of equipment, complete system or installation and when applicable the environmental conditions and any peculiar or specific conditions to which it will be subject. Sufficient detail being given to enable the manufacturer to select the correct materials and components to meet the requirement, and conversely to enable the customer to determine whether the products offered are to his satisfaction.

96.02.651 Specification, System

Document detailing the materials, functional performance and standard of a piece of equipment or complete system or installation to meet the performance specification. The document will contain sufficient details to enable the manufacturer to select and determine the material and components necessary to fulfill the requirements of the customer, and conversely to allow the customer to determine whether the products offered are to his satisfaction.

96.02.660 System Bleeding

Removing pockets of air trapped in the circuit. Usually carried out at low pressure by means of small valves or connections at high points. Removing water in the air curcuit with separator drain-valve.

96.02.661 System Cleaning

The removal of contaminant from all fluid passages and internal spaces to which the working fluid has access.

96.02.662 System Draining

The drawing off of fluid from a system. Usually via a shut-off valve suitably positioned.

96.02.663 System Filling

The act of filling the system with the specified amount of fluid.

96.02.664 System Flushing

Operating the system (often containing a special cleaning fluid or flushing oil) at low pressure to clean the inner passages and cavities in the circuit. The cleaning fluid or flushing oil is replaced by the correct working fluid before the system is put into normal service.

96.02.665 System Maintenance

The process of inspecting or checking a unit or system to ensure it is functioning correctly, or to determine why it is not operating correctly. Repairing or renewing defective or damaged components, adjusting settings and controls to maintain the required or specified performance.

96.02.666 System Startup

The specified sequence of operations for starting a unit or system for the first time or re-starting after maintenance, repair, or long period of shut-down. Includes functional verification.

96.02.667 System Testing

The act of testing the unit or system to ensure that output functions make the correct response to input signals.

SECTION 03 — DIAGRAMS

96.03.125 Diagram, Attached Symbols

A diagram in which all functions and connections to component symbols are shown in the symbols.

96.03.200 Diagram, Combination

A drawing utilizing a combination of graphical, cut-away and pictorial symbols showing interconnected lines.

96.03.300 Diagram Cutaway

A drawing showing principal internal parts of all components, controls and actuating mechanisms, all interconnecting lines and functions of individual components.

96.03.310 Diagram, Detached Symbols

A diagram in which various functions and connections of component symbols are shown by separate symbols in various places on the diagram.

96.03.320 Diagram, Detail Logic

A diagram which depicts all logic functions of a circuit including identification and description of logic components, ports, and connecting flow paths.

96.03.350 Diagram, Fluid Power

A drawing which illustrates pertinent characteristics, element, positions, sizes, interconnection, controls and actuation of components and fluid power circuits.

96.03.375 Diagram, Function

Graphical representation of the sequence of operation and control signals of a fluid power circuit, generally for a complete cycle.

96.03.400 Diagram, Graphic (Schematic)

A drawing using graphic symbols and interconnecting lines, generally drawn according to a standard or other code.

96.03.450 Diagram, Ladder

A diagram in which inputs are located to the left in a vertical column and outputs in a right vertical column. Interconnecting horizontal flow paths and components give the diagram a ladder appearance.

96.03.460 Diagram, Logic Control

A diagram which depicts logic control of power controls and interfaces.

96.03.500　　　Diagram, Pictorial

A drawing showing each component in its actual shape according to the manufacturer's installation.

96.03.550　　　Diagram Power Control

A diagram which depicts all powered devices and interfaces including identification and description of components and their effect on the system.

96.03.600　　　Diagram, Pressure-Time

A graphical presentation of pressure plotted against time for a complete cycle.

SECTION 04 — CIRCUITS

96.04.100　　　Circuit

An arrangement of interconnected components and parts.

96.04.110　　　Circuit, Closed

Circuit in which return fluid is directed to the pump inlet.

96.04.120　　　Circuit, Logic Control

A circuit which gathers and processes information to signal power controls and interfaces.

96.04.130　　　Circuit, Meter-In

A speed control circuit in which the control is achieved by regulating the supply flow to the actuator.

96.04.135　　　Circuit, Meter-Out

A speed control circuit in which the control is achieved by regulating the exhaust flow from the actuator.

96.04.140　　　Circuit, Open

A circuit in which return fluid is directed to the reservoir before recirculation.

96.04.150　　　Circuit, Pilot

A circuit used to control a main circuit or component.

96.04.160　　　Circuit, Power Control

A circuit which directs and regulates fluid power to working devices.

96.04.200　　　Circuit, Pressure Control

Any circuit whose main purpose is to adjust or regulate fluid pressure in the system or any branch of the system.

96.04.250　　　Circuit, Regenerative

A circuit in which pressurized fluid discharged from a component is returned to the system to reduce power input requirements.

96.04.300　　　Circuit, Safety

A circuit which prevents accidental operation, protects against overloads, or otherwise assures safe operation.

96.04.350　　　Circuit, Sequence

A circuit which establishes the order in which two or more phases of a circuit occur.

96.04.400　　　Circuit, Servo

A circuit which is controlled by automatic feedback, i.e., the output of the system is sensed or measured and is compared with the input signal. The difference (error) between the actual output and the input controls the circuit. The controls attempt to minimize the error. The system output may be position, velocity, force, pressure, level, flow rate, or temperature.

96.04.450　　　Circuit, Speed Control

Any circuit where components are arranged to regulate speed of operation.

96.04.500　　　Circuit, Synchronizing

A circuit in which multiple operations are controlled to occur at the same time.

96.04.550　　　Circuit, Unloading

A circuit in which pump volume is returned to reservoir at near zero gage pressure whenever delivery to the system is not required.

GROUP 99 — STANDARDS & CODES

　　Section 01 — General
　　Section 02 — Standards
　　Section 03 — Codes

SECTION 02 — STANDARDS

99.02.600　　　Standard

A document, or an object for physical comparison, for defining product characteristics, products, or processes: prepared by a consensus of a properly constituted group of those substantially affected and having the qualifications to prepare the standard for voluntary use.

ALPHABETICAL INDEX

A	ISO 5598	ANSI
Abrasion		30.04.050
Absolute Filtration Rating (Largest Particle Passed)	5.8.4.2	33.11.010
Absorbent (Absorptive) Filter Medium		33.08.010
Accumulator	5.4	61.01.100
Accumulator, Gas Loaded, Transfer Type	5.4.3.2	61.01.401
Accumulator, Hydropneumatic	5.4.1; 5.4.3.1	61.02.400
Accumulator, Hydropneumatic, Bladder		61.02.420
Accumulator, Hydropneumatic, Diaphragm		61.02.440
Accumulator, Hydropneumatic, Non-Separator		61.02.460
Accumulator, Hydropneumatic Piston		61.02.480
Accumulator, Mechanical		61.02.800
Accumulator, Mechanical, Spring	5.4.2	61.02.840
Accumulator, Mechanical, Weighted	5.4.3	61.02.880
Actuator, Pneumatic/Hydraulic	3.7	81.00.000
Actuator, Rotary		85.01.000
Adaptor (Support Ring)		94.06.100
Adaptor, Female		94.06.125
Adaptor, Male		94.06.150
Adaptor, Pedestal		94.06.175
Additive	10.2.14	10.01.001
Adsorbent (Adsorptive)		33.08.030
Aftercooler		33.02.220
Afterfilter		37.05.050
Agglomerate, Contamination		30.06.030
Air		15.01.100
Air Bleeder	5.2.4.4	31.01.100
Air Breather		31.01.200
Air, Compressed (Pressure)		15.02.100
Air, Contamination of		15.01.102
Air, Dried		15.02.200
Air, Free		04.01.005
Air Inclusion, Filter		33.11.020
Air Inclusion (Quick Disconnect Couplings)		55.01.010
Air Motor		84.01.100
Air, Saturated		15.02.800
Air Standard	2.2.7	04.01.010
Ambient Temperature Range		37.04.125
Amplification	4.0.9	04.01.011
Amplification, Flow	4.0.9.3	04.01.012
Amplification, Power	4.0.9.1	04.01.013
Amplification, Pressure	4.0.9.2	04.01.014
Amplifier, Assembly (See: Fluidic)	8.3.1	72.02.100
Amplifier, Flow	8.3.3	72.02.111
Amplifier, Hydraulic Servovalve	4.7.2.3	73.03.200
Analog		72.01.100
AND 10050, Port		50.04.200
AND Device		70.04.100
Angle, Cylinder		81.05.030
Aniline Point		04.01.020
Annunciator (Indicator)	7.2	91.02.051
Annunciator, Flow	7.2.2	91.02.052
Annunciator, Pressure	7.2.1	91.02.053
Anti-Freeze	10.2.17	30.01.050

Term	Section	Number
Apparent Capacity (α)		33.11.025
Area, Head End, Effective, Cylinder	3.5.1.4	81.01.004
Area, Piston, Effective Cylinder	3.5.1.2	81.01.005
Area, Piston Rod, Cylinder	3.5.1.3	81.01.006
Area, Port	5.4.2.9	50.01.701
Aspect Ratio, Nozzle	4.6.2.23	72.01.105
Assurance Level		04.01.025
Attachment, Piston Rod	3.5.6	81.01.008
Automatic Count		30.06.060
Automatic Drain		37.05.100
Average, Outgoing Quality Limit		30.07.150

B

Term	Section	Number
Back Connected		50.07.200
Background Contamination		30.06.090
Back-Up Ring (See: Ring, Antiextrusion)		94.06.180
Baffle		33.05.010
Base, Filter		33.05.020
Bernoulli's Law		02.01.100
Bi-Directional		33.07.010
Bistable	4.6.1.3	04.01.028
Boil Point (Foam All Over) (Mass Bubble Point) (Open Bubble Point)		33.11.030
Bore, Cylinder	3.5.1.1	81.01.010
Bowl (Shell), Filter		33.05.030
Boyle's Law		02.01.200
Break-Away		55.01.020
Breathing Capacity		21.03.100
Bubble Point (First Bubble Point) (Initial Bubble Point)		33.11.040
Bulk Modulus	10.3.13	04.01.030
Burst Pressure		33.11.050
Burst Pressure Rating		33.11.060

C

Term	Section	Number
Cap, Filter		33.05.040
Cap (Back End) (Blind Head) (Rear End) (Rear Head), Cylinder	3.5.0.2	81.05.180
Capacitance, Fluid	4.6.1.17	72.01.250
Capacitive - Impedance (See: Impedance, Capacitive)		
Capacitor	5.3.1.3	21.01.010
Capacity, Breather	5.3.8	21.03.100
Capacity, Dirt	5.8.4.7	33.11.120
Capacity (Displacement)	3.0.2.1	04.01.043
Capacity, Compressed Air Dryer		37.04.190
Capacity, Cylinder	3.5.1.8	81.01.012
Capacity, Derived (Pump)	3.1.2.1.1	43.01.005
Capacity, Derived (Motor)	3.2.2.2.1.1	82.01.020
Capacity, Effective	3.0.2.1.1	04.01.035
Capacity, Extending, Cylinder	3.5.3.3.1	81.01.013
Capacity, Geometric	3.0.2.1.2	04.01.036
Capacity, Reservoir Breather	5.3.8	21.03.100
Capacity, Reservoir Expansion	5.3.6	21.03.800
Capacity, Reservoir Fluid	5.3.5	21.03.400
Capacity, Retracting, Cylinder	3.5.3.3.2	81.01.014
Cartridge, Filter, (See: Element)		33.05.080
Case (Shell), Filter		33.05.060

Term	Ref	Code
Cavitation	2.2.6.3	03.01.200
Center Tube (Core), Filter		33.09.010
Centrifuge Volume		30.06.120
Channel (Flow Path)		50.01.200
Channel, Control	4.6.2.19	50.01.201
Channel, Output	4.6.2.20	50.01.202
Characteristic, Switching	4.0.8.4	72.01.254
Charles' Law		02.01.300
Circuit	8.5	96.04.100
Circuit, Closed	8.5.11	96.04.110
Circuit, Logic Control		96.04.120
Circuit, Meter In	8.5.7.1	96.04.130
Circuit, Meter Out	8.5.7.2	96.04.135
Circuit, Open	8.5.10	96.04.140
Circuit, Pilot	8.5.1	96.04.150
Circuit, Power Control		96.04.160
Circuit, Pressure Control	8.5.2	96.04.200
Circuit, Regenerative	8.5.4	96.04.250
Circuit, Safety	8.5.3	96.04.300
Circuit, Sequence	8.5.5	96.04.350
Circuit, Servo	8.5.6	96.04.400
Circuit, Speed Control	8.5.7	96.04.450
Circuit, Synchronizing	8.5.8	96.04.500
Circuit, Unloading	8.5.9	96.04.550
Clamp, Pipe (See: Pipe Clamp)		
Clean Fluid		30.07.250
Cleanability		33.11.070
Cleanable	5.8.2.1	33.07.020
Cleanliness Level		30.01.200
Cleanliness, Roll Off		30.01.201
Clevis (Hinge) (Pendulum), Cylinder		81.05.210
Clogging (See Silting)		30.04.500
Coanda Effect	4.6.2.17	03.01.210
Collapse Pressure		33.11.080
Collapse Pressure Rating		33.11.090
Collector (Receiver)	4.6.2.21	72.01.255
Combination (Composite)		33.08.050
Commissioning	9.2.1	96.01.090
Compartment		96.01.100
Compatibility, Seal	10.3.5	94.01.020
Compressed Air Dryer	5.5.7	37.01.100
Compressed Air Dryer, Automatic		37.02.000
Compressed Air Dryer, Demand Cycle		37.02.013
Compressed Air Dryer, Desiccant		37.02.016
Compressed Air Dryer, Deliquescent		37.02.010
Compressed Air Dryer, Fixed Cycle		37.02.400
Compressed Air Dryer, Manual		37.02.500
Compressed Air Dryer, Refrigerated	5.7.1.1	37.02.600
Compressed Air Dryer, Refrigerated, Cycling		37.02.620
Compressed Air Dryer, Refrigerated, Non-Cycling		37.02.660
Compressed Air Dryer, Regenerative		37.02.700
Compressed Air Dryer, Regenerative, Closed System		37.02.710
Compressed Air Dryer, Regenerative, Dual Tower		37.02.720
Compressed Air Dryer, Regenerative, Externally (Connected) Heated		37.02.730
Compressed Air Dryer, Regenerative, Internally Heated		37.02.740
Compressed Air Dryer, Regenerative, Open System		37.02.750

ANSI/B93.2

Term	Ref	Number
Compressed Air Dryer, Semi-Automatic		37.02.800
Compressibility		04.01.040
Compressor		44.01.100
Compressor, Installation	8.1.2	44.01.101
Compressor, Multiple Stage		44.02.300
Compressor, Single Stage		44.02.600
Compressor, Unloading Device	8.1.2.1	44.01.800
Condensation		37.04.200
Condensing Unit		37.05.200
Condition, Acceptable	2.1.12	96.02.050
Condition, Actual	2.1.7	96.02.051
Condition, Continuous Working	2.1.3	96.02.052
Condition, Cyclic Stabilized	2.1.9	96.02.053
Condition Discontinuous (Unstable)	2.1.10	96.02.054
Condition, Instantaneous	2.1.6	96.02.055
Condition, Intermittent	2.1.11	96.02.056
Condition, Limiting	2.1.4	96.02.057
Condition, Operating	2.1.1	96.02.058
Condition, Rated	2.1.2	96.02.059
Condition, Specified	2.1.8	96.02.060
Condition, Steady State	2.1.5	96.02.061
Conditioner, Air	8.4	30.01.202
Conductance, Fluid	4.6.1.13	72.01.257
Conductor		50.02.100
Conduit		50.02.200
Confidence Level		04.01.041
Connect Under Pressure		55.01.030
Connection (See Fitting or Coupling)	5.2.2	51.01.100
Connection, Flange	5.2.2.7	51.02.450
Connection, Rotary	5.2.2.9	56.02.300
Connection, Spherical	5.2.2.13	56.02.598
Connection, Swivel	5.2.2.10	56.02.600
Connection, Telescopic	5.2.2.11	56.02.602
Connector, Electrical		96.01.150
Consecutive Exceptance Number (N)		30.07.300
Contact Time		37.04.250
Contaminant, (See: Contamination)		30.02.100
Contaminant, Artificial		30.02.200
Contaminant, Dissolved Air	10.4.1.5	30.03.050
Contaminant, Dissolved Water		30.03.100
Contaminant, Effective Particle Diameter		30.05.100
Contaminant, Entrained Air	10.4.1.6	30.03.150
Contaminant, Environmental	10.4.1.2	30.02.400
Contaminant, Fiber		30.05.200
Contaminant, Free Air		30.03.200
Contaminant, Free Water	10.4.1.7	30.03.250
Contaminant, Gel		30.03.300
Contaminant, Generated	10.4.1.3	30.02.500
Contaminant, Incompatible Fluids		10.03.450
Contaminant, Liquid	5.5.1.2	30.03.390
Contaminant, Longest Dimension		30.05.300
Contaminant, Magnetic		30.03.400
Contaminant, Mass Index		30.05.390
Contaminant, Micro-Biological		30.03.450
Contaminant, Micrometre (Micron)* *Deprecated		30.05.400
Contaminant, Particle		30.05.500
Contaminant, Second Longest Dimension		30.05.550

Contaminant, Silt	10.4.1.9	30.03.500
Contaminant, Slime		30.03.550
Contaminant, Sludge		30.03.600
Contaminant, Solid	5.5.1.1	30.03.648
Contaminant, Vapor	5.5.1.3	30.03.649
Contaminant, Varnish		30.03.650
Contamination	10.4	30.02.100
Contamination, Abrasion		30.04.050
Contamination, Agglomerate of		30.06.030
Contamination, Automatic Counting of	10.4.2.3	30.06.060
Contamination, Background		30.06.090
Contamination, Centrifuge Volume		30.06.120
Contamination, Classes of	10.4.2.8	30.02.000
Contamination, Corrosion		30.04.100
Contamination, Counting Calibration Factor		30.06.150
Contamination, Decomposition		30.04.150
Contamination, Erosion		30.04.200
Contamination, Filterable Solids		30.06.180
Contamination, Fluid Breakdown		30.04.250
Contamination, Fretting		30.04.300
Contamination, Fretting Corrosion		30.04.350
Contamination, Globe and Circle Reticule		30.06.210
Contamination, Gravimetric Value of		30.06.240
Contamination, Initial	10.4.1.1	30.03.355
Contamination, Level of	10.4.2.7	30.01.300
Contamination, Linear Scale Reticule		30.06.270
Contamination, Mass Index of	10.4.2.1	30.05.390
Contamination, Microscope Gating		30.06.330
Contamination, Microscope Reticule (Microscope Graticule)		30.06.360
Contamination, Microscope Split Image Eyepiece		30.06.390
Contamination, Microscopic		30.06.420
Contamination, Microscopic Filar Eyepiece		30.06.300
Contamination, Multi-Pass Test		30.06.430
Contamination, Non-Combustible Residue		30.06.450
Contamination, Non-Volatile Residue		30.06.480
Contamination, Optical Density		30.06.510
Contamination, Particle Count Blank		30.06.540
Contamination, Oxidation		30.04.400
Contamination, Particle Size Distribution of	10.4.2.6	30.06.570
Contamination, Patch Test		30.06.600
Contamination, Precipitate		30.06.630
Contamination, Raw Count of	10.4.2.4	30.06.660
Contamination, Scoring		30.04.450
Contamination, Scoring Size, Particle		30.06.690
Contamination, Silting	10.4.1.9	30.04.500
Contamination, Total Statistical Count of	10.4.2.5	30.06.720
Contamination, Visual Count of	10.4.2.2	30.06.730
Continuity Equation		02.01.400
Control		70.01.100
Control, Automatic	6.0	70.02.105
Control, Auxiliary	6.8	70.02.106
Control Combination (Combined)	6.6	70.02.110
Control, Cylinder	6.4.3	70.02.115
Control, Detent	6.1.2	70.02.117
Control, Electric	6.5	70.02.120
Control, Emergency	6.8.2	70.02.121
Control, Feedback	6.7.1	70.02.122
Control, Feedback, Mechanical	6.7.1.1	70.02.123

ANSI/B93.2

Term	Section	Number
Control, Feedback, Hydraulic	6.7.1.2	70.02.124
Control, Feedback Electric	6.7.1.3	70.02.125
Control, Feedback Pneumatic	6.7.1.4	70.02.126
Control, Force Motor	6.5.3	70.02.127
Control, Hydraulic	6.4.5	70.02.130
Control, Impulse Generator	7.4.1.2	70.02.133
Control, Latch	6.1.3	70.02.134
Control, Lever	6.2.2	70.02.135
Control		70.01.100
Control, Linkage	6.1.5	70.02.136
Control, Liquid Level		70.02.137
Control, Manual	6.2	70.02.139
Control, Mechanical	6.3	70.02.140
Control, One-Way Trip	6.3.3.4	70.02.141
Control, Over Center	6.1.4	70.02.142
Control, Override	6.8.1	70.02.143
Control, Pedal	6.2.3	70.02.144
Control, Plunger	6.3.1	70.02.145
Control, Pneumatic	6.4.4	70.02.146
Control, Pneumatic Programmer	7.4.2	70.02.147
Control, Pneumatic Programmer Cyclic	8.3.2	70.02.148
Control, Pneumatic Time Delay	7.4.1	70.02.149
Control, Pressure	6.4	70.02.150
Control, Pressure Compensated		70.02.151
Control, Pressure, Direct	6.4.1	70.02.152
Control, Pressure, Indirect	6.4.2	70.02.153
Control, Pressure Pulse Generator	7.4.1.1	70.02.154
Control, Pump		70.02.155
Control, Push-Pull Button	6.2.1	70.02.156
Control, Roller	6.3.3	70.02.157
Control, Roller Lever	6.3.3.1	70.02.158
Control, Roller Plunger	6.3.3.2	70.02.159
Control, Roller Rocker	6.3.3.3	70.02.160
Control, Rotating Shaft	6.1.1	70.02.161
Control, Servo	6.7	70.02.162
Control, Solenoid, Single Acting (One-Way)	6.5.1	70.02.163
Control, Solenoid, Double Acting (Two-Way)	6.5.2	70.02.164
Control, Spring Return (Offset)	6.3.2	70.02.165
Control, Stepping Motor	6.5.5	70.02.166
Control, Torque Motor	6.5.4	70.02.167
Control, Tracer	6.5.6	70.02.168
Control, Treadle	6.2.4	70.02.169
Controller	7.4.5	70.01.101
Controller Temperature		31.01.700
Cooling System	9.1.5	96.02.067
Cooling System, Air	9.1.5.2	96.02.068
Cooling System, Refrigeration	9.1.5.4	96.02.069
Cooling System, Water	9.1.5.1	96.02.070
Cooler	5.6.1	32.01.050
Corrosion		30.04.100
Counting Calibration Factor		30.06.150
Coupling (See Connection or Fitting)		
Coupling, One Valve		55.02.400

Term	Section	Number
Coupling, Quick Disconnect Un-Valved* *Deprecated		55.02.600
Coupling, Quick Disconnect, Valved		55.02.800
Coupling, Quick Release	5.2.2.8	55.01.100
Coupling, Quick Release, Breakaway	5.2.2.8.2	55.02.200
Coupling, Quick Release, Claw Type	5.2.2.8.1	55.02.220
Coupling, Self Sealing	5.2.2.12	55.02.800
Cover, Filter		33.05.070
Crest, Filter		33.09.040
Cushion		63.01.100
Cushion, Cylinder	3.5.0.5	63.02.210
Cushion, Die		63.02.220
Cushion, Hydraulic		63.02.230
Cushion, Hydropneumatic		63.02.240
Cushion, Pressure, Damping		63.01.150
Cushion, Pneumatic		63.02.250
Cycle	2.4.1	96.01.200
Cycle, Adsorbent Dryer		37.04.280
Cycle, Automatic	2.4.1.1	96.01.210
Cycle, Manual	2.4.1.2	96.01.220
Cycle, Semi-Automatic	2.4.1.3	96.01.230
Cycle, Speed of	2.4.1.5	96.01.231
Cycle, Speed Maximum	2.4.1.6	96.01.232
Cycle, Working	2.4.1.4	96.01.233
Cylinder	3.5	81.01.000
Cylinder, Adjustable Stroke		81.02.050
Cylinder, Angle		81.05.030
Cylinder, Bolted		81.02.075
Cylinder, Area, Headend Effective	3.5.1.4	81.01.004
Cylinder, Area, Piston, Effective	3.5.1.2	81.01.005
Cylinder, Area, Piston Rod	3.5.1.3	81.01.006
Cylinder, Bore	3.5.1.1	81.01.010
Cylinder, Cap (Back End) (Blind End) (Blind Head) (Rear End) (Rear Head)	3.5.0.2	81.05.180
Cylinder, Capacity	3.5.1.8	81.01.012
Cylinder, Capacity, Extending	3.5.3.3.1	81.01.013
Cylinder, Capacity, Retracting	3.5.3.3.2	81.01.014
Cylinder, Clevis (Hinge) (Pendulum)		81.05.210
Cylinder, Cushioned		81.02.100
Cylinder, Dashpot	3.8	81.02.101
Cylinder, Diaphragm Type	3.5.4.2	81.02.102
Cylinder, Differential	3.5.4.7	81.02.104
Cylinder, Double Acting	3.5.4.4	81.02.150
Cylinder, Double Rod	3.5.4.6	81.02.200
Cylinder, Dual Stroke		81.02.250
Cylinder, Duplex	3.5.4.8	81.02.252
Cylinder Efficiency, Overall	3.5.3.7	81.05.280
Cylinder Efficiency, Speed	3.5.3.7.2	81.05.281
Cylinder Efficiency, Thrust	3.5.3.7.1	81.05.282
Cylinder End		81.05.240
Cylinder Eye (Hinge) (Pendulum)		81.05.270
Cylinder Flange		81.05.300
Cylinder Force, Piston Rod		81.01.200
Cylinder Force, Actual	3.5.3.2.3	81.01.201
Cylinder Force, Nominal	3.5.3.2.2	81.01.202

ANSI/B93.2

Term	Section	Number
Cylinder Force, Theoretical	3.5.3.2.1	81.01.203
Cylinder Head (Front End) (Front Face) (Front Head) (Rod Head)	3.5.0.1	81.05.450
Cylinder Input Power	3.5.3.6.1	81.01.600
Cylinder Lug (Foot)		81.05.510
Cylinder Mounting	3.5.5	81.15.000
Cylinder Mounting, Centerline		81.15.100
Cylinder Mounting Centerline, Lug	3.5.5.3.2	81.15.150
Cylinder Mounting End		81.15.200
Cylinder Mounting, Both Ends		81.15.204
Cylinder Mounting, Both Ends, Tie Rods Extended	3.5.5.2.4	81.15.203
Cylinder Mounting, Cap End		81.15.212
Cylinder Mounting, Cap End, Circular		81.15.216
Cylinder Mounting, Cap End, Circular Flange		81.15.220
Cylinder Mounting, Cap End, Detachable Clevis		81.15.224
Cylinder Mounting, Cap End, Detachable Eye		81.15.228
Cylinder Mounting, Cap End, Fixed Clevis		81.15.232
Cylinder Mounting, Cap End, Fixed Eye	3.5.5.2.1	81.15.236
Cylinder Mounting, Cap End, Flange	3.5.5.2.2	81.15.239
Cylinder Mounting, Cap End, Rectangular Flange	3.5.5.2	81.15.240
Cylinder Mounting, Cap, End Spherical	3.5.5.3.4	81.15.242
Cylinder Mounting, Cap End, Square		81.15.244
Cylinder Mounting, Cap End, Square Flange		81.15.248
Cylinder Mounting, Cap End, Threaded	3.5.5.2.3	81.15.250
Cylinder Mounting, Cap End, Tie Rods Extended		81.15.252
Cylinder Mounting, Cap End, Trunnion		81.15.256
Cylinder Mounting, Head End		81.15.260
Cylinder Mounting, Head End, Circular		81.15.264
Cylinder Mounting, Head End, Circular Flange		81.15.268
Cylinder Mounting, Head End, Female Rabbet		81.15.272
Cylinder Mounting, Head End, Male Rabbet		81.15.276
Cylinder Mounting, Head End, Rectangular Flange		81.15.280
Cylinder Mounting, Head End, Square		81.15.284
Cylinder Mounting, Head End, Square Flange		81.15.288
Cylinder Mounting, Head End, Tie Rods Extended		81.15.292
Cylinder Mounting, Head End, Threaded	3.5.5.2.5	81.15.290
Cylinder Mounting, Head End, Trunnion		81.15.296
Cylinder Mounting, Fixed		81.15.300
Cylinder Mounting, Intermediate		81.15.400
Cylinder Mounting, Intermediate, Fixed Trunnion		81.15.425
Cylinder Mounting, Intermediate, Moveable Trunnions		81.15.450
Cylinder Mounting, Intermediate, Side Trunnions	3.5.5.3.3	81.15.460
Cylinder Mounting, Clevis	3.5.5.3.1	81.15.232
Cylinder Mounting, Pivot	3.5.5.3	81.15.500
Cylinder Mounting, Side	3.5.5.1	81.15.600
Cylinder Mounting, Side, End Angles	3.5.5.1.1	81.15.615
Cylinder Mounting, Side, End Lugs	3.5.5.1.3	81.15.630
Cylinder Mounting, Side, End Plates		81.15.645
Cylinder Mounting, Side Lugs		81.15.660

Cylinder Mounting, Side, Tapped	3.5.5.1.2	81.15.675
Cylinder Mounting, Side Through Holes		81.15.690
Cylinder Mounting, Side, Trunnion	3.5.5.3.3	81.15.410
Cylinder Mounting, Universal		81.15.700
Cylinder, Multiposition	3.5.4.9	81.02.254
Cylinder, Non-Rotating		81.02.290
Cylinder Output Power	3.5.3.6.2	81.01.601
Cylinder, Piston	3.5.4.1	81.02.300
Cylinder, Plunger (Ram)		81.02.350
Cylinder Pressure	3.5.3.1.1	81.01.610
Cylinder Rabbet		81.05.570
Cylinder, Retractable Stroke		81.02.450
Cylinder, Rotating		81.02.460
Cylinder Side (Base)		81.05.720
Cylinder, Single Acting	3.5.4.3	81.02.500
Cylinder, Single Acting, Gravity Return	3.5.4.3.2	81.02.501
Cylinder, Single Rod	3.5.4.5	81.02.550
Cylinder, Spring Return	3.5.4.3.1	81.02.600
Cylinder Stroke, Extend (Out)	3.5.0.3	81.01.700
Cylinder Stroke, Full	3.5.1.6	81.01.701
Cylinder Stroke, Retract	3.5.0.4	81.01.702
Cylinder Stroke, Working	3.5.1.7	81.01.703
Cylinder, Tandem	3.5.4.10	81.02.650
Cylinder, Telescoping	3.5.4.11	81.02.700
Cylinder, Tie Rod		81.05.750
Cylinder Time, Extend Stroke	3.5.3.6.1	81.01.800
Cylinder Trunnion		81.05.780

D

Damper, Gage Pulsation	7.4.3	91.01.300
Damping Parameter		72.06.050
Darcy's Formula		02.01.500
Dashpot, Cylinder	3.8	81.02.800
Deaerator	5.8.6	34.01.050
Decay		71.07.200
Decay Rate		71.07.300
Decomposition		30.04.150
Decontamination		30.01.400
Demulsibility, Water	10.3.14	11.01.050
Deposited		33.08.070
Depth		33.08.080
Desiccant		37.05.250
Desiccant, Absorbent (Deliquescent)		37.05.255
Desiccant, Adsorbent		37.05.260
Desiccant, Tabular Support		37.05.280
Dew Point		37.04.300
Dew Point, Atmospheric		37.04.310
Dew Point, Pressure		37.04.340
Dew Point, Depression		37.04.350
Diagram, Attached Symbols		96.03.125
Diagram, Combination	2.3.2.1	96.03.200
Diagram, Cutaway	2.3.2.2	96.03.300
Diagram, Detached Symbols		96.03.310
Diagram, Detail Logic		96.03.320
Diagram, Fluid Power	2.3.2	96.03.350
Diagram, Function	2.3.2.7	96.03.375

ANSI/B93.2

Term	Section	Number
Diagram, Graphic (Schematic)	2.3.2.3	96.03.400
Diagram, Ladder		96.03.450
Diagram, Logic Control		96.03.460
Diagram, Pictorial	2.3.2.4	96.03.500
Diagram, Power Control		96.03.550
Diagram, Pressure-Time	2.3.2.6	96.03.600
Digital		72.01.300
Diode, Fluid	4.6.2.26	70.03.050
Dipstick	7.1.4.2	91.01.280
Direction of Rotation, (See: Rotation)		
Dirt Capacity (Dust Capacity) (Contaminant Capacity)	5.8.4.7	33.11.120
Displacement, Volumetric (See Capacity)		04.01.043
Dither	4.7.3.3	73.04.100
Drift	2.4.8	72.06.100
Droop		74.03.100
Dryer, Deliquescent	5.5.7.3	37.02.010
Dryer, Desiccant		37.02.016
Dryer, Regenerative		37.02.700
Dryer, Regenerative, Heat	5.5.7.4.1	37.02.725
Dryer, Regenerative, Heatless	5.5.7.4.2	37.02.745
Dryer, Refrigerant	5.5.7.1	37.02.600
Durometer Hardness		94.01.030

E

Term	Section	Number
Edge		33.08.090
Effective Area	5.8.4.1	33.11.140
Effective Particle Diameter, Contaminant		30.05.100
Efficiency	2.2.1	04.01.046
Efficiency, Curve, Filter	5.8.4.6	33.11.151
Efficiency, Filter	5.8.4.5	33.11.150
Efficiency, Hydro-Mechanical Motor	3.2.2.2.7.2	82.01.030
Efficiency, Mechanical, Pump	3.1.2.8.2	43.01.015
Efficiency, Overall, Cylinder	3.5.3.7	81.05.280
Efficiency, Overall, Motor	3.2.2.2.7.3	82.01.031
Efficiency, Overall, Pump	3.1.2.8.3	43.01.016
Efficiency, Speed, Cylinder	3.5.3.7.2	81.05.281
Efficiency, Thrust, Cylinder	3.5.3.7.1	81.05.282
Efficiency, Volumetric, Motor	3.2.2.2.7.1	82.01.032
Efficiency, Volumetric, Pump	3.1.2.8.1	43.01.017
Effluent		33.04.100
Element, Artificially Loaded Filter		33.10.010
Element, Bi-Directional Filter		33.10.015
Element, Bridging Filter		33.10.020
Element, Burst Filter	5.8.3.3	33.10.030
Element (Cartridge)		33.05.080
Element, Clean Filter		33.10.040
Element, Cleanable Filter	5.8.2.1	33.10.041
Element, Clogged Filter	5.8.3.1	33.10.045
Element, Clogging Filter	5.8.3	33.10.046
Element, Collapsed Filter	5.8.3.2	33.10.050
Element, Contaminated Filter		33.10.060
Element, Dirty Filter		33.10.070

Element, Disposable Filter	5.8.2.2	33.10.074
Element, Effective Filtration Area	5.8.4.1	33.10.076
Element, Extended Area Filter		33.10.078
Element, Fatigued Filter		33.10.080
Element, Full System, Differential Pressure Filter	5.8.2.4	33.10.084
Element, Inside-Out Flow Filter	5.8.2.6	33.10.086
Element, Loaded Filter (See: Element, Clogged)		33.10.090
Element, Magnetic Filter	5.8.2.5	33.10.092
Element Medium		33.09.090
Element, Modular Filter (Plug-In)		33.10.094
Element, Outside-In Flow Filter	5.8.2.7	33.10.096
Element, Pinched Pleat Filter		33.10.100
Element, Plain Filter		33.10.103
Element, Pleated Filter		33.10.106
Element, Primary Filter		33.10.109
Element Removal Clearance		33.04.200
Element, Renewable Filter	5.8.2.3	33.10.112
Element, Reserve Filter		33.10.115
Element, Ruptured Filter		33.10.120
Element, Secondary Filter		33.10.123
Element, Self-Cleaning Filter	5.8.2.1.1	33.10.126
Element, Service Loaded Filter		33.10.130
Element, Two Stage Filter	5.8.2.8	33.10.133
Element, Wash Filter		33.10.136
Emulsion		11.01.100
Emulsion, Oil in Water	10.2.2	11.02.525
Emulsion, Water in Oil	10.2.3	11.02.550
Enclosure		96.01.300
End, Cylinder		81.05.240
End, Cap, Cylinder		33.09.050
End, Cap, Filter		33.09.050
End Load, Filter Element		33.11.160
End Load Rating, Filter Element		33.11.170
End Seal, Filter		33.09.060
Envelope, Pressure Containing		01.01.350
Erosion		30.04.200
Etched		33.08.100
Excluder Bore		94.15.300
Excluder Cavity		94.15.320
Excluder Device		94.06.190
Expansion Device		37.05.400
Expectancy, Life	2.4.3	04.01.047
Extension, Pump Shaft	3.1.3.2	43.01.018
External Support, Filter		33.09.070
Evaporator		37.05.300
Evaporator Back-Pressure Valve		37.05.350
Evaporator Freeze-Up		37.04.400
Eye (Hinge) (Pendulum)		81.05.270

F

Fabrication Integrity		33.11.176
Fan, Cooling	9.1.5.3.2	32.01.070
Fan In Ratio	4.6.3.12	72.01.400
Fan Out Ratio Fluidic Control	4.6.3.11	72.01.500
Fiber, Contaminant		33.05.200
Filler, Ring		94.06.200
Filter	5.8	33.01.000

Filter, Absorbent (Absorptive)	33.08.010	Filter Sintered	33.08.130
Filter Accessories	33.06.000	Filter Surface	33.08.140
Filter Base	33.05.020	Filter Wound	33.08.150
Filter Bowl (Shell)	33.05.030	Filter Woven	33.08.160
Filter Adsorbent (Adsorptive)	33.08.030	Filterable Solids, Contamination	30.06.180
Filter Cap	33.05.040	Filtration Rating, Absolute 5.8.4.2	33.11.010
Filter Case (Shell)	33.05.060	Filtration Rating, Nominal 5.8.4.3	33.11.260
Filter Center Tube (Core)	33.09.010	Filtration Ratio ($\beta\mu$) 5.8.4.4	33.11.182
Filter, Centrifugal 5.8.1.7	33.02.015	Fire Point 10.3.11	04.01.048
Filter Combination (Composite)	33.08.050	Fitting (See Connection) ... 5.2.2	51.01.100
Filter Components	33.05.000	Fitting, Bushing	51.02.050
Filter Crest	33.09.040	Fitting, Cap	51.02.100
Filter Deposited	33.08.070	Fitting, Closure	51.02.150
Filter Depth	33.08.080	Fitting, Compression 5.2.2.4	51.02.200
Filter, Duplex 5.8.1.4	33.02.040	Fitting, Connector	51.02.250
Filter Edge	33.08.090	Fitting, Coupling (See Coupling)	51.02.300
Filter Element (Cartridge) 5.8.2	33.05.080	Fitting, Cross 5.2.2.14	51.02.350
Filter End Cap	33.09.050	Fitting, Elbow 5.2.2.15	51.02.400
Filter End Seal	33.09.060	Fitting, Female Thread 5.2.2.2	51.02.440
Filter Etched	33.08.100	Fitting, Flange 5.2.2.7	51.02.450
Filter External Support	33.09.070	Fitting, Flared 5.2.2.5	51.02.500
Filter Grooving	33.09.080	Fitting, Flared AN	51.02.550
Filter Head	33.05.090	Fitting, Flareless	51.02.620
Filter Housing	33.05.100	Fitting, Male Thread 5.2.2.1	51.02.600
Filter Magnetic Plug	33.06.300	Fitting, Plug	51.02.750
Filter Medium	33.09.090	Fitting, Plug, Dryseal	51.02.760
Filter Non-Woven	33.08.110	Fitting, Plug, Short Pipe Thread	51.02.770
Filter Outer Wrapper	33.09.110	Fitting, Plug, Standard Pipe Thread	51.02.780
Filter Performance	33.11.000	Fitting, Plug, Straight Thread	51.02.790
Filter Pleats (Corrugations)	33.09.110	Fitting, Pneumatic 5.2.2	51.02.795
Filter Precoat	33.08.120		
Filter Root	33.09.120		
Filter Side Seal	33.09.130		

Term	Section	Number
Fitting, Reducer	5.2.2.16	51.02.800
Fitting, Reusable Hose		51.02.650
Fitting, Tailpiece	5.2.2.6	51.02.840
Fitting, Tee	5.2.2.17	51.02.850
Fitting, Threaded Union	5.2.2.3	51.02.860
Fitting, Union	5.2.2.18	51.02.900
Fitting, Welded		51.02.700
Fitting, Wye (Y)	5.2.2.19	51.02.950
Flange, Cylinder		81.05.300
Flange, Hump Mount		43.03.100
Flapper Action, Valve		71.03.200
Flash Point	10.3.10	04.01.050
Flip Flop		70.04.400
Flip, Flop, Double Input		70.04.420
Flip, Flop, Single Input		70.04.440
Flow	2.2.6	03.01.409
Flow, Characteristic Curve, Regulator		74.03.200
Flow Degradation Ratio		04.01.051
Flow Factor	2.2.6.4.4	04.01.052
Flow Fatigue		33.11.190
Flow Input, Motor	3.2.2.2.3	82.01.040
Flow Input, Derived, Motor	3.2.2.2.3.2	82.01.041
Flow Input, Effective, Motor	3.2.2.2.3.3	82.01.042
Flow, Input, Geometric, Motor	3.2.2.2.3.1	82.01.043
Flow, Laminar	2.2.6.1	03.01.410
Flow, Metered		03.01.420
Flow, Minimum Control	4.0.10	72.06.120
Flow, Output, Pump	3.1.2.3	43.01.020
Flow, Output, Derived, Pump	3.1.2.3.2	43.01.021
Flow, Output, Effective, Pump	3.1.2.3.3	43.01.022
Flow, Output, Geometric, Pump	3.1.2.3.1	43.01.023
Flow Passage, Controlled		50.01.610
Flow Path		50.01.625
Flow Path (Gallery), Valves	5.2.3	71.08.200
Flow Rate	2.2.6.4	04.01.060
Flow Rate, Relief	2.2.6.4.3	04.01.061
Flow, Steady State		03.01.430
Flow, Supply Port	2.2.6.4.2	04.01.063
Flow, Turbulent	2.2.6.2	03.01.450
Flow, Unsteady		03.01.460
Flowlines	5.2.1	50.02.100
Flow/Pressure Characteristic, Fluidic	4.0.14	72.06.121
Fluid		10.01.100
Fluid Aeration, Foam	10.3.8	10.04.010
Fluid, Air Release	10.3.4	10.04.011
Fluid, Anti-Corrosive	10.3.22	10.04.012
Fluid, Aqueous	10.2.1	10.02.010
Fluid, Ash Content	10.3.16	10.04.013
Fluid, Auto Ignition Temperature		10.04.016
Fluid Breakdown		30.04.250
Fluid Capacity		21.03.400
Fluid, Chlorinated Hydrocarbon	10.2.6	13.02.090
Fluid Conditioning		30.01.500
Fluid Density	10.3.3	10.04.100
Fluid, Di-Basic Ester	10.2.7	13.02.095
Fluid, Evaporation Deposits	10.3.20	10.04.110
Fluid, Fatty Oil		10.02.100

Fluid, Fire Resistant	10.1.2	10.02.200
Fluid, Friction		04.01.070
Fluid, Halogenated	10.2.8	13.02.100
Fluid, Hydraulic	10.0.1	10.02.300
Fluid Logic	4.6.1.4	01.01.080
Fluid Memory, Off Return		70.01.860
Fluid Memory, Retentive		70.01.870
Fluid Miscibility	10.3.18	10.04.200
Fluid, Newtonian	10.0.2	10.02.350
Fluid, Oil and Water Emulsion		11.01.100
Fluid, Oil-in-Water Emulsion	10.2.2	11.02.525
Fluid, Organic Ester		13.02.200
Fluid, Petroleum	10.1.1	12.01.100
Fluid, Phosphate Ester	10.2.9	13.02.300
Fluid, Phosphate Ester-Base	10.2.10	13.02.400
Fluid, Pneumatic		10.02.400
Fluid Polyglycol	10.2.4	13.02.500
Fluid, Polyglycol Ester	10.2.11	13.02.550
Fluid Power	2.0.0	01.01.100
Fluid Power System (Circuit)	8.0.1	01.01.200
Fluid, Rust Protection	10.3.17	10.04.270
Fluid Sampling, Dynamic		30.07.400
Fluid Sampling, Static		30.07.450
Fluid Signal		70.01.900
Fluid Signal, Maintained		70.01.908
Fluid Signal, Momentary		70.01.910
Fluid Signal, Timed		70.01.914
Fluid, Silicate Ester	10.2.13	13.02.600
Fluid, Silicone	10.2.12	13.02.700
Fluid Stability	10.3.23	10.03.000
Fluid Stability, Chemical		10.03.100
Fluid Stability, Emulsion	10.3.23.2	10.03.150
Fluid Stability, Hydrolytic		10.03.200
Fluid Stability, Oxidation	10.3.23.3	10.03.300
Fluid Stability, Shear	10.3.23.1	10.03.350
Fluid Stability, Thermal		10.03.400
Fluid, Synthetic	10.2.5	13.01.100
Fluid, Vapor Pressure	10.3.19	10.04.280
Fluid, Velocity		37.04.450
Fluid Water Content	10.3.2	10.04.290
Fluid, Waterglycol		14.01.100
Fluid, Water-In-Oil	10.2.3	11.02.550
Fluidic		72.01.600
Fluidic Amplification, Flow		72.03.200
Fluidic Amplification, Power		72.03.400
Fluidic Amplification, Pressure		72.03.600
Fluidic Amplifier (See: Amplifier)		72.02.100
Fluidic Amplifier, Analog	4.6.2.2	72.01.100
Fluidic Amplifier, Closed		72.02.110
Fluidic Amplifier, Digital	4.6.2.3	72.01.115
Fluidic Amplifier, Flow (See: Amplifier, Flow)		72.01.111
Fluidic Amplifier, Impact Modulator	4.6.2.7	72.02.120
Fluidic Amplifier, Momentum	4.6.2.5	72.02.122
Fluidic Amplifier, Open		72.02.125
Fluidic Amplifier, Stream Deflection		72.02.130
Fluidic Amplifier, Turbulence	4.6.2.8	72.02.135

Term	Section	Code
Fluidic Amplifier, Vortex	4.6.2.6	72.02.140
Fluidic Amplifier, Wall Attachment	4.6.2.9	72.02.145
Fluidic Device, Active	4.6.1.24	72.02.050
Fluidic Device, Interface	4.6.2.10	72.02.710
Fluidic Device, Passive	4.6.1.25	72.02.700
Fluidic Device, Power Supply (See: Power Supply)		
Fluidic Device, Rated Flow (See: Rated Flow)		
Fluidic Output Power	4.6.1.22	72.06.325
Fluidics	4.6.1.5	01.01.300
Fluidics, Digital	4.6.1.7	72.01.300
Foot, Pump Mount	3.1.3.1.3	43.03.120
Force, Piston Rod, Cylinder	3.5.3.2	81.01.200
Force, Actual, Cylinder	3.5.3.2.3	81.01.201
Force, Nominal, Cylinder	3.5.3.2.2	81.01.202
Force, Theoretical, Cylinder	3.5.3.2.1	81.01.203
Force Motor (See: Motor, Force)		73.03.100
Frequency Response	2.4.5	04.01.072
Fretting		30.04.300
Fretting Corrosion		30.04.350
Front Connected		50.07.400

G

Term	Section	Code
Gage (Gauge) (See: Instrument)		91.02.050
Gage, Bellows		91.02.100
Gage, Bourdon Tube		91.02.200
Gage, Damper	7.4.3	91.01.300
Gage, Diaphragm		91.02.300
Gage, Fluid Level	7.1.4.3	91.02.400
Gage, Manometer	7.1.1.2	91.02.500
Gage, Piston		91.02.600
Gage, Pitot Tube		91.02.650
Gage, Pressure	7.1.1.1	91.02.700
Gage, Protector	7.4.4	91.01.301
Gage, Reservoir Contents	7.1.4.4	91.02.730
Gage, Vacuum		91.02.800
Gain, Flow	4.6.3.7	72.03.200
Gain, Power	4.6.3.8	72.03.400
Gain, Pressure	4.6.3.6	72.03.600
Gallery (See: Flow Path, Valves)		
Gland		94.06.210
Gland Follower		94.06.250
Globe and Circle Reticule, Contamination		30.06.210
Gravimetric Value, Contamination		30.06.240
Grooving, Filter		33.09.080

H

Term	Section	Code
Hagen Poiseuille Law		02.01.600
Hammer, Liquid	2.2.4.9	04.01.170
Head	2.2.5	04.01.100
Head, Filter		33.05.090
Head, Friction	2.2.5.1	04.01.110
Head (Front End) (Front Face) (Front Head) (Rod Head)	3.5.0.1	81.05.450
Head, Pressure	2.2.5.6	04.01.115
Head, Static	2.2.5.2	04.01.120
Head, Static Discharge	2.2.5.3	04.01.130
Head, Static Suction	2.2.5.4	04.01.140
Head, Total Static (Suction)	2.2.5.5	04.01.150
Head, Velocity	2.2.5.7	04.01.160
Heat Exchanger	5.6	32.01.100

Heater	5.6.2	32.01.101
High Side Components		37.05.440
Hose	5.2.1.2	52.01.100
Hose, Wire Braided		52.02.800
Hot Gas By-Pass Valve		37.05.450
Housing, Filter		33.05.100
Hydraulic Amplifier (See Amplifier, Hydraulic Servovalve)		
Hydraulic Filter, By-Pass (Reserve)		33.02.010
Hydraulic Filter, Centrifugal	5.8.1.7	33.02.015
Hydraulic Filter, Disposable		33.02.020
Hydraulic Filter, Dual		33.02.030
Hydraulic Filter, Duplex	5.8.1.4	33.02.040
Hydraulic Filter, Fill Cap		33.02.050
Hydraulic Filter, Filtered By-Pass	5.8.1.3	33.02.060
Hydraulic Filter, Full Flow	5.8.1.1	33.02.070
Hydraulic Filter, Full Flow with By-Pass	5.8.1.2	33.02.075
Hydraulic Filter, In-Line		33.02.080
Hydraulic Filter, L-Type		33.02.090
Hydraulic Filter, Manifold		33.02.100
Hydraulic Filter, Modular		33.02.110
Hydraulic Filter, Partial Flow	5.8.1.10	33.02.120
Hydraulic Filter, Reservoir (Sump)		33.02.140
Hydraulic Filter, Spin-On	5.8.1.6	33.02.150
Hydraulic Filter, Strainer	5.8.1.9	33.02.160
Hydraulic Filter, T-Type		33.02.170
Hydraulic Filter, Two-Stage	5.8.1.5	33.02.180
Hydraulic Filter, Wash		33.02.190
Hydraulic Filter, Y-Type		33.02.200
Hydraulic Motor (See: Motor, Hydraulic)	3.2.2	82.01.000
Hydraulic Motor, Rated Speed		82.01.125
Hydraulic Motor, Speed Degradation Ration		82.01.150
Hydraulic Power		04.01.200
Hydraulic Pump (See: Pump, Hydraulic)	3.1.1	43.01.100
Hydraulics	2.0.1	01.01.400
Hydrodynamics	2.0.2	01.01.500
Hydrokinetics		01.01.600
Hydropneumatics	2.0.3	01.01.700
Hydrostatics	2.0.4	01.01.800
Hydrostatic Transmission		01.01.801
Hysteresis, Servovalve		73.05.300

I

Impedance, Capacitive, Fluid	4.6.1.15	72.01.650
Impedance, Fluid	4.6.1.12	72.01.651
Impedance, Fluidic Input	4.6.1.20	72.01.652
Impedance, Fluidic Output	4.6.1.23	72.01.653
Impedance, Inductive, Fluid	4.6.1.14	72.01.654
Impingement, Fluid Filter		33.04.300
Improver (See: Viscosity Index)		
Indicator		91.01.400
Indicator, Clogging (Bypass)	5.8.5.2	33.06.210
Indicator, Differential Pressure		33.06.220
Indicator, Filter		33.06.200
Indicator, Pressure		33.06.230

Inductance, Fluid	4.6.1.16	72.01.659
Inductive-Impedance (See: Impedance, Inductive)		
Influent		33.04.400
Inhibitor	10.2.16	10.01.150
Input, Magnetic	5.8.5.3	33.06.300
Inside-Out Flow		33.07.070
Inspection Ration (R)		30.07.500
Installation	9.1.	96.02.150
Installation, Fixed	9.1.1.1	96.02.151
Installation, Integral	9.1.1.4	96.02.152
Installation, Mobile	9.1.1.3	96.02.153
Installation, Pipe	9.1.3.1	96.02.154
Installation, Portable	9.1.1.2	96.02.155
Intensification, Ratio of	3.6.0.3	41.01.099
Intensifier,	3.6	41.01.100
Intensifier, Continuous	3.6.1.6	41.02.100
Intensifier, Double Acting	3.6.1.4	41.02.120
Intensifier, Dual Fluid	3.6.1.2	41.02.125
Intensifier, Primary Fluid	3.6.0.1	41.01.700
Intensifier, Secondary Fluid	3.6.0.2	41.02.130
Intensifier, Single Acting	3.6.1.3	41.02.250
Intensifier, Single Fluid	3.6.1.1	41.02.255
Intensifier, Single Shot	3.6.1.5	41.02.260
Interaction Region, Jet, Fluidic	4.6.2.24	72.06.140
Intercooler		32.02.240
Interface		72.01.700
Instrument (See: Gage)		91.02.050
Instrument, Absolute Pressure Measuring	7.1.1.5	91.02.850
Instrument, Differential Pressure Measuring	7.1.1.6	91.02.852
Instrument, Flow Measuring	7.1.2	91.02.853
Instrument, Liquid Level Measuring	7.1.4	91.02.860
Instrument, Pressure Measuring (See: Gage)	7.1.1	91.02.700

J

Jet	4.6.2.13	72.01.710
Jet Action-Valve		71.03.300
Jet, Attached	4.6.2.18	72.01.711
Jet, Confined	4.6.2.16	72.01.712
Jet, Free	4.6.2.14	72.01.713
Jet, Main Power	4.6.2.15	72.01.714
Joint (See: Connection or Fitting)		56.01.100
Joint, Rotary	5.2.2.9	56.02.300
Joint, Spherical	5.2.2.13	56.02.598
Joint, Swivel	5.2.2.10	56.02.600
Joint, Telescopic	5.2.2.11	56.02.602

L

Lantern Ring (Seal Cage)		94.06.360
Lift	2.2.5.8	04.01.210
Lift, Static Suction		04.01.220
Line	5.2.1	50.02.300
Line, Bleed	5.2.1.10	50.03.050
Line, Drain	5.2.1.9	50.03.100
Line, Exhaust	5.2.1.6	50.03.200
Line, Make-Up	5.2.1.7	50.03.350
Line, Pilot	5.2.1.8	50.03.500

ANSI/B93.2

Term	Ref	Code
Line, Pump Inlet	5.2.1.4	50.03.510
Line, Return	5.2.1.5	50.03.560
Line, Suction		50.03.600
Line, Working or Feed	5.2.1.3	50.03.700
Linear Function	2.4.12	04.01.222
Linear Region	2.4.11	04.01.223
Linear Scale Reticule, Contamination		30.06.270
Linearity		04.01.224
Lines, Joining		50.03.300
Lines, Passing		50.03.400
Load Line	4.6.3.1	72.01.800
Logic Devices	4.6.1.6	70.02.700
Logic Threshold	4.6.3.13	72.01.801
Logical State		70.02.710
Longest Dimension, Contaminant		30.05.300
Losses, Hydrodynamic	3.0.2.2.2	43.01.150
Losses, Mechanical	3.0.2.2.3	43.01.151
Losses, Motor	3.2.2.2.6	82.01.110
Losses, Pump	3.1.2.7	43.01.152
Losses Volumetric	3.0.2.2.1	43.01.153
Low Side Components		37.05.490
Lubricator	5.5.8	36.01.100
Lug (Foot), Cylinder		81.05.510

M

Term	Ref	Code
Magnetic Element (See: Element)	5.8.2.5	33.07.080
Magnetic Plug	5.8.5.3	33.06.300
Maintenance, System	9.2.2; 9.2.2.1	96.01.350
Manifold		57.01.100
Manifold Block, Valve	4.0.2.3	57.01.101
Manifold, Vented		57.02.800
Manometer	7.1.1.2	91.03.300
Manual, Commissioning	9.2.3.4	96.01.351
Manual, Installation	9.2.3.3	96.01.352
Manual, Maintenance	9.2.3.6	96.01.353
Manual, Operating	9.2.3.5	96.01.354
Mass Index, Contaminant		30.05.390
Maximum Excursion		72.06.150
Maximum Inlet Pressure		74.03.300
Mean Filtration Rating		33.11.240
Medium, Filter		33.09.090
Medium, Absorptive Filter Element		33.08.010
Medium, Adsorptive Filter Element		33.08.030
Medium, Combination Filter Element		33.08.050
Medium Deposited Filter Element		33.08.070
Medium, Depth Filter Element		33.08.080
Medium, Edge Filter Element		33.08.090
Medium, Etched Filter Element		33.08.100
Medium, Filter Element		33.09.090
Medium, Non-Woven Filter Element		33.08.110
Medium, Precoat Filter Element		33.08.120
Medium, Sintered Filter Element		33.08.130
Medium, Surface Filter Element		33.08.140
Medium, Wound Filter Element		33.08.150
Medium, Woven Filter Element		33.08.160
Meter, Flow	7.1.2.1	91.03.400
Meter, Flow, Integrating	7.1.2.2	91.03.401
Micrometre (Micron) *Deprecated		30.05.400
Microscope Gating, Comtamination		30.06.330
Microscope Reticule (Microscope Graticule)		30.06.360

Term	Section	Number
Microscope Split Image Eyepiece, Contamination		30.06.390
Microscope, Contaminant		30.06.420
Microscope Filar Eyepiece, Contamination		30.06.300
Migration	10.4.1.8	33.11.250
Migration, Abrasion		33.11.251
Migration, Built-In-Dirt		33.11.252
Migration, Contaminant		33.11.253
Migration, Media		33.11.254
Modular (Plug-In)		33.07.090
Moisture Separator		37.05.500
Monostable	4.6.1.1	04.01.230
Motor, Air	3.2.1	84.01.100
Motor, Air Piston	3.2.1.1	84.02.100
Motor, Assembly, Hydraulic	8.2.1	82.02.620
Motor, Assembly, Pneumatic	8.2.2	84.02.500
Motor, Axial Piston	3.2.2.1.8.2	82.02.010
Motor Capacity, Derived	3.2.2.2.1.1	82.01.020
Motor Displacement	3.2.2.1.1	82.02.050
Motor Efficiency, Hydromechanical	3.2.2.2.7.2	82.01.030
Motor Efficiency, Overall	3.2.2.2.7.3	82.01.031
Motor Efficiency, Volumetric	3.2.2.2.7.1	82.01.032
Motor Fixed Displacement	3.2.2.1.4	82.02.100
Motor Flow, Input	3.2.2.2.3	82.01.040
Motor Flow, Input, Derived	3.2.2.2.3.2	82.01.041
Motor Flow, Input, Effective	3.2.2.2.3.3	82.01.042
Motor Flow, Input, Geometric	3.2.2.2.3.1	82.01.043
Motor, Force (See: Control, Force Motor)		
Motor, Gear	3.2.2.1.6	82.02.150
Motor, Gear, External	3.2.2.1.6.2	82.02.151
Motor, Gear, Internal	3.2.2.1.6.1	82.02.152
Motor, Hydraulic (See: Hydraulic Motor)	3.2.2	82.01.000
Motor, Hydraulic		82.02.175
Motor, Hydraulic Rated Speed		82.01.125
Motor, Hydraulic Stepping	3.2.2.1.11	82.02.176
Motor, Linear (See: Cylinder)	3.2.2.1.10	82.02.200
Motor, Multiple	3.2.2.1.9	82.02.210
Motor Output Power	3.2.2.2.4	82.01.104
Motor, Over-Center	3.2.0.1	82.02.220
Motor Power, Hydraulic	3.2.2.2.4.1	82.01.100
Motor Power, Hydraulic Derived	3.2.2.2.4.3	82.01.101
Motor Power, Hydraulic Effective	3.2.2.2.4.4	82.01.102
Motor Power, Hydraulic Geometric	3.2.2.2.4.2	82.01.103
Motor, Radial Piston	3.2.2.1.8.1	82.01.231
Motor, Reciprocating	3.2.2.1.3	82.02.250
Motor, Reversible	3.2.0.2	82.02.260
Motor, Rotary	3.2.2.1.2	82.02.300
Motor, Rotary Limited		82.02.301
Motor, Torque, Hydraulic	3.2.2.2.5	82.02.310
Motor, Torque, Starting	3.2.2.2.5.1	82.02.311
Motor, Vane Air	3.2.1.2	84.02.101
Motor, Vane, Hydraulic	3.2.2.1.7	82.02.320
Motor, Vane, Hydraulic, Balanced	3.2.2.1.7.2	82.02.321

ANSI/B93.2

Motor, Vane, Hydraulic, Unbalanced	3.2.2.1.7.1	82.02.322
Motor, Variable Displacement	3.2.2.1.5	82.02.600
Mounting (See: Cylinder Mounting)		
Mounting, Flange, Pump	3.1.3.1.1	43.03.100
Mounting, Foot, Pump	3.1.3.1.3	43.03.120
Mounting, Pilot (Spigot), Pump	3.1.3.1.2	43.03.300
Moving Parts Logic		01.01.850
Multi-Pass Test		30.06.430
Muffler (See: Silencer)	5.7.2	76.01.100

N

NAND Device		70.04.700
Net Pressure Drop		33.11.256
Neutralization Number	10.3.6	04.01.250
Newt		04.01.260
Nipple		50.02.400
Noise, Acoustic	4.6.3.14	72.06.199
Noise, Fluidic	4.6.1.19	72.06.200
Nominal Filtration Rating	5.8.4.3	33.11.260
Non-Combustible Residue, Contamination		30.06.450
Non-Volatile Residue, Contamination		30.06.480
Non-Woven		33.08.110
NOR Device		70.04.720
Normal Flow		33.04.500
Normalized Capacity ( )		33.11.265
NOT Device		70.04.740

O

Open Area		33.11.270
Open Area Ratio		33.11.280
Operating Band		72.06.250
Optical Density, Contamination		30.06.510
OR Device		70.04.760
Outer Wrapper, Filter		33.09.100
Output Active, Fluidic	4.6.1.26	72.07.600
Output, Inactive or Passive, Fluidic	4.6.1.27	72.07.601
Output Stage (See: Stage, Output)		73.03.300
Outside-In Flow		33.07.100
Oxidation		33.04.400

P

Pack		94.01.080
Packing (See: Seal)		94.02.200
Packing, Coil		94.02.220
Packing, U		94.02.240
Packing, V		94.02.260
Packing, W		94.02.280
Panel		96.01.600
Panel, Control	9.1.4.3	96.01.610
Panel, Mounting		96.01.620
Particle, Contaminant	10.4.1.4	30.05.500
Particle Count Blank, Contamination		30.06.540
Particle Size Distribution, Contamination		30.06.570
Parts List	9.2.3.7	96.01.630
Parts List, Spares	9.2.3.7.1	96.01.631
Pascal's Law		02.01.700

Term	Ref	Code
Passage		50.01.600
Patch Test, Contamination		30.06.600
Period, Cooling		37.04.560
Period, Drying (Sorption)		37.04.570
Period, Heating		37.04.580
Period, Regeneration (Desorption)		37.04.590
Permeability		33.11.300
Petroleum Fluid		12.01.100
Phase	2.4.2	96.01.700
Phase, Dwell	2.4.2.1	96.01.710
Phase, Feed		96.01.720
Phase, Neutral	2.4.2.3	96.01.730
Phase, Rapid Advance	2.4.2.4	96.01.740
Phase, Rapid Return	2.4.2.5	96.01.750
Phase, Working	2.4.2.2	96.01.770
Pilot, Pump (Spigot)		43.03.300
Pipe, Port		53.01.100
Pipeline, (See: Flowline)	5.2.1	50.02.100
Pipe Clamp	9.1.3.1.1	53.01.500
Pipe Thread		50.06.400
Pipe Thread, Dryseal		50.06.420
Pipe Thread, Tapered		50.06.440
Piston Rod	3.5.0.9	81.05.550
Plain		33.07.110
Plain "O" Ring, Port		50.04.600
Pleated (Corrugated)		33.07.120
Pleats (Corrugations), Filter		33.09.110
Pneumatics	2.0.5	01.01.900
Poise		04.01.270
Pore Size Distribution		33.11.310
Porosity (Voil Fraction)		33.11.320
Port	5.2.4	50.01.700
Port, AND 10050		50.04.200
Port, Area of	5.2.4.9	50.01.701
Port, Bias		50.05.030
Port, Bleed	5.2.4.7	50.05.050
Port, Control	5.2.4.1	50.05.100
Port, Cylinder		50.05.150
Port, Differential Pressure		50.05.180
Port, Discharge		50.05.200
Port, Drain	5.2.4.6	50.05.250
Port, Drain, Airline	5.2.4.8	50.05.251
Port, Exhaust	5.2.4.4	50.05.300
Port, Fill		50.05.350
Port, Flanged	5.2.4.11.2	50.05.360
Port, Inlet	5.2.4.2	50.05.400
Port, Manifold	5.2.4.10	50.05.416
Port, Outlet (Output)	5.2.4.3	50.05.430
Port, Pipe		50.04.400
Port, Plain "O" Ring		50.04.600
Port, Pressure		50.05.450
Port, SAE J514		50.04.800
Port, Suction		50.05.600
Port, Supply		50.05.620
Port, Take-Off Point (Power P-T-O)		50.04.850
Port, Tank (Reservoir) (Return)	5.2.4.5	50.05.650
Port, Threaded	5.2.4.11.1	50.06.000
Port, Vent		50.05.700
Port-to-Port Dimension		50.07.600
Pour Point	10.3.9	04.01.280
Power, Capacity, Fluidic		72.06.300

ANSI/B93.2

Term	Section	Number
Power Consumption	2.4.4	04.01.282
Power, Cylinder Output	3.5.3.6.2	81.01.601
Power, Fluidic Output	4.6.1 22	72.06.325
Power, Hydraulic, Motor	3.2.2.2.4.1	82.01.100
Power, Hydraulic, Motor, Derived	3.2.2.2.4.3	82.01.101
Power, Hydraulic, Motor, Effective	3.2.2.2.4.4	82.01.102
Power, Hydraulic, Motor, Geometric	3.2.2.2.4.2	82.01.103
Power, Hydraulic, Pump	3.1.2.4	43.01.170
Power, Hydraulic, Pump, Derived	3.1.2.4.2	43.01.171
Power, Hydraulic, Pump, Effective	3.1.2.4.3	43.01.172
Power, Hydraulic, Pump, Geometric	3.1.2.4.1	43.01.173
Power Input, Cylinder	3.5.3.6.1	81.01.600
Power Input, Pump	3.1.2.5	43.01.174
Power, Motor Output	3.2.2.2.4	82.01.175
Power, Pump, Installed	3.1.2.5.3	43.01.104
Power, Required, Pump	3.1.2.5.2	43.01.176
Power Supply, Fluid	5.1.1	04.01.283
Power Supply, Fluidic Devices		72.06.350
Power, Switching	4.0.8.3	
Power Unit	8.1.1	42.01.100
Power Unit, Hydraulic		42.02.400
Precipitate, Contamination		30.06.630
Precipitation Number		04.01.290
Precoat		33.08.120
Precooler		32.02.260
Precooler Reheater		37.05.550
Prefilter		37.05.600
Pressure	2.2.4	04.01.300
Pressure, Absolute	2.2.4.1.1	04.01.310
Pressure, Atmospheric		04.01.320
Pressure, Back	2.2.4.11	04.01.330
Pressure, Breakloose	2.2.4.12	04.01.340
Pressure, Burst	2.2.4.13	04.01.360
Pressure, Charge (Boost)	2.2.4.14	04.01.370
Pressure, Control Range	2.2.4.7	04.01.375
Pressure, Cracking	2.2.4.15	04.01.380
Pressure, Cyclic Test		04.01.385
Pressure, Cylinder	3.5.3.1.1	81.01.610
Pressure, Damping, Cushion	3.5.3.1.3	63.01.150
Pressure, Dead Head	4.6.3.2	43.01.180
Pressure, Differential (Drop)	2.2.4.6	04.01.390
Pressure, Flow (See: Flow, Pressure)		
Pressure Gage	2.2.4.1	04.01.400
Pressure, Induced	3.5.3.1.2	04.01.412
Pressure, Inlet	2.2.4.4	04.01.414
Pressure, Intensified	3.5.3.1.4	04.01.415
Pressure, Loss (Ψ)		33.11.330
Pressure, Maximum Inlet		04.01.417
Pressure, Nominal		04.01.419
Pressure, Operating	2.2.4.17	04.01.420
Pressure, Outlet	2.2.4.5	04.01.425
Pressure, Overrange Rating		04.01.427
Pressure, Override	2.2.4.26	04.01.430
Pressure, Peak	2.2.4.16	04.01.435
Pressure, Pilot	2.2.4.22	04.01.440
Pressure, Precharge	2.2.4.23	04.01.450

ANSI/B93.2

Term	Section	Code
Pressure, Proof	2.2.4.17	04.01.460
Pressure, Rated	2.1.2	04.01.470
Pressure, Rated Fatigue		04.01.473
Pressure, Rated Static		04.01.476
Pressure, Recovery	4.6.3.3	04.01.472
Pressure, Regulation	2.2.4.7	04.01.474
Pressure, Regulation Characteristic	4.0.13	74.03.500
Pressure, Regulation Steady State Supply	4.0.11	73.03.501
Pressure, Residual	4.6.3.10	04.01.477
Pressure, Shock	2.2.4.19	04.01.480
Pressure, Shock Wave	2.2.4.8	04.01.481
Pressure, Static	2.2.4.20	04.01.490
Pressure, Static Test		04.01.495
Pressure, Suction	2.2.4.2.4	04.01.500
Pressure, Supply	2.2.4.4	04.01.505
Pressure, Surge	2.2.4.21	04.01.510
Pressure, System		04.01.520
Pressure, Vapor		04.01.530
Pressure Vessel		62.01.100
Pressure, Working	2.2.4.2	04.01.540
Pressure, Working, Range	2.2.4.3	04.01.541
Primary, Fluid Intensifier	3.6.0.1	41.01.700
Prime Mover	9.1.2	96.02.600
Programmer, Pneumatic (See: Control)		
Protector, Gauge (See: Gage Damper)	7.4.3	91.01.300
Protector, Gauge (See: Gage Protector)	7.4.4	91.01.301
Pump, Centrifugal	3.1.1.1	43.02.100
Pump, Centrifugal, Diffuser (Concentric)		43.02.130
Pump, Centrifugal, Peripheral		43.02.145
Pump, Centrifugal, Volute (Spiral)		43.02.175
Pump, Fixed Displacement	3.1.1.2.1	43.02.200
Pump, Gear	3.1.1.5	43.02.250
Pump, Gear, External	3.1.1.5.1	43.02.251
Pump, Gear, Fixed Clearance	3.1.1.5.3	43.02.252
Pump, Gear, Internal	3.1.1.5.2	43.02.253
Pump, Gear, with Pressure Loading	3.1.1.5.4	43.02.254
Pump, Hand	3.1.1.9	43.02.300
Pump, Hand, Double Acting	3.1.1.9.2	43.02.301
Pump, Hand, Single Acting	3.1.1.9.1	43.02.302
Pump, Hydraulic	3.1.1	43.01.100
Pump, Motor	3.3	43.02.410
Pump, Multistage	3.1.1.10	43.02.350
Pump, Piston	3.1.1.8	43.02.396
Pump, Piston, Angled	3.1.1.8.3	43.02.397
Pump, Piston, Axial	3.1.1.8.2	43.02.398
Pump, Piston, Inline	3.1.1.8.4	43.02.399
Pump, Piston, Radial	3.1.1.8.1	43.02.400
Pump, Reciprocating, Duplex		43.02.450
Pump, Reciprocating, Single Piston		43.02.500
Pump, Screw	3.1.1.6	43.02.550
Pump, Vane	3.1.1.7	43.02.600
Pump, Vane, Balanced	3.1.1.7.2	43.02.601
Pump, Vane, Unbalanced	3.1.1.7.1	43.02.602
Pump, Variable Displacement	3.1.1.2.2	43.02.650
Purge Flow		37.04.600
Purge Flow, Co-Current		37.04.620
Purge Flow, Countercurrent		37.04.630

Q

Quick Disconnect, Breakaway		55.02.200
Quick Disconnect, Claw Type		55.02.220
Quick Disconnect, Coupling		55.01.100
Quick Disconnect, One Valve		55.02.400
Quick Disconnect, Un-Valved Deprecated		55.02.600
Quick Disconnect, Valved	5.2.2.12	55.02.800

R

Rabbet, Cylinder		81.05.570
Radial Cavity		64.15.560
Radiator	9.1.5.3.1	32.02.270
Rated Flow		82.01.041
Rated Flow, Fluidics		72.06.400
Rated Speed, Motor		82.01.125
Ratio, Fan In Fluidic Device (See: Fan In)	4.6.3.12	72.01.400
Ratio, Fan Out, Fluidic Device (See: Fan Out)	4.6.3.11	72.01.500
Ratio, Flow Degradation Ratio		04.01.051
Ratio, Series, Fluidic Device	4.6.3.11.1	72.01.900
Ratio, Signal to Noise, Fluidic Device	4.6.3.15	72.01.920
Raw Count, Contamination		30.06.660
Receiver (See: Collector, Reservoir)	4.6.2.21	72.01.255
Recorder, Flow	7.1.2.4	91.04.100
Recorder, Pressure	7.1.1.4	91.04.200
Recovery, Flow Rate	4.6.3.4	72.06.410
Recovery, Power	4.6.3.5	72.06.420
Reduced Pressure Range		74.03.400
Refrigerant		37.05.650
Refrigerant Gauge		37.05.700
Refrigerant Compressor		37.05.800
Refrigerant Compressor, Hermetic		37.05.830
Refrigerant Compressor, Open-Type		37.05.840
Regeneration (Reactivation)		37.04.650
Regeneration, Atmospheric		37.04.660
Regeneration, Heat		37.04.670
Regeneration, Heatless		37.04.680
Regeneration, Pressure		37.04.690
Regulation Characteristic Curve		74.03.500
Regulator, Air Line Pressure		74.02.100
Regulator, Constant Bleed Air Line Pressure		74.02.120
Regulator, Pressure Relieving Air Line		74.02.160
Relief Characteristic Curve		74.03.600
Relief Flow Rate	2.2.6.4.3	04.01.061
Repeatability	2.4.6	91.01.700
Repressurization		37.04.700
Required Cleanliness Level (CL)		30.07.800
Reserve		33.07.150
Reserve Capacity		21.03.800
Reservoir (See: Receiver)	5.3; 5.3.1	21.01.100
Reservoir, Atmospheric	5.3.2	21.02.100
Reservoir, Hydraulic		21.02.200
Reservoir, Non Integral		21.02.300
Reservoir, Pressure Sealed	5.3.3	21.02.400
Reservoir, Sealed	5.3.4	21.02.500
Reservoir, Top Mounted		21.02.600
Residual Dirt Capacity		33.11.350
Resistance, Fluid	4.6.1.11	

Term	Section	Number
Response, Frequency	2.4.5	04.01.072
Response Time (Fluidic)	2.2.9.5	72.07.700
Restrictor	4.6.2.12	70.03.100
Restrictor, Choke		70.03.130
Restrictor, Orifice		70.03.160
Reversal, Position		96.01.910
Reversal, Pressure		96.01.920
Reverse Flow		33.04.800
Reyn		04.01.610
Reynolds Number		02.01.800
Ring, Anti-Extrusion	5.9.2.16	94.06.180
Ring, O (See: Seal)	5.9.2.11	94.02.410
Ring, Piston		94.02.420
Ring, Scraper		94.02.430
Ring, U (See: Seal)	5.9.2.11	94.02.440
Ring, V (See: Seal)	5.9.2.11	94.02.450
Ring, W (See: Seal)	5.9.2.11	94.02.460
Ring, Wiper		94.02.470
Ripple	2.4.9	72.06.450
Rise Rate		72.07.800
Root, Filter		33.09.129
Rotary Actuator		85.01.000
Rotation	2.2.2	04.01.611
Rotation, Anti-Clockwise	2.2.2.2	04.01.612
Rotation, Clockwise	2.2.2.1	04.01.613
Rotational Frequency	3.1.2.2	04.01.614

S

Term	Section	Number
SAE, Port		50.04.800
Sampler, Turbulent		30.07.900
Saponification, Value of	10.3.15	11.01.400
Schmitt Trigger		70.03.800
Scoring		30.04.450
Scoring Size, Particle, Contamination		30.06.690
Seal, (Sealing Device) (Ring)	5.9	94.01.100
Seal, Axial	5.9.2.3	94.02.603
Seal, Butyl	5.9.3.5.2	94.02.604
Seal, Chevron	5.9.2.14	94.02.605
Seal, Composite	5.9.3.1	94.02.606
Seal, Cork	5.9.3.4	94.02.607
Seal, Cup	5.9.2.12	94.02.608
Seal, Diaphragm (Flat Diaphragm)		94.02.609
Seal, Dished Diaphragm		94.02.610
Seal, Dynamic	5.9.2.1	94.02.612
Seal, Elastomer	5.9.3.5	94.02.613
Seal, Ethylene Propylene	5.9.3.5.8	94.02.617
Seal, Flange (Hat)	5.9.2.13	94.02.621
Seal, Fluorinated (Viton)	5.9.3.5.5	94.02.622
Seal Lip	5.9.2.15	94.02.630
Seal, Leather	5.9.3.3	94.02.631
Seal Lubricant		94.02.633
Seal, Mechanical	5.9.1.1	94.02.636
Seal, Nitrile	5.9.3.5.1	94.02.638
Seal, Oil		94.02.639
Seal, O Ring	5.9.2.11	94.02.642
Seal, Packing	5.9.1	94.02.200
Seal, Piston	5.9.2.8	94.02.645
Seal, Polyacrylic	5.9.3.5.9	94.02.646
Seal, Polyamide (Nylon)	5.9.3.5.10	94.02.647
Seal, Polychloroprene	5.9.3.5.4	94.02.648

Term	Section	Number
Seal, Polytetrafluoroethylene (PTFE)	5.9.3.5.6	94.02.649
Seal, Polyurethane	5.9.3.5.7	94.02.650
Seal, Pressure Actuated		94.02.651
Seal, Radial	5.9.2.4	94.02.654
Seal, Ram (Plunger)		94.02.657
Seal, Rod (Shaft, Stem)	5.9.2.9	94.02.660
Seal, Rotary	5.9.2.5	94.02.663
Seal, Silicone	5.9.3.5.3	94.02.665
Seal, Sliding	5.9.2.7	94.02.672
Seal, □ (Square) Ring	5.9.2.11	94.02.674
Seal, Static	5.9.2.2	94.02.675
Seal, U Ring	5.9.2.11	94.02.676
Seal, V Ring	5.9.2.11	94.02.677
Seal, W Ring	5.9.2.11	94.02.678
Seal, Water		94.02.681
Seal, Wiper	5.9.2.10	94.02.684
Seal, X Ring	5.9.2.11	94.02.685
Sealing Device	5.9	94.01.100
Sealing Action, Valve		71.03.500
Second Longest Dimension Contaminant		30.05.550
Secondary		33.07.170
Secondary, Fluid, Intensifier	3.6.0.2	41.02.130
Self Cleaning		33.07.180
Sensing Device		70.02.880
Sensor		70.02.900
Sensor, Instrumentation	4.6.2.11	91.05.000
Separator	5.5.3	34.01.100
Separator, Adsorbent		34.02.100
Separator, Centrifugal		34.02.200
Separator, Coalescing (See: 5.5.5; 5.8.1.8)		34.02.300
Separator, Electrostatic		34.02.400
Separator, Magnetic		34.02.500
Separator, Oil Remover	5.5.5	34.02.550
Separator, Two Phase		34.02.600
Separator, Vacuum		34.02.900
Separator, Water Trap	5.5.4	34.02.910
Servovalve	4.7	73.01.100
Servovalve Amplitude Ratio	4.7.5.2	73.06.200
Servovalve Control Flow	4.7.4.1	73.05.050
Servovalve Control Flow, Loaded	4.7.4.4	73.05.060
Servovalve Control Flow, No-Load	4.7.4.3	73.05.070
Servovalve, Electrohydraulic	4.7.1.1	73.02.200
Servovalve, Electrohydraulic Flow Control	4.7.1.3	73.02.300
Servovalve Flow Curve	4.7.4.5	73.05.100
Servovalve Flow Curve, Normal	4.7.4.6	73.05.130
Servovalve Flow Gain	4.7.4.7	73.05.150
Servovalve Flow Gain, No-Load		73.05.160
Servovalve Flow Gain, Normal		73.05.170
Servovalve Flow Limit	4.7.4.9	73.05.200
Servovalve Flow Linearity	4.7.4.16	73.05.450
Servovalve Flow Saturation Region		73.05.250
Servovalve, Four-Way	4.7.1.5	73.02.400
Servovalve Frequency Response	4.7.5.1	73.06.500
Servovalve Hysteresis	4.7.4.10	73.05.300

Term	Section	Code
Servovalve Internal Leakage	4.7.4.11	73.05.350
Servovalve Lap	4.7.4.13	73.05.400
Servovalve Lap, Over	4.7.4.13.2	73.05.410
Servovalve Lap, Under	4.7.4.13.3	73.05.420
Servovalve Lap, Zero	4.7.4.13.1	73.05.430
Servovalve, Mechanical Hydraulic	4.7.1.2	73.02.450
Servovalve, Null	4.7.4.18	73.05.500
Servovalve Null Bias	4.7.4.19	73.05.510
Servovalve Null Leakage	4.7.4.12	73.05.405
Servovalve Null Pressure	4.7.4.20	73.05.520
Servovalve Null Region	4.7.4.21	73.05.530
Servovalve Null Shaft	4.7.4.22	73.05.540
Servovalve Phase Lag	4.7.5.3	73.06.800
Servovalve Pressure Control	4.7.1.4	73.02.460
Servovalve Pressure Drop	4.7.4.26	73.05.550
Servovalve Pressure Drop, Load	4.7.4.17	73.05.570
Servovalve Pressure Gain	4.7.4.27	73.05.600
Servovalve Rated Flow	4.7.4.2	73.05.650
Servovalve Resolution	4.7.4.28	73.05.660
Servovalve Symmetry	4.7.4.8	73.05.700
Servovalve, Three-Way	4.7.1.6	73.02.500
Servovalve Threshold	4.7.4.29	73.05.750
Servovalve, Torque Motor	6.5.4	73.03.500
Servovalve, Two-Way	4.7.1.7	73.02.600
Shaft Extension	3.1.3.2	43.01.018
Shear Action, Valve		71.03.700
Shell		94.06.400
Shock Wave	2.2.4.8	03.01.600
Side (Base) Cylinder		81.05.720
Side Seal, Filter		33.09.130
Silencer		76.01.600
Silt (See: Contaminant)	10.4.1.9	30.03.500
Silting (Clogging)	5.8.3	30.04.500
Sintered		33.08.130
Slime (See: Contaminant)		30.03.550
Slip	3.2.2.2.2.2	82.01.200
Sloughing Off		33.11.360
Sludge (See: Contaminant)		30.03.600
Specific Gravity, Liquid		04.01.620
Specification, Performance	9.2.3.1	96.02.650
Specification, System	9.2.3.2	96.02.651
Speed (See: Rotational Frequency)		
Speed Degradation Ration		82.01.150
Spillage		55.01.080
Spigot (See: Mounting)		
Splitter	4.6.2.25	72.01.930
Spring, Expander		94.06.510
Spring, Fringer (Lug)		94.06.520
Spring, Garter		94.06.530
Spring, Spreader		94.06.560
Spring, Wave (Marcel) (Wave Washer)		94.06.570
Stage, Servovalve	4.7.2.4	73.03.400
Stage, Output, Servovalve	4.7.2.5	73.03.300
Standard		99.02.600
Start-Up Time		72.06.500
Starvation		33.04.900
Steady State Pressure Regulation		72.06.550

Term	Ref	Code
Stoke		04.01.630
Stop, Positive Position		96.01.810
Stop, Positive Safety		96.01.820
Stroke, Extended (Out), Cylinder	3.5.0.3	81.01.700
Stroke, Full, Cylinder	3.5.1.6	81.01.701
Stroke, Retract (In), Cylinder	3.5.0.4	81.01.702
Stroke, Working, Cylinder	3.5.1.7	81.01.703
Stuffing Box		94.06.600
Subplate, Valve (Sub-Base) (Backplate) (Manifold)	4.0.2	58.01.100
Subplate, (Sub-Base), Ganged Valve	4.0.2.2	58.01.101
Subplate, Multiple Valve (Sub-Base)	4.0.2.1	58.01.102
Superfical Bed Velocity		37.04.750
Surface		33.08.140
Surface Tension	10.3.21	04.01.640
Surge	2.2.4.10	03.01.800
Switch, Float		75.02.400
Switch, Flow	7.3.2	75.02.500
Switch, Liquid Level	7.3.3	75.02.600
Switch, Pressure	7.3.1	75.02.700
Switch, Pressure Differential	7.3.4	75.02.750
Symbol, Combination	2.3.4	05.02.100
Symbol, Cutaway	2.3.2	05.02.200
Symbol, Fluid Power		05.01.100
Symbol, Graphical (Schematic)	2.3.1	05.02.300
Symbol, Pictorial	2.3.3	05.02.400
Synthetic Fluid (See: Fluid)		13.01.100
Synthetic Fluid, Chlorinated Hydrocarbon	10.2.6	13.02.090
Synthetic Fluid, Di-Base Ester	10.2.7	13.02.095
Synthetic Fluid, Halogenated	10.2.8	13.02.100
Synthetic Fluid, Phosphate Ester	10.2.9	13.02.300
Synthetic Fluid, Phosphate Ester Base	10.2.10	13.02.400
Synthetic Fluid, Polyglycol	10.2.11	13.02.500
Synthetic Fluid, Polyglycol Ester		13.02.550
Synthetic Fluid, Silicate Ester	10.2.13	13.02.600
Synthetic Fluid, Silicone	10.2.12	13.02.700
System Bleeding	9.2.1.4	96.02.660
System Cleaning	9.2.1.1	96.02.661
System Draining	9.2.1.5	96.02.662
System Filling	9.2.1.3	96.02.663
System, Fluid Power	8.0.1	96.01.850
System Flushing	9.2.1.2	96.02.664
System Maintenance	9.2.2	96.02.665
System Startup	9.2.1.6	96.02.666
System Testing	9.2.1.7	96.02.667

T

Term	Ref	Code
Take-Off Point (Power, P-T-O), Port	5.2.2.20	50.04.850
Tank (See: Reservoir 5.3)		22.01.100
Tank, Air-Oil		22.02.100
Tank, Vacuum		22.02.900
Temperature, Ambient	2.2.3.5	04.01.641
Temperature, Auto Ignition	10.3.12	10.04.016
Temperature Controller	5.6.3	32.01.700

Term	Section	Number
Temperature, Equipment	2.2.3.1	04.01.642
Temperature, Fluid	2.2.3.2	04.01.643
Temperature, Inlet	2.2.3.6	04.01.644
Temperature, Outlet	2.2.3.7	04.01.645
Temperature, Range	2.2.3.3	04.01.646
Terminal Pressure Drop		33.11.365
Test Duration Factor		04.01.642
Threshold, Logic (See: Logic Threshold)	4.6.3.13	72.01.801
Tie Rod, Cylinder		81.05.750
Time	2.2.9	01.01.910
Time, Actuated	2.2.9.4.1	01.01.911
Time, Extended Stroke, Cylinder	3.5.3.8.1	81.01.800
Time, Fall	2.2.9.3	01.01.912
Time, Operated	2.2.9.4	01.01.913
Time, Released	2.2.9.4.2	01.01.914
Time, Relative Duty	2.2.9.4.3	01.01.915
Time Response	3.5.3.8.3	01.01.916
Time, Retract Stroke, Cylinder	3.5.3.8.2	81.01.802
Time, Rise	2.2.9.2	01.01.918
Time, Start Up	2.2.9.1	01.01.919
Time, Transient Recovery	4.0.13	72.06.600
Toricelli's Theorem		02.01.900
Torque	3.0.2.3	01.01.930
Torque, Motor	3.2.2.2.5	82.02.310
Torque, Pump	3.1.2.6	43.01.300
Torque, Derived	3.0.2.3.1	01.01.931
Torque, Effective	3.0.2.3.3	01.01.932
Torque, Geometric	3.0.2.3.2	01.01.933
Torque Motor, (Servovalve)	6.5.4	73.03.500
Torque, Motor, Starting	3.2.2.2.5.1	82.02.311
Torr		04.01.647
Tortuosity		33.11.370
Total Area		33.11.380
Total Statistical Count, Contamination		30.06.720
Transient Recovery Time (See: Time, Transient Recovery)	4.0.12	72.06.600
Transducer, Flow	7.1.2.3	71.03.100
Transducer, Pressure	7.1.1.3	71.03.600
Transmission, Hydrostatic		01.01.801
Trunnion, Cylinder		81.05.780
Tube, Rigid or Semi-Rigid	5.2.1.1	54.01.100
Two-Stage		33.07.230

U

Term	Section	Number
Unistable	4.6.1.2	04.01.648
Unloading		33.11.390
Unloading Device, Compressor	8.1.2.1	44.01.800

V

Term	Section	Number
Vacuum	2.2.4.25	04.01.650
Vacuum Pump		45.01.100
Valve		71.01.100
Valve Actuator		71.06.000
Valve Actuator, Manual	4.3.1.8	71.06.200
Valve, Actuator, Mechanical	4.3.1.6	71.06.400

ANSI/B93.2

Term	Ref	Code
Valve Actuator, Pilot		71.06.600
Valve Actuator, Pilot, Barrier		71.06.610
Valve Actuator, Pilot, Differential Area		71.06.620
Valve Actuator, Pilot, Differential Pressure		71.06.630
Valve Actuator, Pilot, External		71.06.640
Valve Actuator, Pilot, Internal		71.06.650
Valve Actuator, Pilot, Solenoid Controlled		71.06.660
Valve Actuator, Solenoid		71.06.800
Valve, Air		71.02.050
Valve, Ball, Seat Action		71.02.055
Valve, Ball, Shear Action		71.02.056
Valve, Butterfly	4.5.1.8	71.02.060
Valve, Cartridge	4.0.4.4	71.02.070
Valve, Diaphragm Type	4.3.1.2	71.02.080
Valve, Directional Control	4.1	71.02.100
Valve, Directional Control Check	4.2	71.02.110
Valve, Directional Control, Check, Springloaded	4.2.1.1	71.02.111
Valve, Directional Control, Check, Pilot Operated	4.2.1.2	71.02.112
Valve, Directional Control, Check, Cushioned	4.2.1.3	71.02.113
Valve, Directional Control, Four-Way		71.02.130
Valve, Directional Control, Selector		71.02.140
Valve, Directional Control, Straightway		71.02.160
Valve, Directional Control, Three-Way		71.02.170
Valve, Directly Operated	4.3.1.4	71.02.172
Valve, Disc (Globe)	4.5.1.3	71.02.174
Valve, Disc (Swing)		71.02.175
Valve, Filter Bypass	5.8.5.1	33.02.010
Valve, Flapper Action		71.03.200
Valve, Flow Combining	4.4.1.8	71.02.190
Valve Flow Condition		71.05.000
Valve Flow Condition, Closed		71.05.100
Valve Flow Condition, Float	4.1.2.11	71.05.200
Valve Flow Condition, Hold		71.05.300
Valve Flow Condition, Open		71.05.400
Valve Flow Condition, Regenerative		71.05.500
Valve Flow Condition, Tandem	4.1.2.10	71.05.600
Valve, Flow Control	4.4	71.02.200
Valve, Flow Control, Adjustable Restrictor	4.4.1.2	71.02.202
Valve, Flow Control, Bypass	4.4.1.6	71.02.201
Valve, Flow Control, Deceleration	4.4.1.3	71.02.210
Valve, Flow Control, Fixed, Restrictor	4.4.1.1	71.02.203
Valve, Flow Control, One-Way Restrictor	4.4.1.4	71.02.216
Valve, Flow Control, Pressure Compensated	4.4.1.5	71.02.220
Valve, Flow Control, Pressure and Temperature Compensated		71.02.230
Valve, Flow Dividing		71.02.250
Valve, Flow Dividing, Pressure Compensated	4.4.1.7	71.02.280
Valve, Four Position		71.04.600
Valve, Gate	4.5.1.5	71.02.300
Valve, Gate, Spreader		71.02.301
Valve, Gate, Wedge		71.02.302
Valve, Globe (See: Valve, Disc)	4.5.1.3	71.02.320
Valve, Hydraulic		71.02.350
Valve, Jet Action		71.03.300

Term	Ref	Number
Valve, Manually Operated (See: Valve Actuator, Manual)		
Valve, Mechanically Controlled (See: Valve Actuator, Mechanical)		
Valve, Monoblock	4.0.4.1	71.01.500
Valve Mounting		71.07.000
Valve Mounting, Base		71.07.200
Valve, Mounting, Base, Gang	4.0.4.3	71.07.201
Valve Mounting, Line		71.07.400
Valve Mounting, Manifold		71.07.600
Valve Mounting, Sub-Plate	4.0.4.2	71.07.800
Valve, Needle	4.5.1.4	71.02.380
Valve, Pilot	4.0.4.5	71.02.400
Valve, Pilot Operated	4.3.1.5; 4.3.1.7	71.02.401
Valve, Pinch	4.5.1.7	71.02.410
Valve, Piston Type	4.3.1.3	71.02.420
Valve, Plug	4.5.1.1	71.02.430
Valve, Plug, Cylinder	4.5.1.1.2	71.02.431
Valve, Plug, Shear Action		71.02.432
Valve, Plug, Spherical	4.5.1.1.3	71.02.433
Valve, Plug, Tapered	4.5.1.1.1	71.02.434
Valve, Pneumatic		71.02.450
Valve, Poppet	4.1.1.2	71.02.460
Valve Position	4.1.2.1	71.04.000
Valve Position, Actuated	4.1.2.5	71.04.002
Valve Position, Center	4.1.2.4	71.04.100
Valve Position, Closed (See: Valve, Flow Condition, Closed)		
Valve Position, Detent		71.04.200
Valve Position, Float (See: Valve, Flow Condition, Float)		
Valve Position, Initial	4.1.2.3	71.04.220
Valve Position, Intermediate (Transit)	4.1.2.6	71.04.222
Valve Position, Middle (See: Valve Position, Center)		71.04.100
Valve, Position, Normal (At Rest)	4.1.2.2	71.04.300
Valve Position, Offset		71.04.400
Valve Position, Open (See: Valve, Flow Condition, Open)		
Valve Position, Return		71.04.500
Valve, Power Control		71.02.470
Valve, Prefill	4.2.1.4	71.02.500
Valve, Pressure Control	4.3	71.02.550
Valve, Pressure Control, Counterbalance	4.3.0.3.3	71.02.560
Valve, Pressure Control, Decompression	4.3.0.3.4	71.02.570
Valve, Pressure Control, Load Dividing		71.02.580
Valve, Pressure Control, Reducing	4.3.0.3	71.02.590
Valve, Pressure Control, Reducing and Relieving	4.3.0.1	71.02.591
Valve, Pressure Control, Relief		71.02.600
Valve, Pressure Control, Relief, Safety		71.02.620
Valve, Pressure Control, Sequence	4.3.0.2	71.02.700
Valve, Pressure Control, Unloading	4.3.0.4	71.02.630
Valve, Pressure Proportioning	4.3.0.3.1	71.02.635
Valve, Pressure Sensing		71.02.640
Valve, Priority		71.02.650
Valve, Quick Exhaust	4.2.3	71.02.655
Valve, Relay	4.3.0.3.1	71.02.640
Valve, Relay, Free Floating		71.02.662
Valve, Relay, One Shot		71.02.664

Entry	Reference	Code
Valve, Relay, Time Delay After Exhausting A Control Point		71.02.666
Valve, Relay, Time Delay After Pressurizing A Control Point		71.02.668
Valve, Relay, Time Delay		71.02.670
Valve, Relay, Time Delay, Detented		71.02.672
Valve, Relay, Time Delay, Detented, Delayed Action		71.02.674
Valve, Relay, Time Delayed, Detented, Delayed Reset		71.02.676
Valve, Rotary Selector		71.02.680
Valve, Seating Action		71.03.500
Valve, Separator Drain	5.5.6	71.02.690
Valve, Sequence (See: Valve Pressure, Control)		
Valve, Servo (See: Servovalve)		
Valve, Shear Action		71.03.700
Valve, Shut Off	4.5	71.02.750
Valve, Shut Off, Automatic	4.2.4	71.02.751
Valve, Shut Off, Sliding	4.5.1.2	71.02.752
Valve, Shut Off, Sliding, Flat	4.5.1.2.1	71.02.753
Valve, Shut Off, Sliding, Spool	4.5.1.2.2	71.02.754
Valve Shuttle	4.2.2	71.02.800
Valve Shuttle, High Pressure	4.2.2.1	71.02.801
Valve Shuttle, Low Pressure	4.2.2.2	71.02.802
Valve, Slide (Was: 71.03.750)	4.1.1.1	71.02.810
Valve, Slide, Linear (Was: 71.03.760)	4.1.1.1.1	71.02.811
Valve, Slide, Rotary (Was: 71.03.770)		71.02.812
Valve, Spool (Was: 71.03.740)	4.1.1.1.2	71.02.820
Valve, Surge Damping		71.02.850
Valve, Swing (See: Valve, Disc)		
Valve, Three Position		71.04.700
Valve, Three Position, Closed Center		71.04.701
Valve, Three Position, Open Center		71.04.702
Valve, Time Delay		71.02.900
Valve, Two Position		71.04.800
Valve, Unloading (See: Valve, Pressure Control)		
Variability Factor		04.01.657
Vent	4.6.2.22	31.01.600
Viscosity	10.3.1	04.01.700
Viscosity, Absolute	10.3.1.2	04.01.710
Viscosity Index	10.3.1.3	04.01.800
Viscosity Index, Improver	10.2.15	04.01.801
Viscosity, Kinematic	10.3.1.1	04.01.720
Viscosity, SAE Number		04.01.730
Viscosity, SUS		04.01.740
Visual Count, Contamination		30.06.730
Volume, Fluidic Control	4.6.1.21	72.01.950
Vortex	4.6.1.10	04.01.810

W

Entry	Code
Wash	33.07.240
Water	16.01.100
Water Glycol Fluid	14.01.100
Wound	33.08.150
Woven	33.08.160

NFPA Recommended Standard
T2.10.1M-1978

AN INDUSTRY STANDARD FOR FLUID POWER

Metric Units

For Fluid Power Applications

Approved as NFPA Recommended Standard
1 March 1978

published by
NATIONAL FLUID POWER ASSOCIATION, INC.

3333 N. Mayfair Road / Milwaukee, WI 53222 / 414-778-3344 / TLX 26898

NFPA/T2.10.1M

FOREWORD

This Foreword is not part of NFPA Recommended Standard Metric Units for Fluid Power Applications, NFPA/T2.10.1-1978.

In 1975, NFPA reactivated its Metrication Coordinating Committee to deal with the increasing concern with the application of the metric language to fluid power components. The Metrication Coordinating Committee immediately recognized the desirability of a metric practice standard which is tailored for the fluid power industry.

A Title, Scope and Purpose was prepared and submitted to the NFPA Technical Board. The TSP was approved on 21 August 1975 and Project Group meetings were held on 1 October 1975, 18 November 1975, and 6 January 1976. Although the Project Group reached consensus at its 6 January 1976 meeting, a number of editorial details delayed the preparation of the General Review Draft until 5 January 1977. General Review was conducted immediately thereafter. General Review comments were answered on 17 February 1977, and the project was approved for ballot by the Technical Board on 11 May 1977. Headquarters prepared the Ballot Draft on 20 June 1977.

A ballot of all NFPA Member Companies was conducted on 4 August 1977 and resulted in 3 negative ballots. Two of the three negatives were based on editorial objections which were easily resolved by Chairman Johnston. The remaining negative was philosophical in as much as it opposed the move toward metrication without close government coordination. Chairman Johnston was able to resolve this objection through additional discussion about the purpose of the document.

On 18 January 1978 the NFPA Technical Board unaminously recommended that this document be approved as an NFPA Recommended Standard. The NFPA Board of Directors granted approval to NFPA/T2.10.1-1978 on 1 March 1978.

NFPA/T2.10.1M

PROJECT GROUP MEMBERS WHO DEVELOPED THIS STANDARD

Johnston, N. B.	Project Chairman	International Harvester Company
Moyer, Don	Coordinating Committee Chairman	International Harvester Company
Schaeffer, Cliff	Coordinating Committee Vice Chairman	J I Case
Murphy, Ray	Coordinating Committee Secretary	Sundstrand Corporation
Chenoweth, Robert	Technical Auditor	Abex Corporation
Luecke, John R.	Director of National Technical Services	National Fluid Power Association

Fisher, J. L.	Bellows International
Mueller, J	The Weatherhead Co.
Ricker, E. C.	Abex Corporation
Schaeffer, C. O.	J I Case

REFERENCES

1. American National Standard Glossary of Terms for Fluid Power, ANSI/B93.2-1971, and Supplements thereto. (ISO/DP 5598)

2. SI units and recommendations for the use of their multiples and of certain other units, ISO 1000-1973.

3. American National Standard for Metric Practice, ANSI/ASTM E 380-76.

4. The International System of Units (SI), NBS Special Publication 330. (1977 Edition)

5. General principles concerning quantities, units and symbols, ISO 31/0-1974, ISO 31/1-1965, ISO 31/2-1958, ISO 31/3-1960, ISO 31/4-1960 and ISO 31/5-1965.

6. American National Standard Glossary of Terms Concerning Letter Symbols, ANSI/Y10.1-1972.

7. American National Standard Letter Symbols for Hydraulics, ANSI/Y10.2-1958.

8. Conditioning atmosphere - Test atmosphere - Reference atmosphere - Definitions, ISO/R 558-1967.

9. Standard atmospheres for conditioning and/or testing - Specifications, ISO 554-1976.

10. American National Standard Method for Conversion of Kinematic Viscosity to Saybolt Universal Viscosity or to Saybolt Furol Viscosity, ANSI/Z11.129-1975.

NFPA/T2.10.1M

METRIC UNITS FOR FLUID POWER APPLICATIONS

INTRODUCTION

In fluid power systems, power is transmitted and controlled through a fluid (liquid or gas) under pressure within an enclosed circuit. In fluid power technology, the use of metric units of measurement is increasing as result of the needs for international communication and in anticipation of the gradual obsolesence of customary U.S. units. In fact, measurement practice all over the world is in transition. Engineers who currently use, or are planning to use, metric units are confronted with the struggle between the noncoherent technical metric units such as the kilogram-force and the calorie, and the modern metric units (SI*) such as the newton and the joule.

The intent of this recommended standard is to provide the fluid power industry with guidance in the use of metric units.

1. SCOPE

To include quantities, metric units, their symbols, and their conversion factors for use in fluid power applications.

2. PURPOSE

To facilitate communication between manufacturer and user by providing a list of metric units for use in fluid power applications.

3. TERMS AND DEFINITIONS

See Reference No. 1 for definitions of terms used.

4. THE METRIC SYSTEM

4.1 General

The "metric system" of measurement used in this recommended standard is defined as SI with a limited number of additional units which are not formally recognized as SI units. SI is a coherent system of units created and maintained by the General Conference of Weights and Measures (CGPM), an international treaty organization. There are precise rules for the use of SI which must be followed in order to obtain the advantages of its simplicity and convenience. See Reference Nos. 2, 3 and 4 for more information about SI.

---

*Le Système International d'Unités - The International System of Units - abbreviated "SI" in all languages.

4.2 Differences from Technical Metric System

SI uses many of the same units as the technical metric system but also has many differences. The greatest difference is the departure of SI from the gravimetric system of metric units. SI is an absolute system and has distinct units for force and mass. The unit for force, the newton, is not related to gravity but is defined in terms of its acceleration of unit mass, the kilogram.

4.3 Mass and Force

Considerable confusion exists in the use of the term **weight** as a quantity to mean either **force** or **mass**. In commercial and everyday use, the term weight nearly always means mass; thus, when one speaks of a person's weight, the quantity referred to is mass. This nontechnical use of the term weight in everyday life will probably persist. In science the technology, the term **weight of body** has usually meant the force that, if applied to the body, would give it an acceleration equal to the local acceleration of free fall. The adjective "local" in the phrase "local acceleration of free fall" has usually meant a location on the surface of the earth: in this context the "local acceleration of free fall" has the symbol g (sometimes referred to as "acceleration of gravity" with observed values of g differing by over 0.5% at various points on the earth's surface. The use of **force of gravity** (mass times acceleration of gravity) instead of **weight** with this meaning is recommended. Because of the dual use of the term weight as a quantity, this term should be avoided in technical practice except under circumstances in which its meaning is completely clear. When the term is used, it is important to know whether mass or force is intended and to use SI units properly by using kilograms for mass or newtons for force.

4.4 Spelling of Metre and Litre

The Fluid Power Industry prefers the spellings metre and litre and encourages worldwide uniformity in the English spelling of the units of the International System. However, the widespread use of spellings meter and liter in the U.S. is recognized.

5. METRIC UNITS FOR FLUID POWER

5.1 General

In the past, specific units to be used in fluid power technology were selected from the many available customary U.S. units. For example, "GPM" was selected to describe flow; but "barrels per hour" could have been chosen. The same choices exist in the metric system. Table 1 presents a list of recommended metric units for use in the fluid power industry.

5.2 Conversion Factors

For those units not listed for fluid power applications use conversion factors in Reference No. 3.

TABLE 1 - Preferred metric units for fluid power

QUANTITIES						Conversion (Note 1)			
Name	Symbol	Typical Applications	Customary U.S. Units	Metric Units	U.S. Unit	Multiply By	To Get Metric	Multiply By	To Get U.S.
Acceleration	a	system damper, shock absorber	foot per second squared	metre per second squared	ft/sec^2	0.3048*	m/s^2	3.281	ft/sec^2
	g	gravity							
Angle, Plane	α, β etc.	swing arc, actuator rotation, conductor direction	degree minute second	degree minute second	° ′ ″	1.0* 1.0* 1.0*	° ′ ″	1.0* 1.0* 1.0*	° ′ ″
Area	A	orifice,	square inch	square millimetre	in^2	645.2	mm^2	0.001550	in^2
		heat exchanger	square inch	square centimetre	in^2	6.452	cm^2	0.1550	in^2
			square foot	square metre	ft^2	0.09290	m^2	10.76	ft^2
Conductivity, Thermal	λ	component material heating and cooling calculations	British thermal unit per hour foot degree Fahrenheit	watt per metre kelvin	$\dfrac{Btu}{hr \cdot ft \cdot °F}$	1.731	$\dfrac{W}{m \cdot K}$ (Note 10)	0.5778	$\dfrac{Btu}{hr \cdot ft \cdot °F}$
Cubic Expansion, Coefficient (Note 9)	γ	closed system pressure changes	per degree Fahrenheit	per degree Celsius	$\dfrac{1}{°F}$	1.8*	$\dfrac{1}{°C}$	$\dfrac{1}{1.8}$*	$\dfrac{1}{°F}$
Current, Electric	I	component operation rating	ampere	ampere	A	1.0*	A	1.0*	A

(continued)

NFPA/T2.10.1M

TABLE 1 - Preferred metric units for fluid power, continued

QUANTITIES						Conversion (Note 1)			
Name	Symbol	Typical Applications	Customary U.S. Units	Metric Units	U.S. Unit	Multiply By	To Get Metric	Multiply By	To Get U.S.
Density	ϱ	hydraulic fluids	pound per gallon	kilogram per litre	lb/gal	0.1198	kg/L (Note 2)	8.345	lb/gal
		gases	pound per cubic foot	kilogram per cubic metre	lb/ft^3	16.02	kg/m^3	0.06243	lb/ft^3
Displacement (Unit discharge) (Note 3)	q	pumps, motors, rotary actuators, intensifiers, cylinder, air and hand pumps	cubic inch	cubic centimetre	in^3	16.39	cm^3	0.06102	in^3
		(pneumatic) gallon	gallon	litre	gal	3.785	L (Note 2)	0.2642	gal
		(hydraulic) cubic inch	cubic inch	millilitre	in^3	16.39	mL Note 2)	0.06102	in^3
Energy	E	heat energy	British thermal unit	kilojoule	Btu	1.055	kJ	0.9478	Btu
Flow Rate, Heat	Φ	heat exchangers	British thermal unit per minute	watt	Btu/min	17.58	W	0.05687	Btu/min
Flow Rate, Mass	q_m	process pumps	pound per minute	gram per second	lb/min	7.560	g/s	0.1323	lb/min
			pound per second	kilogram per second	lb/s	0.4536	kg/s	2.205	lb/s
Flow Rate, Volume (Note 4)	q_v	pneumatic	cubic foot per minute	cubic decimetre per second	ft^3/min (cfm)	0.4719	dm^3/s	2.119	ft^3/min (cfm)
			cubic inch per minute	cubic centimetre per second	in^3/min (cim)	0.2731	cm^3/s	3.661	in^3/min (cim)
		hydraulic	gallon per minute	litre per minute	gal/min (gpm)	3.785	L/min (Note 2)	0.2642	gal/min (gpm)
			cubic inch per minute	millilitre per minute	in^3/min (cim)	16.39	mL/min (Note 2)	0.06102	in^3/min (cim)

(continued)

NFPA/T2.10.1M

TABLE 1 - Preferred metric units for fluid power, continued

QUANTITIES						Conversion (Note 1)			
Name	Symbol	Typical Applications	Customary U.S. Units	Metric Units	U.S. Unit	Multiply By	To Get Metric	Multiply By	To Get U.S.
Force	F	actuator thrust, spring force	pound	newton	lb	4.448	N	0.2248	lb
Force per Length	F/l	spring rate	pound per inch	newton per millimetre	lb/in	0.1751	N/mm	5.710	lb/in
Frequency, (Cycle)	f	acoustic vibration, electrical period, pressure pulsation	hertz (cycle per second)	hertz	Hz (cps)	1.0*	Hz	1.0*	Hz (cps)
		mechanical oscillation	cycle per minute	reciprocal minute	cpm	1.0*	1/min (Note 5)	1.0*	cpm
Frequency, (**R**otational)	n	rotational speed of pumps, motors, compressors and transmissions	revolution per minute	reciprocal minute	rpm	1.0*	1/min (Note 5)	1.0*	rpm
Heat	Q	quantity of heat	British thermal unit	kilojoule	Btu	1.055	kJ	0.9478	Btu
Heat Capacity, Specific	c	hydraulic fluid temperature change	British thermal unit per pound degree Fahrenheit	kilojoule per kilogram kelvin	$\frac{Btu}{lb \cdot °F}$	4.187	$\frac{kJ}{kg \cdot K}$ (Note 10)	0.2388	$\frac{Btu}{lb \cdot °F}$

(continued)

NFPA/T2.10.1M

TABLE 1 - Preferred metric units for fluid power, continued

QUANTITIES			Customary U.S. Units	Metric Units	U.S. Unit	Conversion (Note 1)			
Name	Symbol	Typical Applications				Multiply By	To Get Metric	Multiply By	To Get U.S.
Heat Transfer, Coefficient of	h	heating and cooling, thermal conductance	British thermal unit per hour square foot degree Fahrenheit	watt per square metre kelvin	$\frac{Btu}{hr \cdot ft^2 \cdot °F}$	5.678	$\frac{W}{m^2 \cdot K}$ (Note 10)	0.1761	$\frac{Btu}{hr \cdot ft^2 \cdot °F}$
Inertia, Moment of	I	links, rocker arms, walking beams	pound foot squared	kilogram metre squared	$lb \cdot ft^2$	0.04214	$kg \cdot m^2$	23.73	$lb \cdot ft^2$
			pound inch squared	kilogram metre squared	$kg \cdot m^2$	0.0002926	$kg \cdot m^2$	3417.3	$lb \cdot in^2$
Length	l	bore and stroke, orifice diameter	inch	millimetre	in	25.4*	mm	0.03937	in
		component location	foot	metre	ft	0.3048*	m	3.281	ft
		filter rating, particle size	micron	micrometre	μ	1.0*	μm	1.0*	μ
		surface finish	microinch	micrometre	μin	0.0254*	μm	39.37	μin
Linear Expansion, Coefficient of (Note 8)	α	shrink fit, shaft, gears	per degree Fahrenheit	per degree Celsius	$\frac{1}{°F}$	1.8*	$\frac{1}{°C}$	$\frac{1}{1.8}$*	$\frac{1}{°F}$
Mass	m	component weight	pound	kilogram	lb	0.4536	kg	2.205	lb
Modulus, Bulk	K	system fluid compression	pound per square inch	megapascal	lb/in^2 (psi)	0.006895	MPa	145.0	lb/in^2 (psi)

(continued)

NFPA/T2.10.1M

TABLE 1 - Preferred metric units for fluid power, continued

QUANTITIES			Customary U.S. Units	Metric Units	U.S. Unit	Conversion (Note 1)			
Name	Symbol	Typical Applications				Multiply By	To Get Metric	Multiply By	To Get U.S.
Momentum	p	shock absorbers	pound foot per second	kilogram metre per second	$\dfrac{\text{lb·ft}}{\text{sec}}$	0.1383	kg·m/s	7.233	lb·ft/sec
Potential, Electric	V	component	volt	volt	V	1.0*	V	1.0*	V
Power	P	input and output of hydraulic and pneumatic systems	horsepower (550 ft·lb/sec)	kilowatt	hp	0.7457	kW	1.341	hp
		heat exchangers	British thermal unit per minute	watt	Btu/min	17.58	W	0.05687	Btu/min
Pressure	p	component and system rating, atmospheric pressure	pound per square inch	bar (Note 6)	lb/in^2 (psi)	0.06895	bar	14.50	lb/in^2 (psi)
				kilopascal (Note 6)	lb/in^2 (psi)	6.895	kPa	0.1450	lb/in^2 (psi)
Stress, Normal	σ	normal stress, shear stress, strength of materials	pound per square inch	megapascal	lb/in^2 (psi)	0.006895	MPa	145.0	lb/in^2 (psi)
Shear	τ								
Temperature, Customary	θ	thermal operating limits of components, dryer dew-point	degree Fahrenheit	degree Celsius	°F	$t_{°C} = \dfrac{(t_{°F}-32)}{1.8}$*	°C	$t_{°F} = 1.8(t_{°C})+32$*	°F

(continued)

NFPA/T2.10.1M

TABLE 1 - Preferred metric units for fluid power, continued

QUANTITIES						Conversion (Note 1)			
Name	Symbol	Typical Applications	Customary U.S. Units	Metric Units	U.S. Unit	Multiply By	To Get Metric	Multiply By	To Get U.S.
Temperature, Thermodynamic (absolute)	T	pneumatic system calculations	degree Rankine	kelvin	°R	1/1.8*	K	1.8*	°R
Time	t	component response, system cycle, maintenance cycle	second	second	sec	1.0*	s	1.0*	sec
			minute	minute	min	1.0*	min	1.0*	min
			hour	hour	hr	1.0*	h	1.0*	hr
Torque	T	motor, rotary actuator	pound foot	newton metre	lb·ft	1.356	N·m	0.7376	lb·ft
		output, bolt tightening	pound inch	newton metre	lb·in	0.1130	N·m	8.851	lb·in
Velocity, linear	v	fluid flow	foot per second	metre per second	ft/sec	0.3048*	m/s	3.281	ft/sec
		cylinder actuator	inch per second	millimetre per second	in/sec	25.4*	mm/s	0.03937	in/sec
Viscosity, Dynamic	η	system fluid properties	centipoise	millipascal second	cP	1.0*	mPa·s	1.0*	cP
Viscosity, Kinematic	v	system fluid properties	centistokes (Note 7)	square millimetre per second	cSt	1.0*	mm²/s	1.0*	cSt
Volume	V	lubricators, accumulators, reservoir capacity (pneumatics)	cubic foot	cubic decimetre	ft³	28.32	dm³	0.0353	ft³
		(pneumatics)	cubic inch	cubic centimetre	in³	16.39	cm³	0.06102	in³
		(hydraulics)	gallon	litre	gal	3.785	L (Note 2)	0.2642	gal
		(hydraulics)	ounce	millilitre	oz	29.57	mL (Note 2)	0.03381	oz
Work	W	cylinders, linear actuators	foot pound	joule	ft·lb	1.356	J	0.7376	ft·lb

See Notes to Table 1, which follows

NOTES TO TABLE 1

1. An asterisk (*) after the conversion factor indicates that it is exact and that all subsequent digits are zeros. All other conversion factors have been rounded to four significant digits. See Reference No. 3 if more exact conversion factors are required.

2. The international symbol for litre is the lowercase "l" which can easily be confused with the numeral "1". Accordingly, the symbol "L" is recommended for U.S. fluid power use.

3. Indicate displacement of a rotary device as "per revolution" and of a non-rotary device as "per cycle".

4. For gases, this quantity is frequently expressed as free gas at Standard Reference Atmosphere as defined in Reference No. 8 and specified in Reference No. 9. In such cases, the abbreviation "ANR" is to follow the expression of the quantity, "q_v(ANR)" or the unit, "m^3/min (ANR)". See Reference 3, for more detailed information of the attachment of letters to a unit symbol.

5. Mechanical oscillations are normally expressed in cycles per unit time and rotational frequency in revolutions per unit time. Since "cycle" and "revolution" are not units, they do not have internationally recognized symbols. Therefore, they are normally expressed by abbreviations which are different in various languages. In English the symbology for mechanical oscillations is c/min and for rotational frequency, r/min.

6. The bar and kilopascal are given equal status as pressure units. At this time the domestic fluid power industry does not agree on one preferred unit. The pascal is the SI unit for pressure and a major segment of U.S. industry has accepted the multiple, kilopascal (kPa), as the preferred unit. On the other hand, the majority of the international fluid power industry has accepted the bar as the metric pressure unit.

    The bar is recognized by the EEC as an acceptable metric unit and is shown in ISO 1000 for use in specialized fields. Conversely, the bar is considered by both the International Committee on Weights and Measures and the NBS as a unit to be used for a limited time only. Further, the bar has been deprecated by Canada, ANMC, and some U.S. standards organizations and is illegal in some countries.

7. Viscosity is frequently expressed in SUS (Saybolt Universal Seconds). SUS is the time in seconds for 60 mL of fluid to flow through a standard orifice at a specified temperature. Conversion between kinematic viscosity, mm^2/s (centistokes) and SUS can be made by reference to tables in Reference No. 10.

8. The linear expansion coefficient is a ratio, not a unit, and is expressed in customary U.S. units such as in/in and in metric units such as mm/mm per unit temperature change.

9. The cubic expansion coefficient is a ratio, not a unit, and is expressed in customary U.S. units such as in^3/in^3 and in metric units such as cm^3/cm^3 per unit temperature change.

10. In these expressions K indicates temperature interval. Therefore K may be replaced with °C if desired without changing the value or affecting the conversion factor.

NFPA/T2.10.1M

6. **IDENTIFICATION STATEMENT**

Use the following statement in catalogs and sales literature when electing to comply with this voluntary standard:

"Metric units are selected and used in accordance with NFPA/T2.10.1-1978."

7. **KEY WORDS**

The following Key Words, useful in indexes and in information retrieval systems, are suggested for this recommended standard:

fluid power
metric conversion
metric units, fluid power
quantities
SI units
units, metric

NFPA Recommended Standard
T2.10.2M-1977

AN INDUSTRY STANDARD FOR FLUID POWER

Survey on Metric Language
Usage by the U.S. Fluid
Power Industry

Approved for Release on
1 December 1977

---

published by
NATIONAL FLUID POWER ASSOCIATION, INC.
3333 N. Mayfair Road / Milwaukee, WI 53222 / 414-778-3344 / TLX 26898

EXECUTIVE SUMMARY

In December 1976 the National Fluid Power Association conducted a survey of its membership to assess the present and planned use of the metric language by the U.S. fluid power industry. The survey examined two general areas: (1) Product Description and (2) In-Plant Use. Questionnaires were sent to 162 NFPA member companies. On 8 February 1977, 127 questionnaires had been returned. Because of the high rate of return, the Association did not conduct a follow-up.

After reviewing the questionnaire results, the project group responsible for this study reached the following conclusions:

1. Many fluid power manufacturers appear to either be using or considering the use of dual dimensioning. Fewer manufacturers have chosen to use only metric units.

2. A significant number of manufacturers are using or are planning to use metric units in product descriptions; however, a very limited number of manufacturers reported the present or planned in-plant use of metric units.

3. Three-quarters of those responding indicated that they are either using or considering the use of metric units.

4. One-half of the companies responding are using metric units in some aspect of their operation.

5. One-quarter of those responding are neither using nor considering the use of metric units.

6. The open-ended comments varied, but seemed to center around the following ideas:

    6.1 Manufacturers are waiting for customer demand.

    6.2 Manufacturers are moving cautiously.

    6.3 Manufacturers are waiting for the establishment of firm metric standards before initiating metric production.

7. This survey did not produce sufficient evidence to identify any trends in the use of metric language by the U. S. fluid power industry.

NFPA/T2.10.2M

INTRODUCTION

This report summarizes the findings of a study conducted to assess the present and planned use of the metric language by the U.S. fluid power industry. The purpose of this study is to provide NFPA member companies and other interested organizations with an industry profile describing the use of the metric language.

In December 1976, 162 self-administered questionnaires were mailed to member companies Official Representatives. An Official Representative is that individual selected to serve as spokesman for an NFPA member company. An introductory letter and business reply envelope accompanied the questionnaire. (See Appendix X). The introductory questionnaire was developed by a Project Group of the NFPA Metrication Coordinating Committee. The Project Group was chaired by F. Flick (Miller Fluid Power) and included A. Komarek (Miller Fluid Power), R. Murphy (Sundstrand), D. Moyer (International Harvester), R. Dix (IIT), R. Berg (WABCO) and J. Luecke (NFPA).

A total of 127 questionnaires were returned before the cut-off date of 31 January 1977. Because of the high response rate, no attempts were made to contact non-respondents. Data relating to non-respondents of questions 1 thru 9 is tabulated on pages 8, 9, 10 and 11 of the Statistical Summary. (Appendix Y)

For ease of reporting and interpretation, all of the quantitative responses have been converted into percentages. All percentages were calculated on the basis of the number of responses to the individual question, and have been rounded off to the nearest tenth of a percent. Consequently, the percentages may not always add up to 100%.

The NFPA Metrication Coordinating Committee reviewed the final draft on 10 November 1977 and recommended its release with some minor modifications. The document was circulated to the NFPA Technical Board on 14 November 1977. The Technical Board review resulted in one comment that recommended an amplification in the Executive Summary. This change was made and the document was submitted to the NFPA Board of Directors for approval to release the document as an "Information Report". The NFPA Board of Directors granted approval at its 1 December 1977 meeting.

NFPA/T2.10.2M

SUMMARY OF RESULTS

Past Use of the Metric Language

Product Descriptions

Approximately one-fifth (21.3%) of those responding reported the use of metric units alone (no customary U. S. units) for **product descriptions** in catalogs, product drawings, spec sheets, brochures, etc. **at the special request of the customer. On a voluntary basis,** nearly one-fourth (23.6%) reported the use of metric units alone outside the U. S. and only six percent (6.3%) reported their use in the U. S. The highest increment in the use of metric units alone, either voluntarily or at customer request, was reported in 1970 (8.7%) and 1975 (8.5%).

Nearly thirty-eight percent (37.8%) of those responding reported the use of **both metric units and customary U. S. units** for product descriptions at the customer's request. A similar number (38.6%) reported the voluntary use of metric and customary U. S. units outside the U. S. and nearly forty percent (39.6%) used both measurement systems in the U. S. The use of both systems of measurement started to increase appreciably in 1970 (10.3%) and reached a high point in 1975 (28.4%).

In-Plant Use

With regard to the use of metric units on drawings for in-plant use, approximately four percent (3.9%) reported the exclusive use of metric units on drawings and one-fifth (20.5%) reported the use of both metric and customary U. S. units on drawings. Most in-plant use of metric units (either alone or with customary U. S. equivalents) has taken place since 1970.

Less than two percent (1.6%) reported the use of metric equipment as a single measurement system for in-plant use, while one tenth (10.2%) reported the use of both metric and customary U. S. equipment for in-plant measuring systems. In the latter case, customary U. S. measurement equipment is gradually being replaced by metric measurement equipment.

NFPA/T2.10.2M

Anticipated Use of the Metric Language

Product Description

Only three percent (3.1%) have definite plans to start using metric units for product descriptions **at the special request of the customer.** Nearly seventeen percent (16.5%) of those reporting indicated that they had plans for using metric units for product descriptions, **but no definite schedule.** Close to half of the respondents (47.2%) reported no plans for the use of metric units on product descriptions when requested by the customers.

Regarding the **voluntary use of metric units alone** for product descriptions, only three percent (3.1%) reported definite plans for the sole use of metric units **outside the U. S.** and less than one percent (0.8%) indicated definite plans for the voluntary use of metric units on product descriptions within the U. S. More than thirteen percent (13.4%) indicated that they plan to use metric units alone for product descriptions outside the U. S., but lacked definite plans to do so. Approximately half of those responding (50.4%) reported no plans for the sole use of metric units on product descriptions outside the U. S.

Within the U. S. nearly nineteen percent (18.9%) of those responding indicated that they had plans for the voluntary use of metric units on product descriptions, but did not have a definite schedule. More than sixty percent (61.4%) reported no plans for the voluntary use of metric units on product descriptions within the U. S.

The use of **both metric and customary U. S. units** on product descriptions is anticipated by approximately seven percent (7.1%) of the respondents **at the request of the customer.** Eight percent (7.9%) of those responding reported plans, but no schedule, for dealing with requests of product descriptions in both metric and customary U. S. units. One-fourth (27.6%) of the respondents indicated no plans for the use of metric and customary U. S. units on product descriptions, when requested by the customer.

Outside the U. S. four percent (3.9%) of the respondents plan to **voluntarily** use both metric and customary U. S. units on product descriptions. Another thirteen percent (13.4%) of the respondents indicated similar plans, but no definite schedule. More than thirty percent (31.5%) of those responding reported no plans for the voluntary use of metric and customary U. S. units on product descriptions outside the U. S.

Within the U. S. eight percent (7.9%) of the respondents indicated definite plans to voluntarily use metric and customary U. S. units on product descriptions. Another sixteen percent (15.7%) reported plans, but no schedule, and the remaining thirty percent (29.9%) reported no plans for the voluntary use of metric and customary U. S. units on product descriptions in the U. S.

In-Plant Use

For **in-plant use,** slightly more than two percent (2.4%) reported definite plans for the **sole use of metric units** on drawings. Approximately seventeen percent (17.3%) reported plans for the in-plant use of metric units on drawings, but no schedule. Just over seventy percent (70.8%) of those responding reported no plans for the use of metric units on drawings for in-plant use.

The planned use of **both metric and customary U. S.** units on drawings for in-plant use is anticipated by approximately four percent (3.9%) of the respondents. Another twenty-two percent (22.0%) of the respondents indicated plans, but no schedule, for the use of both metric and customary U. S. units on drawings for in-plant use; while forty-seven percent (47.2%) have made no such plans.

Slightly more than two percent (2.4%) reported they planned **exclusive use of metric measuring equipment.** Seventeen percent of those responding reported plans, but no schedule, to use only metric measuring equipment, and the remaining three-quarters (72.4%) indicated no plans for the sole use of metric measuring equipment.

The planned use of **both metric and customary U. S. measuring equipment** (during a gradual conversion to metric) was reported by nearly eight percent (7.9%) of the respondents. Twenty-nine percent (29.1%) of the respondents indicated plans, but no schedule, for the gradual conversion of customary U. S. measuring equipment to metric measuring equipment; and the remaining forty-seven percent (47.2%) indicated no plans in this area.

Metric Language Impact

Those repondents now using metric units assessed the impact of their metrication efforts as follows:

1. Customers: twenty-seven percent (26.8%) reported a minor impact.

2. Data Processing: Twenty-three percent (22.8%) considered the impact to be minor.

3. Personnel (employees): Twenty percent (20.4%) considered the impact to be minor.

4. Product Engineering: Approximately seventeen percent (16.5%) indicated a minor impact.

5. Manufacturing: Nearly nineteen percent (18.9%) considered the impact to be minor.

6. Suppliers: Again, nineteen percent (18.9%) considered the impact to be minor.

Those respondents who plan to use metric units expect metrication to have the following impact:

1. Customers: Fifteen percent (14.9%) expect metrication to have some impact on customers.

2. Data Processing: Fifteen percent (14.9%) anticipate a minor impact.

3. Personnel (employees): Fourteen percent (14.2%) expect a major impact on personnel.

4. Product Engineering: The response to this item was bi-modal. Twelve percent (11.8%) expect a major impact and thirteen percent (12.6%) expect some impact on this group.

5. Manufacturing: Nearly nineteen percent (18.9%) expect a major impact.

6. Suppliers: Approximately sixteen percent (15.7%) expect a major impact.

NOTE: With the noted exception, the above responses are modal. For full information on these items, see Appendix Y, pages AY-1 - AY-11.

CONCLUSIONS

It is significant that in 1976 - the year in which the Metric Conversion Act was signed into law by President Ford – many U. S. fluid power manufacturers had already embraced the metric language in some aspect of their operation, e.g. product descriptions, in-plant use, with or without customary U. S. units, within and outside the United States. It is equally significant to note that U. S. fluid power manufacturers are moving with caution and are waiting for a demand to arise from the marketplace. This approach appears to be consistent with other U. S. industries that manufacture capital equipment.

With regard to the impact, experienced and anticipated, of the metric language, the fluid power industry also appears consistent with other capital equipment industries. Those who are using the metric language find its impact to be of minor importance, while those who are planning its use expect a considerable impact.

NOTES

APPENDIX X TO

NFPA/T2.10.2M-1977

METRIC LANGUAGE USAGE QUESTIONNAIRE

IN U. S. PLANTS

NFPA/T2.10.2M

National Fluid Power Association

3333 N. Mayfair Road - Milwaukee, Wisconsin 53222
Phone (414) 259-0990   TLX 26898

Date: 1 December 1976
Brief Title: Voluntary Metric Language Usage
Project No. T2.10.2

TO: NFPA Official Representatives

   To the Attention of Mr. _____

The NFPA Metrication Coordinating Committee is conducting a study to assess the present and planned use of the metric language by the fluid power industry. The purpose of this study is to provide NFPA member companies and other interested organizations with an industry profile describing the use of the metric language. This study will not establish a timetable, nor will it inquire about corporate plans regarding dimensional changes of products.

The need for this study was recognized when President Ford signed the Metric Conversion Act of 1975 which declares "that the policy of the United States shall be to coordinate and plan the increasing use of the metric system in the United States and to establish a United States Metric Board to coordinate the voluntary conversion to the metric system." The Metric Conversion Act defines "metric system" as the "International System of Units (SI)". After careful study, the Committee concluded that the main thrust of the Metric Conversion Act is directed at the <u>use of the metric language, or SI</u>, and, therefore, it would be helpful if fluid power manufacturers and users had an indication of the trends in the use of the metric language.

This report is intended to provide the fluid power industry with data on the present and planned use of the metric language, thus helping the industry to successfully and economically deal with the problems inherent in metrication.

Please complete the attached questionnaire and return it to NFPA's accountant, W. A. Knorr & Co., S.C., Certified Public Accountants, P.O. Box 265, Cedarburg, Wisconsin 53012 by <u>31 January 1977</u>.   A pre-addressed envelope is included for your convenience. The information you provide will be summarized into a final report by the accountant. This procedure will insure the confidentiality of your information. Complimentary copies of the final report will be provided to those NFPA member companies which participate in the study. Ninety days after the distribution of the complimentary copies, the final report will be released to non-participating NFPA member companies and to the general public.

Thank you for your cooperation.

Very truly yours,

John R. Luecke
John R. Luecke
Director of National Technical Services

ODE NO. _____   NFPA/T2.10.2M

METRIC LANGUAGE USAGE QUESTIONNAIRE
IN U.S. PLANTS

Introduction

This study is intended to assess the present and planned use of the metric language in the fluid power industry in U.S. plants. It will not be used to establish an industry timetable, nor will it deal with dimensional changes of products. Please respond to this questionnaire by providing the appropriate year date for items 1-6 and information as required in items 7-10. Indicate which respresentative(s) answered the questionnaire by checking the appropriate box(es). One questionnaire per company.

☐ Official Representative   ☐ Technical Representative

            ☐ Marketing Representative

Please return by 31 January 1977

	Used Since	Plan to Start Using	Plans But No Schedule	No Definite Plans
The use of metric units alone (no customary U.S. units) for product descriptions in catalogues, product drawings, spec sheets, brochures, etc.				
1.1 Used at special request of the customer.	19___	19___	___	___
1.2.1 outside the United States	19___	19___	___	___
1.2.2 in the United States	19___	19___	___	___
The use of both metric units and customary U.S. units for product descriptions in catalogues, product drawings, spec sheets, brochures, etc.				
2.1 Used at special request of the customer.	19___	19___	___	___
2.2.1 outside the United States	19___	19___	___	___
2.2.2 in the United States	19___	19___	___	___

continued. . .

NFPA/T2.10.2M

METRIC LANGUAGE USAGE QUESTIONNAIRE
IN U.S. PLANTS (continued)

	Used Since	Plan to Start Using	Plans But No Schedule	No Definite Plans
3. The use of metric units alone (no customary U.S. units) on drawings for in-plant use.	19___	19___	___	___
4. The use of both metric units and customary U.S. units on the same drawing for in-plant use.	19___	19___	___	___
5. The use of metric unit measuring equipment as a single measuring system for in-plant use.	19___	19___	___	___
6. The use of both metric unit and customary U.S. unit measuring equipment (a combination of metric and inch languages) during a gradual conversion in which customary U.S. unit instruments are replaced, as needed, by metric instruments.	19___	19___	___	___

7. If your company <u>now uses metric units</u>, please describe its impact on the following groups by placing an "x" in the appropriate space:

    Customers   Major ___ ___ ___ ___ ___ Minor

    Data Processing   Major ___ ___ ___ ___ ___ Minor

    Personnel (employees)   Major ___ ___ ___ ___ ___ Minor

    Product Engineering   Major ___ ___ ___ ___ ___ Minor

    Manufacturing   Major ___ ___ ___ ___ ___ Minor

    Suppliers   Major ___ ___ ___ ___ ___ Minor

    Other _____ Major ___ ___ ___ ___ ___ Minor
        (please specify)

(continued)

NFPA/T2:10.2M

METRIC LANGUAGE USAGE QUESTIONNAIRE
IN U.S. PLANTS (concluded)

8.0 If your company <u>plans to use metric units</u>, what do you expect its impact to be on the following groups: Mark an "x" in the appropriate space.

Customers	Major __ __ __ __ __	Minor
Data Processing	Major __ __ __ __ __	Minor
Personnel (employees)	Major __ __ __ __ __	Minor
Product Engineering	Major __ __ __ __ __	Minor
Manufacturing	Major __ __ __ __ __	Minor
Suppliers	Major __ __ __ __ __	Minor
Other _____ (please specify)	Major __ __ __ __ __	Minor

9.0 Does your company have a formal policy towards metric unit conversion.

Yes _____   No _____

10.0 General Comment, if any:

NOTES

NFPA/T2.10.2M

APPENDIX Y TO

NFPA/T2.10.2M-1977

STATISTICAL SUMMARY

NFPA/T2.10.2M

February 8, 1977

National Fluid Power Association
3333 N. Mayfair Road, Suite 311
Milwaukee, WI 53222

Gentlemen

We have reviewed the National Fluid Power Association Metric Language Usage In U.S. Plants questionnaire submitted by 127 member companies and tabulated the information in accordance with your instructions.

Individual company data has been kept under our control and in the strictest confidence. No data that can be specifically identified with a reporting member will be made available to the association's members, it's officers, or to any other persons. After a limited period of time, the completed questionnaires will be destroyed.

Very truly yours

W.A. KNORR & CO., s.c.

METRIC LANGUAGE USAGE IN U.S. PLANTS

RESPONDING COMPANIES

6, 8, 9, 10, 12, 13, 14, 18, 20, 21, 22, 23, 24, 25, 28, 33, 34, 36, 43, 46, 49, 50, 52, 53, 54, 59, 60, 62, 65, 67, 68, 69, 70, 77, 78, 82, 83, 87, 88, 89, 90, 92, 94, 99, 100, 101, 108, 115, 116, 118, 119, 123, 124, 125, 133, 138, 139, 140, 145, 151, 152, 154, 155, 158, 159, 160, 161, 162, 164, 168, 170, 173, 175, 177, 181, 183, 193, 199, 208, 211, 216, 223, 224, 226, 231, 232, 238, 239, 244, 245, 246, 248, 249, 250, 251, 254, 256, 257, 260, 263, 264, 266, 268, 269, 272, 274, 275, 276, 277, 279, 282, 285, 288, 289, 290, 292, 293, 294, 296, 301, 302, 304, 306, 307, 308, 309, 310

One of the responding companies replied "We are magazine publishers and this doesn't apply." Another responded "As an educational institution IIT does not manufacture a product in the fluid power area. Thus, I am not supplying data, but returning the questionnnaire blank."

NFPA/T2.10.2M

METRIC LANGUAGE USAGE IN U.S. PLANTS

USED SINCE

PERCENT OF COMPANIES RESPONDING

	Prior to 1950	1950 to 1955	1955 to 1960	1960 to 1965	1965	1966	1967	1968	1969	1970	1971	1972	1973	1974	1975	1976
1. The use of metric units alone (no customary U.S. units) for product drawings, spec sheets, brochures, etc.																
1.1 Used at special request of the customer 21.3%*					.8	.8		.8	.8	3.9	1.6	2.4	.8	2.4	3.9	1.5
1.2 Used voluntarily:																
1.2.1 outside the United States 23.6%*	1.6	1.6	.8	3.1	1.6	.8	1.5		.8	2.4	.8	1.6	.8	1.5	3.1	1.6
1.2.2 in the United States 6.3%*								.8		2.4					1.5	1.6
2. The use of both metric units and customary U.S. units for product descriptions in catalogues, product drawings, spec sheets, brochures, etc.																
2.1 Used at special request of the customer 37.8%*	.8				.8	1.6	.8	.8	2.3	5.5	2.4	2.4	3.1	6.3	6.3	4.7
2.2 Used voluntarily:																
2.2.1 outside the United States 38.6%*	.8	2.4	3.1	.8	3.1	.8	.8	.8	.8	2.4	.8	3.9	2.4	5.5	7.9	3.9
2.2.2 in the United States 39.4%*			.8		.8			.8	.8	2.4	.8	3.9	4.7	4.7	14.2	5.5
3. The use of metric units alone (no customary U.S. units on the same drawing for in-plant use. 3.9%*								.8						.8	1.5	
4. The use of both metric units and customary U.S. units on the same drawing for in-plant use. 20.5%*				.8			.8	.8	.8	1.6	.8	1.6	2.4	3.1	3.1	4.7

NFPA/T2.10.2M

METRIC LANGUAGE USAGE IN U.S. PLANTS

USED SINCE

	Prior to 1950	1950 to 1955	1955 to 1960	1960 to 1965	1965	1966	1967	1968	1969	1970	1971	1972	1973	1974	1975	1976
5. The use of metric unit measuring equipment as a single measuring system for in-plant use. 1.6%*																.8
6. The use of both metric unit and customary U.S. unit measuring equipment (a combination of metric and inch languages) during a gradual conversion in which customary U.S. unit instruments are replaced, as needed, by metric instruments 10.2%*					.8			.8		.8		.8	2.4	.8	3.1	1.5

PERCENT OF COMPANIES RESPONDING

* Percent of companies responding to question.

NFPA/T2.10.2M

METRIC LANGUAGE USAGE IN U.S. PLANTS

PLAN TO START USING
PERCENT OF COMPANIES RESPONDING

	1976	1977	1978	1979	1980	1982	1985	1987
1. The use of metric units alone (no customary U.S. units) for product descriptions in catalogues, product drawings, spec sheets, brochures, etc.								
1.1 Used at special request of the customer. 3.1%*		2.3			.8			
1.2 Used voluntarily:								
1.2.1 outside the United States 3.1%*		2.3				.8		
1.2.2 in the United States .8%*								.8
2. The use of both metric units and customary U.S. units for product descriptions in catalogues, product drawings, spec sheets, brochures, etc.								
2.1 Used at special request of the customer. 7.1%*	.8	4.7	1.6					
2.2 Used voluntarily:								
2.2.1 outside the United States 3.9%*	.8	2.3	.8					
2.2.2 in the United States 7.9%*		5.5	2.4					
3. The use of metric units alone (no customary U.S. units) on drawings for in-plant use. 2.4%*			1.6		.8			
4. The use of both metric units and customary U.S. units on the same drawing for in-plant use. 3.9%*	.8	1.6	1.5					
5. The use of metric unit measuring equipment as a single measuring system for in-plant use. 2.4%*					1.6		.8	

METRIC LANGUAGE USAGE IN U.S. PLANTS

	PLAN TO START USING						
1976	1977	1978	1979	1980	1982	1985	1987
PERCENT OF COMPANIES RESPONDING							
.8	3.1	2.4	.8	.8			

6. The use of both metric unit and customary U.S. unit measuring equipment (a combination of metric and inch languages) during a gradual conversion in which customary U.S. unit instruments are replaced, as needed, by metric instruments. 7.9%*

* Percent of companies responding to question.

NFPA/T2.10.2M

METRIC LANGUAGE USAGE IN U.S. PLANTS

	Plans But No Schedule	No Definite Plans	Did Not Answer
	PERCENT OF COMPANIES		
1. The use of metric units alone (no customary U.S. units) for product descriptions in catalogues, product drawings, spec sheets, brochures, etc.	16.5	47.2	11.9
1.1 Used at special request of the customer			
1.2 Used voluntarily:			
1.2.1 outside the United States	13.4	50.4	9.5
1.2.2 in the United States	18.9	61.4	12.6
2. The use of both metric units and customary U.S. units for product descriptions in catalogues, product drawings, spec sheets, brochures, etc.			
2.1 Used at special request of the customer	7.9	27.6	19.6
2.2 Used voluntarily:			
2.2.1 outside the United States	13.4	31.5	12.6
2.2.2 in the United States	15.7	29.9	7.1
3. The use of metric units alone (no customary U.S. units) on drawings for in-plant use	17.3	70.8	5.6
4. The use of both metric units and customary U.S. units on the same drawing for in-plant use	22.0	47.2	6.4

NFPA/T2.10.2M

METRIC LANGUAGE USAGE IN U.S. PLANTS

	Plans But No Schedule	No Definite Plans	Did Not Answer
	PERCENT OF COMPANIES		
5. The use of metric unit measuring equipment as a single measuring system of in-plant use	16.5	72.4	7.1
6. The use of both metric unit and customary U.S. unit measuring equipment (a combination of metric and inch languages) during a gradual conversion in which customary U.S. unit instruments are replaced, as needed, by metric instruments	29.1	47.2	5.6

METRIC LANGUAGE USAGE IN U.S. PLANTS

7. The Metrication Impact By Companies Now Using Metric Units On The Following Groups

	Major				Minor	Did Not Answer
Customers	.8		5.5	3.1	26.8	31.5
Data Processing	.8	2.4	3.1	2.4	22.8	36.2
Personnel (Employees)	2.4	1.6	6.3	3.1	20.4	33.9
Product Engineering	3.9	1.6	8.7	3.1	16.5	33.9
Manufacturing	4.7	2.4	3.2	3.9	18.9	34.6
Suppliers	1.6	2.4	4.7	4.7	18.9	35.4
Other (Specify):						
Quality Control Testing	.8					
Research and Testing		.8		.8		
Licensee	.8					
Cost	.8					
U.S. Sales		.8				
Instruments and Lab Equipment						
No Discription				.8	2.4	

67.7% of the responding companies were eligible to respond to this section.

1.6% of the responding companies not presently using metric units responded to this section, therefore, were not included in the results.

METRIC LANGUAGE USAGE IN U.S. PLANTS

8. The Metrication Impact By Companies Planning To Use Metric Units On The Following Groups

	Major		Minor	Did Not Answer
Customers	6.3	4.7	14.9	5.5	7.1	20.5

	Major			Minor	Did Not Answer
Customers	6.3	4.7	14.9	5.5	7.1	20.5
Data Processing	7.1	3.9	3.2	6.3	14.9	23.6
Personnel (Employees)	14.2	5.5	11.0	3.9	3.1	21.3
Product Engineering	11.8	4.7	12.6	4.7	3.9	21.3
Manufacturing	18.9	9.4	7.1	2.4	1.6	19.6
Suppliers	15.7	5.5	10.2	4.7	2.4	20.5
Other (Specify):						
Licensee	.8					
Purchasing and Stores		.8				
Management		.8				
Sales/Distribution		.8				
Research and Testing					.8	
No Description		.8	1.6	1.6	.8	

59.0% of the responding companies were eligible to respond to this section.

Approximately 17.3% of the responding companies not planning to use metric units responded to this section, therefore, were not included in the results.

9. Of the responding companies, 20.5% have a formal policy towards metric unit conversion, 75.6% do not, and 3.9% did not respond to this section.

NOTES

NFPA Recommended Standard
T3.9.13-1982

AN INDUSTRY STANDARD FOR FLUID POWER

Hydraulic fluid power -

Pumps and motors -

Glossary

Approved as NFPA Recommended Standard
2 June 1982

---

published by
NATIONAL FLUID POWER ASSOCIATION, INC.
3333 N. Mayfair Road / Milwaukee, WI 53222 / 414-778-3344 / TLX 26898

NFPA/T3.9.13

FOREWORD

This Foreword is not part of NFPA Recommended Standard - Hydraulic fluid power - Pumps and motors - Glossary, NFPA/T3.9.13-1982.

On 17 March 1970, the NFPA Pump & Motor Section agreed to undertake the definition of terms specifically related to hydraulic fluid power pumps, motors and hydrostatic transmissions. This project was assigned Project Number T3.9.13.

The Pump & Motor Section appointed a Project Group on 28 September 1971. Draft No. 1 was presented to the Section for review at the 22 March 1972 meeting. Comments received were incorporated into Draft No. 2, prepared on 22 January 1973.

After much inaction, on 9 November 1977, proposed terms and definitions were reviewed and changes recommended were incorporated into Draft No. 3. The Project Group met on 19 April 1978 to discuss and review Draft No. 3. On 8 August 1979, Draft No. 4 was developed. It was reviewed at the 18 September 1979 Project Group meeting. On 26 September 1979, Draft No. 5 was compiled and incorporated all of the changes made at the 18 September meeting (It is the first draft which combines Draft No. 4 of the pump terms and Draft No. 2 of the motor terms).

At the request of Project Chairman Clay Thorson, Draft No. 5 was sent to members of the T3.9 Section for review and comment. These comments were discussed at the 14 April 1980 Project Group meeting and it was recommended that Chairman Thorson incorporate these changes into the document and submit it to Headquarters for General Review. The General Review Draft was prepared by NFPA Technical Staff on 30 July 1980.

All comments concerning the General Review Draft were resolved and the document was sent to the Technical Board for approval to ballot. On 11 February 1981, the Technical Board approved the document for ballot. The Project Group Chairman forwarded it to Headquarters.

Headquarters Technical Staff prepared the document for ballot on 29 May 1981. Ballot closed on 29 June 1981 with one negative comment and several editorial suggestions. These were reviewed and resolved.

On 24 February 1982, T3.9.13 was granted approval by the Technical Board. It was forwarded to and granted final approval by the Board of Directors on 10 March 1982.

Project Group Members who developed this standard:

Clay Thorson*
Project Chairman
Eaton Corporation

Al Mills
Section Chairman
Continental Hydraulic Division

Cliff Wobig
Section Vice Chairman
Sundstrand Corporation

Jack Wilcox
Section Secretary
Abex Corporation

Ed Saloum
Technical Auditor
Snap-Tite, Inc.

James C. White*
Director of Technical Services
National Fluid Power Association

G. Berman
Ross Gear Division of TRW

J. Johnson
W.H. Nichols Co.

P. Mardosa
Dayton T. Brown, Inc.

R. Ohnesorge
Gresen Mfg. Co./Dana Corp.

E. Ratkay
Commercial Shearing Inc.

E. Whitacre
International Harvester Co.

ksg

---

*Company affiliation has changed.

NFPA/T3.9.13

Hydraulic fluid power - Pumps and motors - Glossary

0 INTRODUCTION

In hydraulic fluid power systems, power is transmitted and controlled thru a liquid under pressure within an enclosed circuit. Hydraulic pumps are used to convert mechanical power into hydraulic fluid power. Hydraulic motors are used to convert hydraulic fluid power into mechanical power.

1 SCOPE AND FIELD OF APPLICATION

1.1 This NFPA Recommended Standard includes the following for technical terms applicable to hydraulic fluid power pumps and motors:

— Definitions;

— Numerical classifications.

1.2 This NFPA Recommended Standard provides:

— A unified glossary for the pump and motor segment of the fluid power industry, educational programs and users of fluid power;

— A convenient reference for technicians.

1.3 This NFPA Recommended Standard is intended to:

— Clarify terms and definitions for beginners;

— Simplify technical communications;

— Reduce interpretation errors;

— Encourage individuals to expand their working vocabulary.

1.4 This NFPA Recommended Standard promotes:

— A common understanding of pump and motor terms;

— Greater use of hydraulic fluid power pumps and motors.

2 REFERENCES

ANSI/B93.2-1971 and Supplement ANSI/B93.2A-1978, American National Standard Glossary of Terms for Fluid Power.

ISO 5598, Fluid power systems and components - Vocabulary.

ANSI/B93.27-1973 (R1979), American National Standard Method of Testing and Presenting Basic Performance Data for Positive Displacement Hydraulic Fluid Power Pumps and Motors.

3 TERMS AND DEFINITIONS

For definitions of other terms, see ANSI/B93.2 and Supplement ANSI/B93.2A.

NOTE: The terms set forth herein are presented in format for direct insertion in Section 43 and 82 of ANSI/B93.2 and ANSI/B93.2A at its next revision.

NFPA/T3.9.13

GROUP 43 — HYDRAULIC PUMPS

Section 01 — General
Section 02 — Types of Hydraulic Pumps
Section 03 — Hydraulic Pump — Centrifugal
Section 04 — Hydraulic Pump — Gear
Section 05 — Hydraulic Pump — Screw
Section 06 — Hydraulic Pump — Piston
Section 07 — Hydraulic Pump — Vane
Section 08 — Pump Controls
Section 09 — Pump Installation Features
Section 10 — Pump Performance

SECTION 01 — GENERAL

43.01.000 Capacity, Derived (Displacement, Derived)

Volume displaced at defined minimum working pressure calculated from two measurements at different speeds.

43.01.050 Compensation

The automatic regulation of a variable.

43.01.100 Displacement, Positive

A fluid transfer principle where fluid enters a chamber as a close fitted rigid member withdraws, and where it exits, as the member reenters the chamber.

43.01.150 Displacement, Volumetric (Capacity, Geometric)

Volume displaced per stroke of a cylinder or per revolution of a pump, calculated geometrically without reference to tolerances, clearances or deformation.

SECTION 02 — TYPES OF HYDRAULIC PUMPS

43.01.250 Pump, Hydraulic

A device which converts mechanical power into hydraulic fluid power.

43.02.000 Charge

Any hydraulic pump whose primary function is to elevate inlet pressure to another pump.

43.02.050 Circulating

Any hydraulic pump whose primary function is to circulate flow (for such things as cooling, filtration and lubrication).

43.02.100 Displacement, Fixed

A hydraulic pump in which the displacement per cycle cannot be varied.

43.02.150 Displacement, Variable

A hydraulic pump in which the volume displaced per cycle can be varied.

43.02.200 Drive, Tandem

Two or more hydraulic pumps driven directly from a common drive or pump shaft.

Synonym: Double pump, dual, triple, piggyback.

43.02.250 Hand

A hand operated hydraulic pump.

43.02.300 Multi-State

Two or more pumps hydraulically in series.

43.02.350 Over center

A variable positive displacement hydraulic pump wherein the inlet/outlet relationship is reversible with unidirectional input shaft rotation.

SECTION 03 — HYDRAULIC PUMP — CENTRIFUGAL

43.03.000 Centrifugal

A hydraulic pump which produces fluid velocity and converts it to pressure head.

43.03.050 Centrifugal, Diffuser (Concentric)

A centrifugal hydraulic pump in which fluid enters at the center of the impeller, is accelerated radially, and leaves through vanes arranged to provide a gradually enlarged flow passage.

43.03.100 Centrifugal, Peripheral

A centrifugal hydraulic pump in which fluid enters, follows, and leaves the periphery of the impeller.

43.03.150 Centrifugal, Volute (Spiral)

A centrifugal hydraulic pump in which fluid enters at the center of the impeller, is accelerated radially, and leaves through a gradually enlarging flow passage.

SECTION 04 — HYDRAULIC PUMP — GEAR

43.04.000 Gear

Pump in which two or more gears act in engagement as pumping members.

43.04.050 Gear, External

Pump with two or more external gears.

43.04.100 Gear, Internal

Pump with an internal gear engagement with one or more external gears.

43.04.150 Lobed

A hydraulic pump in which the intermeshing of the lobes is sustained by driving the lobes externally and in synchronization.

SECTION 05 — HYDRAULIC PUMP — SCREW

43.05.000 Screw

A hydraulic pump having one or more screws rotating in a housing.

SECTION 06 — HYDRAULIC PUMP — PISTON

43.06.000 Duplex, Reciprocating

A hydraulic pump having two reciprocating pistons.

43.06.050 Piston

Pump in which the fluid volume is displaced by one or more reciprocating pistons.

43.06.100 Piston, Angled

Axial piston pump in which the drive shaft is at an angle to the common axis of the pistons.

43.06.150 Piston, Axial

Pump having several pistons with mutually parallel axes which are arranged around and parallel to a common axis.

43.06.200 Piston, Inline

Pump having several pistons with mutually parallel axes arranged on a common plane.

43.06.250 Piston, Radial

Pump having several pistons arranged to operate radially.

43.06.300 Piston, Single, Reciprocating

A hydraulic pump having a single reciprocating piston.

SECTION 07 — HYDRAULIC PUMP — VANE

43.07.000 Vane

A hydraulic pump having multiple radial vanes within a supporting rotor.

43.07.050 Vane, Balanced

Pump in which the transverse forces on the rotor are balanced.

43.07.100 Vane, Unbalanced

Pump in which the transverse forces on the rotor are not balanced.

SECTION 08 — PUMP CONTROLS

43.08.000 Control, Pump

A control applied to a positive displacement variable delivery pump to adjust the volumetric output or direction of flow.

SECTION 09 — PUMP INSTALLATION FEATURES

43.09.000 Flange

Mounted by a flange with the supporting face at right angles to the driving shaft.

43.09.050 Foot

Mounting with the supporting face parallel to the driving shaft.

43.09.100 Pilot (Spigot)

A guide for accurately locating the pump to the driving force.

43.09.150 Port, Control

A port which provides passage for a control signal.

43.09.200 Port, Drain

A port for removal of fluid from a component, open to atmosphere, or connected to an unrestricted line to the reservoir.

43.09.250 Port, Inlet

A port which provides a passage for the influent.

43.09.300 Port, Outlet (Output)

A port which provides a passage for the effluent.

43.09.350 Rotation

The direction of rotation is normally quoted as viewed looking at the shaft end. In dubious cases, provide a sketch.

NFPA/T3.9.13

SECTION 10 — PUMP PERFORMANCE

43.10.000   Efficiency, Mechanical

Ratio of derived torque (See ANSI/B93.27 for a quantitative definition based on measured data.) to absorbed torque.

43.10.050   Efficiency, Overall

Ratio of the effective power to absorbed power.

43.10.100   Efficiency, Volumetric

Ratio of the effective output flow to the derived output flow.

43.10.150   Pressure, Rated

The operating pressure of a hydraulic pump recommended by the pump manufacturer when operating the pump under specified conditions of temperature, fluid, and drive speed.

GROUP 82 — HYDRAULIC MOTORS

Section 01 — General
Section 02 — Types of Hydraulic Motors
Section 03 — Motor — Gear
Section 04 — Motor — Piston
Section 05 — Motor — Vane
Section 06 — Motor Controls
Section 07 — Motor Installation Features
Section 08 — Motor Performance

SECTION 01 — GENERAL

82.01.000   Capacity, Derived (Displacement, Derived)

Volume absorbed at defined minimum working pressure obtained from measurements at two different speeds.

82.01.050   Compensation

The automatic regulation of a variable.

82.01.100   Displacement, Positive

A fluid transfer principle where fluid enters a chamber as a close fitted rigid member withdraws, and where it exits as the member reenters the chamber.

82.01.150   Displacement, Volumetric (Capacity, Geometric)

Volume displaced per stroke of a cylinder or per revolution of a motor shaft, calculated geometrically without reference to tolerances, clearances or deformation.

82.01.200   Motor, Hydraulic

A device which converts hydraulic fluid power into mechanical power and motion.

SECTION 02 — TYPES OF HYDRAULIC MOTORS

82.02.000   Displacement, Fixed

A hydraulic motor in which the displacement per unit of output motion cannot be varied.

82.02.050   Displacement, Variable

A hydraulic motor in which the displacement per unit of output motion can be varied.

82.02.100   Linear

A fluid power cylinder providing reciprocating motion.

82.02.150   Low Speed, High Torque

A hydraulic motor with optimum torque characteristics in the low speed range.

82.02.200   Rotary

A hydraulic motor capable of continuous rotary motion.

82.02.250   Rotary, Limited

A hydraulic rotary motor having limited motion.

SECTION 03 — MOTOR — GEAR

82.03.000   Abutment, Rotary

A hydraulic motor in which the intermeshing of the lobes is sustained by driving the lobes externally in synchronization.

82.03.050   Gear

A motor in which two or more gears act in arrangement as working members.

82.03.100   Gear, External

A motor having two or more external gears.

82.03.150   Gear, Internal

A motor with an internal gear in engagement with one or more external gears.

82.03.200   Gerotor

A hydraulic motor where the inner gear has teeth on its outside periphery and the outer gear has teeth on the inside periphery where the inner gear has fewer teeth and both gears rotate or one gear orbits.

82.03.250 Gerotor, Roller

A motor where the outer gear teeth are rolls (rollers) which are free to rotate in close fitted pockets.

82.03.300 Internal Gear, Crescent

A hydraulic motor where the inner gear has teeth on its outside periphery and the outer gear has its teeth on the inside periphery where the inner gear has less teeth with a fixed crescent-shaped element between the rotating gears.

82.03.350 Screw

A hydraulic motor which has meshing screws, which form consecutive, isolated helical chambers within a close fitted housing.

SECTION 04 — MOTOR — PISTON

82.04.000 Piston

A hydraulic motor which uses the reciprocal action of one or more pistons as its displacement principle.

82.04.050 Piston, Axial

A motor having several pistons with mutually parallel axes which are arranged around and parallel to a common axis.

82.04.100 Piston, Bent Axis

An axial piston motor whose output shaft axis is at an angle to the common axes of the pistons.

82.04.150 Piston, Inline Axial

An axial piston motor whose output shaft axis is parallel to the common axis of the pistons.

82.04.200 Piston, Radial

A motor having several pistons arranged to operate radially.

82.04.250 Piston, Radial Eccentric Crank

A radial piston motor in which the reciprocating pistons cause an eccentric crank to rotate.

82.04.300 Piston, Radial Rotary Barrel

A radial piston motor in which the reciprocating pistons cause the barrel to rotate.

82.04.350 Piston, Radial Rotary Housing

A radial piston motor in which the reciprocating pistons cause the housing to rotate.

SECTION 05 — MOTOR — VANE

82.05.000 Vane

A motor in which the fluid under pressure, acting on a set of radial vanes, causes rotation of an internal member.

82.05.050 Vane, Balanced

A motor in which the transverse forces on the rotor are balanced.

82.05.100 Vane, Unbalanced

A motor in which the transverse forces, acting on the rotor, are not balanced.

SECTION 06 — MOTOR CONTROLS

82.06.000 Control, Motor

A control applied to a variable displacement motor to adjust the speed or direction of rotation or direction of motion.

SECTION 07 — MOTOR INSTALLATION FEATURES

82.07.000 Flange

Mounted by a flange with the supporting face at right angles to the driving shaft.

82.07.050 Foot

Mounting with the supporting face parallel to the driving shaft.

82.07.100 Pilot (Spigot)

A guide for accurately locating the motor to the driven load.

82.07.150 Port, Control

A port which provides passage for a control signal.

82.07.200 Port, Drain

A port for removal of fluid from a component, open to atmosphere, or connected to an unrestricted line to the reservoir.

82.07.250 Port, Inlet

A port which provides a passage for the influent.

82.07.300 Port Outlet

A port which provides a passage for the effluent.

NFPA/T3.9.13

82.07.350   Rotation

The direction of rotation is normally quoted as viewed looking at the shaft end. In dubious cases, provide a sketch.

SECTION 08 — MOTOR PERFORMANCE

82.08.000   Efficiency, Hydromechanical

Ratio of the effective torque and derived torque.

82.08.050   Efficiency, Overall

Ratio of the output power to the effective hydraulic power.

82.08.100   Efficiency, Volumetric

Ratio of the derived speed to effective speed.

82.08.150   Pressure, Rated

The operating pressure of a hyraulic motor recommended by the motor manufacturer when operating the motor under specified conditions of temperature, fluid and speed.

4   IDENTIFICATION STATEMENT

Use the following statement in catalogs and sales literature when electing to conform with this voluntary standard:

"Listing and definitions of terms applicable to hydraulic fluid power pumps and motors conform to NFPA/T3.9.13-1982, **Hydraulic fluid power - Pumps and motors - Glossary.**"

ALPHABETICAL INDEX

C

	ISO 5598	NFPA*/ANSI
Capacity, Derived (Displacement, (Motor) Derived)	3.2.2.2.1.1	82.01.000*
Capacity, Derived (Displacement, (Pump) Derived)	3.1.2.1.1	43.01.000*
Control, Motor		82.06.000*
Control, Pump		43.08.000*
Compensation (Motor)		82.01.050*
Compensation (Pump)		43.01.050*

D

	ISO 5598	NFPA*/ANSI
Displacement, Positive (Motor)		82.01.100*
Displacement, Positive (Pump)		43.01.100*
Displacement, Volumetric (Capacity, Geometric) (Motor)		82.01.150*
Displacement, Volumetric (Capacity, Geometric) (Pump)		43.01.150*

E

	ISO 5598	NFPA*/ANSI
Efficiency, Hydromechanical (Motor)	3.2.2.2.7.2	82.08.000*
Efficiency, Mechanical (Pump)	3.1.2.8.2	43.10.000*
Efficiency, Overall (Motor)	3.2.2.2.7.3	82.08.050*
Efficiency, Overall (Pump)	3.1.2.8.3	43.10.050*
Efficiency, Volumetric (Motor)	3.2.2.2.7.1	82.08.100*
Efficiency, Volumetric (Pump)	3.1.2.8.1	43.10.100*

F

	ISO 5598	NFPA*/ANSI
Flange, Motor Mounting		82.07.000*
Flange, Pump Mounting	3.1.3.1.1	43.09.000*
Foot, Motor Mounting		82.07.050*
Foot, Pump Mounting	3.1.3.1.3	43.09.050*

H

	ISO 5598	NFPA*/ANSI
Hydraulic, Motor		82.01.200*

M

	ISO 5598	NFPA*/ANSI
Motor, Abutment Rotary		82.03.000*
Motor, Displacement, Fixed	3.2.2.1.4	82.02.000* 82.02.100
Motor, Displacement, Positive		82.01.100*
Motor, Displacement, Variable	3.2.2.1.5	82.02.050* 82.02.600
Motor, Gear	3.2.2.1.6	82.03.050*
Motor, Gear, External	3.2.2.1.6.2	82.03.100*
Motor, Gear, Internal	3.2.2.1.6.1	82.03.150*
Motor, Gerotor		82.03.200*
Motor, Gerotor, Roller		82.03.250*
Motor, Internal Gear, Crescent		82.03.300*
Motor, Linear		82.02.100* 82.02.200
Motor, Low Speed, High Torque		82.02.150*
Motor, Piston		82.04.000*
Motor, Piston, Axial	3.2.2.1.8.2	82.04.050*
Motor, Piston, Bent Axis		82.04.100*
Motor, Piston Inline Axial		82.04.150*

NFPA/T3.9.13

	ISO 5598	NFPA*/ANSI
Motor, Piston, Radial	3.2.2.1.8.1	82.04.200*
Motor, Piston Radial, Eccentric Crank		82.04.250*
Motor Piston, Radial Rotary Barrel		82.04.300*
Motor Piston, Radial Rotary Housing		82.04.350*
Motor Gerotor, Roller		82.03.250*
Motor, Rotary	3.2.2.1.2	82.02.200* 82.02.300
Motor Rotary, Limited		82.02.250* 82.02.350
Motor, Screw		82.03.350*
Motor, Vane	3.2.2.1.7	82.05.000*
Motor, Vane, Balanced	3.2.2.1.7.2	82.05.050*
Motor, Vane, Unbalanced	3.2.2.1.7.1	82.05.100*

P

	ISO 5598	NFPA*/ANSI
Pilot (Spigot) Motor		82.07.100*
Pilot (Spigot), Pump		43.09.100*
Port, Control, Motor	5.2.4.1	82.07.150* 50.05.100
Port, Control Pump	5.2.4.1	43.09.150* 50.05.100
Port, Drain, Motor	5.2.4.6	82.07.200* 50.05.250
Port, Drain, Pump	5.2.4.6	43.09.200* 50.05.250
Port, Inlet, Motor	5.2.4.2	82.07.250* 50.05.400
Port, Inlet, Pump	5.2.4.2	43.09.250* 50.05.400
Port, Outlet, Motor	5.2.4.3	82.07.300* 50.05.430
Port, Outlet (Output), Pump	5.2.4.3	43.09.300* 50.05.430
Pump, Centrifugal	3.1.1.1	43.03.000* 43.02.100
Pump, Centrifugal, Diffuser (Concentric)		43.03.050* 43.02.130
Pump, Centrifugal, Peripheral		43.03.100* 43.02.145
Pump, Centrifugal, Volute (Spiral)		43.03.150* 43.02.175
Pump, Charge		43.02.000*
Pump, Circulating		43.02.050*
Pump, Displacement, Fixed		43.02.100*
Pump, Displacement, Variable	3.1.1.2.2	43.02.150* 43.02.650
Pump, Drive, Tandem		43.02.200*
Pump, Duplex, Reciprocating		43.06.000*
Pump, Gear	3.1.1.5	43.04.000*
Pump, Gear, External	3.1.1.5.1	43.04.050*
Pump, Gear, Internal	3.1.1.5.2	43.04.100*
Pump, Hand	3.1.1.9	43.02.250* 43.02.300
Pump, Hydraulic	3.1.1	43.01.250* 43.01.100
Pump, Lobed		43.04.150*
Pump, Multi-Stage		43.02.300*

	ISO 5598	NFPA*/ANSI
Pump, Over Center		43.02.350*
Pump, Piston	3.1.1.8	43.06.050*
Pump, Piston, Angled	3.1.1.8.3	43.06.100*
Pump, Piston, Axial	3.1.1.8.2	43.06.150*
Pump, Piston, Inline	3.1.1.8.4	43.06.200*
Pump, Piston, Radial	3.1.1.8.1	43.06.250*
Pump, Piston, Single, Reciprocating		43.06.300*
Pump, Screw	3.1.1.6	43.05.000* 43.02.550
Pump, Tandem Drive		43.02.200*
Pump, Vane	3.1.1.7	43.07.000* 43.02.600
Pump, Vane, Balanced	3.1.1.7.2	43.07.050*
Pump, Vane, Unbalanced	3.1.1.7.1	43.07.100*
Pump, Variable Displacement	3.1.1.2.2	43.02.150* 43.02.650
Pressure, Rated, Motor	2.1.2	82.08.150* 04.01.470
Pressure, Rated, Pump	2.1.2	43.10.150* 04.01.470

R

	ISO 5598	NFPA*/ANSI
Rotation, Motor	2.2.2	82.07.350*
Rotation, Pump	2.2.2	43.09.350*

NOTES

International Standard  1000

INTERNATIONAL ORGANIZATION FOR STANDARDIZATION•МЕЖДУНАРОДНАЯ ОРГАНИЗАЦИЯ ПО СТАНДАРТИЗАЦИИ•ORGANISATION INTERNATIONALE DE NORMALISATION

SI units and recommendations for the use of their multiples and of certain other units

Unités SI et recommandations pour l'emploi de leurs multiples et de certaines autres unités

Second edition — 1981-02-15

Notice: The copyright on all ISO International Standards and other priced publications is the property of the International Organization For Standardization (ISO). As the U.S. member body of ISO, the American National Standards Institute has an exclusive license to distribute and sell ISO International Standards and other priced publications of ISO within the United States of America.

ISO has granted exclusive permission to the American National Standards Institute to reprint ISO International Standard 1000-1981 (E) for distribution and sale within the United States of America.

UDC 53.081 : 003.62 : 004.1

Ref. No. ISO 1000-1981 (E)

Descriptors : units of measurement, metric system, multiples, international system of units, utilisation.

ISO 1000

Foreword

ISO (the International Organization for Standardization) is a worldwide federation of national standards institutes (ISO member bodies). The work of developing International Standards is carried out through ISO technical committees. Every member body interested in a subject for which a technical committee has been set up has the right to be represented on that committee. International organizations, governmental and non-governmental, in liaison with ISO, also take part in the work.

Draft International Standards adopted by the technical committees are circulated to the member bodies for approval before their acceptance as International Standards by the ISO Council.

International Standard ISO 1000 was developed by Technical Committee ISO/TC 12, *Quantities, units, symbols, conversion factors and conversion tables*.

This second edition was submitted directly to the ISO Council, in accordance with clause 5.10.1 of part 1 of the Directives for the technical work of ISO. It cancels and replaces the first edition (i.e. ISO 1000-1973), which had been approved by the member bodies of the following countries:

Austria	Hungary	Romania
Belgium	India	Sri Lanka
Brazil	Iran	Sweden
Bulgaria	Ireland	Switzerland
Canada	Israel	Thailand
Chile	Italy	Turkey
Denmark	Japan	United Kingdom
Egypt, Arab Rep. of	Netherlands	USA
Finland	New Zealand	USSR
France	Norway	
Germany, F.R.	Portugal	

The member bodies of the following countries had expressed disapproval of the document on technical grounds:

Australia
Czechoslovakia
South Africa, Rep. of

This International Standard cancels and replaces ISO Standard 1000-1973 (E)

Prices

This publication may be ordered from the American National Standards Institute at the following prices:

Single copy	$5.00
10–49 copies	4.50
50–99 copies	4.00
100–499 copies	3.50
500–999 copies	3.00
1000 and over — Special quotation on request	

American National Standards Institute
1430 Broadway, New York, N.Y. 10018

© International Organization for Standardization, 1981

ISO 1000

SI units and recommendations for the use of their multiples and of certain other units

1 Scope and field of application

This International Standard

a) describes the International System of Units[1] (in clauses 2 and 3);

b) recommends selected decimal multiples and sub-multiples of the SI units for general use and gives certain other units which may be used with the International System of Units (in clauses 4 and 5, and annex A);

c) defines base and supplementary SI units (in annex B).

2 SI Units

The name Système International d'Unités (International System of Units), with the international abbreviation SI, was adopted by the 11th Conférence Générale des Poids et Mesures in 1960.

This system includes three classes of units:

— base units

— supplementary units

— derived units,

which together form the coherent system of SI units.

---

[1] Full information about the International System of Units is given in a publication from the International Bureau of Weights and Measures: *Le Système International d'Unités* (authorized English translations have been published in the United Kingdom through the National Physical Laboratory, and in the United States of America through the National Bureau of Standards).

2.1 Base units

The International System of Units is founded on the seven base units listed in table 1.

Table 1

Quantity	Name of base SI unit	Symbol
length	metre	m
mass	kilogram	kg
time	second	s
electric current	ampere	A
thermodynamic temperature	kelvin	K
amount of substance	mole	mol
luminous intensity	candela	cd

For the definitions of the base units and the supplementary units, see annex B.

2.2 Supplementary units

The Conférence Générale des Poids et Mesures has not classified certain units of the International System under either base units or derived units.

These units, listed in table 2, are called "supplementary units" and may be regarded either as base units or as derived units.[1]

Table 2

Quantity	Name of supplementary SI unit	Symbol
plane angle	radian	rad
solid angle	steradian	sr

2.3 Derived units

Derived units are expressed algebraically in terms of base units and/or supplementary units. Their symbols are obtained by means of the mathematical signs of multiplication and division; for example, the SI unit for velocity is metre per second (m/s) and the SI unit for angular velocity is radian per second (rad/s).

For some of the derived SI units, special names and symbols exist; those approved by the Conférence Générale des Poids et Mesures are listed in tables 3 and 4.

It may sometimes be advantageous to express derived units in terms of other derived units having special names; for example, the SI unit for electric dipole moment is usually expressed as C·m instead of A·s·m.

Table 3

Quantity	Special name of derived SI unit	Symbol	Expressed in terms of base or supplementary SI units or in terms of other derived SI units
frequency	hertz	Hz	$1\ Hz = 1\ s^{-1}$
force	newton	N	$1\ N = 1\ kg \cdot m/s^2$
pressure, stress	pascal	Pa	$1\ Pa = 1\ N/m^2$
energy, work, quantity of heat	joule	J	$1\ J = 1\ N \cdot m$
power	watt	W	$1\ W = 1\ J/s$
electric charge, quantity of electricity	coulomb	C	$1\ C = 1\ A \cdot s$
electric potential, potential difference, tension, electromotive force	volt	V	$1\ V = 1\ J/C$
electric capacitance	farad	F	$1\ F = 1\ C/V$
electric resistance	ohm	Ω	$1\ \Omega = 1\ V/A$
electric conductance	siemens	S	$1\ S = 1\ \Omega^{-1}$
flux of magnetic induction, magnetic flux	weber	Wb	$1\ Wb = 1\ V \cdot s$
magnetic flux density, magnetic induction	tesla	T	$1\ T = 1\ Wb/m^2$
inductance	henry	H	$1\ H = 1\ Wb/A$
Celsius temperature	degree Celsius	°C	$1\ °C = 1\ K$ [2]
luminous flux	lumen	lm	$1\ lm = 1\ cd \cdot sr$
illuminance	lux	lx	$1\ lx = 1\ lm/m^2$

Table 4 — Derived SI units with special names accepted for the sake of safeguarding human health

Quantity	Special name of derived SI unit	Symbol	Expressed in terms of base units or derived SI units
activity (of a radionuclide)	becquerel	Bq	$1\ Bq = 1\ s^{-1}$
absorbed dose, specific energy imparted, kerma, absorbed dose index	gray	Gy	$1\ Gy = 1\ J/kg$
dose equivalent	sievert	Sv	$1\ Sv = 1\ J/kg$

---

1) However, in October 1980 the International Committee of Weights and Measures decided to interpret the class of supplementary units in the International System as a class of dimensionless derived units for which the General Conference of Weights and Measures leaves open the possibility of using these or not in expressions of derived units of the International System.

2) For the use of degree Celsius (°C), see note 2 under the definition of kelvin in annex B.

3 Multiples of SI units

The prefixes given in table 5 (SI prefixes) are used to form names and symbols of multiples (decimal multiples and sub-multiples) of the SI units.

The symbol of a prefix is considered to be combined with the single unit symbol[1] to which it is directly attached, forming with it a new symbol (for a decimal multiple or sub-multiple) which can be raised to a positive or negative power, and which can be combined with other unit symbols to form symbols for compound units.

Examples

$1 \text{ cm}^3 = (10^{-2} \text{ m})^3 = 10^{-6} \text{ m}^3$

$1 \mu\text{s}^{-1} = (10^{-6} \text{ s})^{-1} = 10^{6} \text{ s}^{-1}$

$1 \text{ mm}^2/\text{s} = (10^{-3} \text{ m})^2/\text{s} = 10^{-6} \text{ m}^2/\text{s}$

Compound prefixes shall not be used; for example, write nm (nanometre), never mµm.

NOTE — Because the name of the base unit for mass, kilogram, contains the name of the SI prefix "kilo", the names of the decimal multiples and sub-multiples of the unit of mass are formed by adding the prefixes to the word "gram", e.g. miligram (mg) instead of microkilogram (µkg).

Table 5

Factor	Prefix	Symbol
10^{18}	exa	E
10^{15}	peta	P
10^{12}	tera	T
10^{9}	giga	G
10^{6}	mega	M
10^{3}	kilo	k
10^{2}	hecto	h
10	deca	da
10^{-1}	deci	d
10^{-2}	centi	c
10^{-3}	milli	m
10^{-6}	micro	µ
10^{-9}	nano	n
10^{-12}	pico	p
10^{-15}	femto	f
10^{-18}	atto	a

4 Use of the SI units and their multiples

4.1 The choice of the appropriate multiple (decimal multiple or sub-multiple) of an SI unit is governed by convenience, the multiple chosen for a particular application being the one which will lead to numerical values within a practical range.

4.2 The multiple can usually be chosen so that the numerical values will be between 0,1 and 1000.

Examples

$1,2 \times 10^4$ N	can be written as	12 kN
0,003 96 m	can be written as	3,94 mm
1401 Pa	can be written as	1,401 kPa
$3,1 \times 10^{-8}$ s	can be written as	31 ns

However, in a table of values for the same quantity or in a discussion of such values within a given context, it will generally be better to use the same multiple for all items, even when some of the numerical values will be outside the range 0,1 to 1000. For certain quantities in particular applications, the same multiple is customarily used; for example, the millimetre is used for dimensions in most mechanical engineering drawings.

4.3 It is recommended that only one prefix be used in forming a multiple of a compound SI unit.

4.4 Errors in calculations can be avoided more easily if all quantities are expressed in SI units, prefixes being replaced by powers of 10.

4.5 Rules for writing unit symbols

4.5.1 Unit symbols should be printed in roman (upright) type (irrespective of the type used in the rest of the text), should remain unaltered in the plural, should be written without a final full stop (period) except for normal punctuation, e.g. at the end of a sentence, and should be placed after the complete numerical value in the expression for a quantity, leaving a space between the numerical value and the unit symbol.

Unit symbols should generally be written in lower case letters except that the first letter is written in upper case when the name of the unit is derived from a proper name.

Examples

m	metre
s	second
A	ampere
Wb	weber

---

[1] In this case, the term "unit symbol" means only a symbol for a base unit, a derived unit with a special name or a supplementary unit; see, however, the note about the base unit kilogram.

4.5.2 When a compound unit is formed by multiplication of two or more units, this may be indicated in one of the following ways:

N·m   N.m   N m

NOTE — The last form may also be written without a space, provided that special care is taken when the symbol for one of the units is the same as the symbol for a prefix, e.g. mN means millinewton, not metre newton.

When a compound unit is formed by dividing one unit by another, this may be indicated in one of the following ways:

$\frac{m}{s}$, m/s or by writing the product of m and s^{-1}, for example $m \cdot s^{-1}$.

In no case should more than one solidus (as in m/s) on the same line be included in such a combination unless parentheses be inserted to avoid all ambiguity. In complicated cases, negative powers or parentheses should be used.

5 Non SI units which may be used together with the SI units and their multiples

5.1 There are certain units outside the SI which are recognized by the Comité International des Poids et Mesures (CIPM) as having to be retained because of their practical importance (table 6) or for use in specialized fields (table 7).

5.2 Prefixes given in table 5 may be attached to many of the units given in tables 6 and 7; for example, millilitre, ml; megaelectronvolt, MeV. See also annex A, column 6.

5.3 In a limited number of cases, compound units are formed with the units given in tables 6 and 7 together with SI units and their multiples; for example, kg/h; km/h. See also annex A, columns 5 and 6.

Table 6

Quantity	Name of unit	Unit symbol	Definition
time	minute	min	1 min = 60 s
	hour	h	1 h = 60 min
	day	d	1 d = 24 h
plane angle	degree	°	1° = (π/180) rad
	minute	'	1' = (1/60) °
	second	''	1'' = (1/60) '
volume	litre	l, L[1]	1 l = 1 dm^3
mass	tonne	t	1 t = 10^3 kg

[1] The two symbols for litre are on an equal footing. The CIPM will, however, before the 18th CGPM make a survey on the development of the use the two symbols in order to see if one of the two may be suppressed. [16th CGPM (1979), Resolution 6]

Table 7

Quantity	Name of unit	Unit symbol	Definition
energy	electronvolt	eV	1 electronvolt is the kinetic energy acquired by an electron in passing through a potential difference of 1 volt in vacuum; 1 eV = 1,602 19 × 10^{-19} J (approximately)
mass of an atom	atomic mass unit	u	1 (unified) atomic mass unit is equal to the fraction 1/12 of the mass of an atom of the nuclide ^{12}C; 1 u = 1,660 53 × 10^{-27} kg (approximately)
length	astronomic unit	AU[1]	1 AU = 149 597,870 × 10^6 m (adopted value in System of Astronomic Constants, 1979)
	parsec	pc	1 parsec is the distance at which 1 astronomic unit subtends an angle of 1 second of arc; 1 pc = 206 265 AU = 30 857 × 10^{12} m (approximately)
pressure of fluid	bar[2]	bar	1 bar = 10^5 Pa

[1] The unit has no international symbol; AU is the abbreviation of the English name; the abbreviation of the French name is UA.

[2] The bar is not mentioned by CIPM in this group of units; in many countries, however, there are special requirements for this unit.

Annex A

Examples of decimal multiples and sub-multiples of SI units and of some other units which may be used

For a number of commonly used quantities, examples of decimal multiples and sub-multiples of SI units, as well as of some other units which may be used, are given in this annex. It is suggested that the selection shown, while not intended to be restrictive, will none the less prove helpful in presenting values of quantities in an identical manner in similar contexts within the various sectors of technology. For some needs (for example, in applications in science and education), it is recognized that greater freedom will be required in the choice of decimal multiples and sub-multiples of SI units than is exemplified in the list which follows.

NOTE — Factors for conversion to SI units from the other units listed are given in the relevant parts of ISO 31.

Item No. in ISO 31	Quantity	SI unit	Selection of multiples of the SI unit	Units outside the SI which are nevertheless recognized by the CIPM as having to be retained either because of their practical importance or because of their use in specialized fields		Remarks, and information about units used in special fields
				Units	Multiples of units given in column 5	
(1)	(2)	(3)	(4)	(5)	(6)	(7)
Part 1: **Space and time**						
1-1.1	angle (plane angle)	rad (radian)	mrad µrad	° (degree) ′ (minute) ″ (second)		If the radian is not used, the units degree or grade (or gon) may be used. Decimal subdivisions of degree are preferable to minute and second for most applications. grade (g) or gon, $1^g = 1 \text{ gon} = \frac{\pi}{200}$ rad
1-2.1	solid angle	sr (steradian)				
1-3.1...7	length	m (metre)	km cm mm µm nm pm fm			1 international nautical mile = 1 852 m
1-4.1	area	m²	km² dm² cm² mm²			ha (hectare), 1 ha = 10^4 m² a (are), 1 a = 10^2 m²
1-5.1	volume	m³	dm³ cm³ mm³	l, L[1] (litre)	hl 1 hl = 10^{-1} m³ cl 1 cl = 10^{-5} m³ ml 1 ml = 10^{-6} m³ = 1 cm³	In 1964, the Conférence Générale des Poids et Mesures declared that the name litre (l) may be used as a special name for the cubic decimetre (dm³) and advised against the use of the name litre for high-precision measurements.

[1] See table 6.

ISO 1000

Item No. in ISO 31	Quantity	SI unit	Selection of multiples of the SI unit	Units outside the SI which are nevertheless recognized by the CIPM as having to be retained either because of their practical importance or because of their use in specialized fields		Remarks, and information about units used in special fields
				Units	Multiples of units given in column 5	
(1)	(2)	(3)	(4)	(5)	(6)	(7)
1-6.1	time	s (second)	ks ms µs ns	d (day) h (hour) min (minute)		Other units such as week, month and year (a) are in common use.
1-7.1	angular velocity	rad/s				
1-9.1	velocity	m/s		km/h $1 \text{ km/h} = \frac{1}{3,6} \text{ m/s}$		1 knot = 0,514 444 m/s
1-10.1	acceleration	m/s²				
Part 2: **Periodic and related phenomena**						
2-3.1	frequency	Hz (hertz)	THz GHz MHz kHz			
2-3.2	rotational frequency	s^{-1}		min^{-1}		The designations "revolution per minute" (r/min) and "revolution per second" (r/s) are widely used for rotational frequency in specifications on rotating machinery.[1]
Part 3: **Mechanics**						
3-1.1	mass	kg (kilogram)	Mg g m µg	t (tonne)		
3-2.1	density (mass density)	kg/m³	Mg/m³ or kg/dm³ or g/cm³	t/m³ or kg/l	g/ml g/l	For litre, see item 1-5.1.
3-5.1	linear density	kg/m	mg/m			1 tex = 10^{-6} kg/m The unit tex is used for textile filaments.
3-7.1	momentum	kg·m/s				
3-8.1	moment of momentum, angular momentum	kg·m²/s				
3-9.1	moment of inertia	kg·m²				
3-10.1	force	N (newton)	MN kN mN µN			

---

[1] See also IEC Publication 27-1 (1971).

ISO 1000

Item No. in ISO 31	Quantity	SI unit	Selection of multiples of the SI unit	Units outside the SI which are nevertheless recognized by the CIPM as having to be retained either because of their practical importance or because of their use in specialized fields		Remarks, and information about units used in special fields
				Units	Multiples of units given in column 5	
(1)	(2)	(3)	(4)	(5)	(6)	(7)
3-12.1	moment of force	N·m	MN·m kN·m mN·m μN·m			
3-13.1	pressure	Pa (pascal)	GPa MPa kPa mPa μPa	bar[1]	mbar μbar	1 bar = 10^5 Pa
3-13.2	normal stress	Pa or N/m^2	GPa MPa or N/mm^2 kPa			
3-21.1	viscosity (dynamic)	Pa·s	mPa·s			P (poise)[2] 1 cP = 1 mPa·s
3-22.1	kinematic viscosity	m^2/s	mm^2/s			St (stokes)[2] 1 cSt = 1 mm^2/s
3-23.1	surface tension	N/m	mN/m			
3-24.1	energy, work	J (joule)	EJ PJ TJ GJ MJ kJ mJ	eV (electronvolt)	GeV MeV keV	The units W·h, kW·h, MW·h, GW·h and TW·h are used in the field of consumption of electrical energy. The units keV, MeV and GeV are used in atomic and nuclear physics and in accelerator technology.
3-25.1	power	W (watt)	GW MW kW mW μW			
Part 4: **Heat**						
4-1.1	thermodynamic temperature	K (kelvin)				
4-2.1	Celsius temperature	°C (degree Celsius)[3]				The Celsius temperature t is equal to the difference $t = T - T_0$ between two thermodynamic temperatures T and T_0, where $T_0 = 273,15$ K.
4-1.1 4-2.1	temperature interval	K				For temperature interval, °C may be used instead of K.

1) For the bar, see 5.1 and table 7.

2) Poise and stokes are special names for CGS units. They and their multiples should not be used together with SI units.

3) For the definition and the use of degree Celsius (°C), see note 2 under the definition of kelvin in annex B.

ISO 1000

Item No. in ISO 31	Quantity	SI unit	Selection of multiples of the SI unit	Units outside the SI which are nevertheless recognized by the CIPM as having to be retained either because of their practical importance or because of their use in specialized fields		Remarks, and information about units used in special fields
				Units	Multiples of units given in column 5	
(1)	(2)	(3)	(4)	(5)	(6)	(7)
4-3.1	linear expansion coefficient	K^{-1}				For degree Celsius, see footnote 3), page 7.
4-6.1	heat, quantity of heat	J	EJ PJ TJ GJ MJ kJ mJ			
4-7.1	heat flow rate	W	kW			
4-9.1	thermal conductivity	$W/(m \cdot K)$				For degree Celsius, see footnote 3), page 7.
4-10.1	coefficient of heat transfer	$W/(m^2 \cdot K)$				For degree Celsius, see footnote 3), page 7.
4-14.1	heat capacity	J/K	kJ/K			For degree Celsius, see footnote 3), page 7.
4-15.1	specific heat capacity	$J/(kg \cdot K)$	$kJ/(kg \cdot K)$			For degree Celsius, see footnote 3), page 7.
4-17.1	entropy	J/K	kJ/K			
4-18.1	specific entropy	$J/(kg \cdot K)$	$kJ/(kg \cdot K)$			
4-20.1	specific internal energy	J/kg	MJ/kg kJ/kg			
—	specific latent heat	J/kg	MJ/kg kJ/kg			
Part 5: Electricity and magnetism						
5-1.1	electric current	A (ampere)	kA mA µA nA pA			
5-2.1	electric charge, quantity of electricity	C (coulomb)	kC µC nC pC			1 A·h = 3,6 kC
5-3.1	volume density of charge, charge density	C/m^3	C/mm^3 MC/m^3 or C/cm^3 kC/m^3 mC/m^3 $µC/m^3$			

ISO 1000

Item No. in ISO 31	Quantity	SI unit	Selection of multiples of the SI unit	Units outside the SI which are nevertheless recognized by the CIPM as having to be retained either because of their practical importance or because of their use in specialized fields		Remarks, and information about units used in special fields
				Units	Multiples of units given in column 5	
(1)	(2)	(3)	(4)	(5)	(6)	(7)
5-4.1	surface density of charge	C/m^2	MC/m^2 or C/mm^2 C/cm^2 kC/m^2 mC/m^2 $\mu C/m^2$			
5-5.1	electric field strength	V/m	MV/m kV/m or V/mm V/cm mV/m $\mu V/m$			
5-6.1	electric potential	V (volt)	M/V kV mV μV			
5-6.2	potential difference (tension)					
5-6.3	electromotive force					
5-7.1	electric flux density, displacement	C/m^2	C/cm^2 kC/m^2 mC/m^2 $\mu C/m^2$			
5-8.1	electric flux, (flux of displacement)	C	MC kC mC			
5-9.1	capacitance	F (farad)	mF μF nF pF			
5-10.1	permittivity	F/m	$\mu F/m$ nF/m pF/m			
5-13.1	electric polarization	C/m^2	C/cm^2 kC/m^2 mC/m^2 $\mu C/m^2$			
5-14.1	electric dipole moment	$C \cdot m$				
5-15.1	current density	A/m^2	MA/m^2 or A/mm^2 A/cm^2 kA/m^2			

ISO 1000

Item No. in ISO 31	Quantity	SI unit	Selection of multiples of the SI unit	Units outside the SI which are nevertheless recognized by the CIPM as having to be retained either because of their practical importance or because of their use in specialized fields		Remarks, and information about units used in special fields
				Units	Multiples of units given in column 5	
(1)	(2)	(3)	(4)	(5)	(6)	(7)
5-16.1	linear current density	A/m	kA/m or A/mm A/cm			
5-17.1	magnetic field strength	A/m	kA/m or A/mm A/cm			
5-18.1	magnetic potential difference	A	kA mA			
5-19.1	magnetic flux density, magnetic induction	T (tesla)	mT µT nT			
5-20.1	magnetic flux	Wb (weber)	mWb			
5-21.1	magnetic vector potential	Wb/m	kWb/m or Wb/mm			
5-22.1	self inductance	H (henry)	mH µH nH pH			
5-22.2	mutual inductance					
5-24.1	permeability	H/m	µH/m nH/m			
5-27.1	electromagnetic moment, (magnetic moment)	A·m²				
5-28.1	magnetization	A/m	kA/m or A/mm			
5-29.1	magnetic polarization	T	mT			
(IEC Pub. 27, item 86)	magnetic dipole moment	N·m²/A or Wb·m				
5-33.1	resistance (to direct current)	Ω (ohm)	GΩ MΩ kΩ mΩ µΩ			
5-34.1	conductance (to direct current)	S (siemens)	kS mS µS			

ISO 1000

Item No. in ISO 31	Quantity	SI unit	Selection of multiples of the SI unit	Units outside the SI which are nevertheless recognized by the CIPM as having to be retained either because of their practical importance or because of their use in specialized fields		Remarks, and information about units used in special fields
				Units	Multiples of units given in column 5	
(1)	(2)	(3)	(4)	(5)	(6)	(7)
5-35.1	resistivity	$\Omega \cdot m$	$G\Omega \cdot m$ $M\Omega \cdot m$ $k\Omega \cdot m$ $\Omega \cdot cm$ $m\Omega \cdot m$ $\mu\Omega \cdot m$ $n\Omega \cdot m$			$\mu\Omega \cdot cm = 10^{-8} \Omega \cdot m$ $\dfrac{\Omega \cdot mm^2}{m} = 10^{-6} \Omega \cdot m = \mu\Omega \cdot m$ are also used.
5-36.1	conductivity	S/m	MS/m kS/m			
5-37.1	reluctance	H^{-1}				
5-38.1	permeance	H				
5-41.1	impedance (complex impedance)	Ω	$M\Omega$ $k\Omega$ $m\Omega$			
5-41.2	modulus of impedance, (impedance)					
5-41.3	reactance					
5-41.4	resistance					
5-43.1	admittance (complex admittance)	S	kS mS μS			
5-43.2	modulus of admittance, (admittance)					
5-43.3	susceptance					
5-43.4	conductance					
5-44.1	power	W	TW GW MW kW mW μW nW			In electric power technology, active power is expressed in watts (W), apparent power in voltamperes (V·A) and reactive power in vars (var).
Part 6: Light and related electromagnetic radiations						
6-3.1	wavelength	m	μm nm pm			Å (ångström), $1\,\text{Å} = 10^{-10}\,m = 0{,}1\,nm = 10^{-4}\,\mu m$
6-6.1	radiant energy	J				
6-9.1	radiant power, radiant energy flux	W				
6-11.1	radiant intensity	W/sr				
6-12.1	radiance	$W/(sr \cdot m^2)$				
6-13.1	radiant exitance	W/m^2				

ISO 1000

Item No. in ISO 31	Quantity	SI unit	Selection of multiples of the SI unit	Units outside the SI which are nevertheless recognized by the CIPM as having to be retained either because of their practical importance or because of their use in specialized fields		Remarks, and information about units used in special fields
				Units	Multiples of units given in column 5	
(1)	(2)	(3)	(4)	(5)	(6)	(7)
6-14.1	irradiance	W/m^2				
6-19.1	luminous intensity	cd (candela)				
6-20.1	luminous flux	lm (lumen)				
6-21.1	quantity of light	lm·s				1 lm·h = 3 600 lm·s
6-22.1	luminance	cd/m^2				
6-23.1	luminous exitance	lm/m^2				
6-24.1	illuminance	lx (lux)				
6-25.1	light exposure	lx·s				
6-26.1	luminous efficacy	lm/W				

Part 7: Acoustic

Item No. in ISO 31	Quantity	SI unit	Selection of multiples of the SI unit	Units	Multiples of units given in column 5	Remarks
7-1.1	period periodic time	s	ms µs			
7-2.1	frequency	Hz	MHz kHz			
7-5.1	wavelength	m	mm			
7-7.1	density (mass density)	kg/m^3				
7-8.1	static pressure	Pa				
7-8.2	(instantaneous) sound pressure		mPa µPa			
7-10.1	(instantaneous) sound particle velocity	m/s	mm/s			
7-12.1	(instantaneous) volume flow rate, volume velocity	m^3/s				
7-13.1	velocity of sound	m/s				
7-15.1	sound energy flux, sound power	W	kW mW µW pW			
7-16.1	sound intensity	W/m^2	mW/m^2 $µW/m^2$ pW/m^2			

ISO 1000

Item No. in ISO 31	Quantity	SI unit	Selection of multiples of the SI unit	Units outside the SI which are nevertheless recognized by the CIPM as having to be retained either because of their practical importance or because of their use in specialized fields		Remarks, and information about units used in special fields
				Units	Multiples of units given in column 5	
(1)	(2)	(3)	(4)	(5)	(6)	(7)
7-17.2	specific acoustic impedance	Pa·s/m				
7-18.1	acoustic impedance	Pa·s/m^3				
7-19.1	mechanical impedance	N·s/m				
7-20.1	sound pressure level					dB (decibel)
7-21.1	sound power level					dB
7-27.1	sound reduction index, sound transmission loss					dB
7-28.1	equivalent absorption area of a surface or object	m^2				
7-29.1	reverberation time	s				
Part 8: Physical chemistry and molecular physics						
8-3.1	amount of substance	mol (mole)	kmol mmol μmol			
8-5.1	molar mass	kg/mol	g/mol			
8-6.1	molar volume	m^3/mol	dm^3/mol cm^3/mol	l/mol		For litre, see item 1-5.1.
8-7.1	molar internal energy	J/mol	kJ/mol			
8-8.1	molar heat capacity	J/(mol·K)				
8-9.1	molar entropy	J/(mol·K)				
8-13.1	concentration of substance B, amount-of-substance concentration of substance B	mol/m^3	mol/dm^3 or kmol/m^3	mol/l		For litre, see item 1-5.1.
8-16.1	molality of solute substance B	mol/kg	mmol/kg			
8-38.1	diffusion coefficient	m^2/s				
8-40.1	thermal diffusion coefficient	m^2/s				

Annex B

Definitions of the base units and supplementary units of the International System of Units

Base units

metre: The metre is the length equal to 1 650 763,73 wavelengths in vacuum of the radiation corresponding to the transition between the levels $2p_{10}$ and $5d_5$ of the krypton-86 atom.

[11th CGPM (1960), Resolution 6]

kilogram: The kilogram is the unit of mass; it is equal to the mass of the international prototype of the kilogram.

[1st CGPM (1889) and 3rd CGPM (1901)]

second: The second is the duration of 9 192 631 770 periods of the radiation corresponding to the transition between the two hyperfine levels of the ground state of the caesium-133 atom.

[13th CGPM (1967), Resolution 1]

ampere: The ampere is that constant electric current which, if maintained in two straight parallel conductors of infinite length, of negligible circular cross-section, and placed 1 metre apart in vacuum, would produce between these conductors a force equal to 2×10^{-7} newton per metre of length.

[CIPM (1946), Resolution 2 approved by the 9th CGPM (1948)]

kelvin: The kelvin, unit of thermodynamic temperature, is the fraction 1/273,16 of the thermodynamic temperature of the triple point of water.

[13th CGPM (1967), Resolution 4]

NOTES

1 The 13th CGPM (1967, Resolution 3) also decided that the unit kelvin and its symbol K should be used to express an interval or a difference of temperature.

2 In addition to the thermodynamic temperature (symbol T), expressed in kelvins, use is also made of Celsius temperature (symbol t) defined by the equation $t = T - T_0$, where $T_0 = 273,15$ K by definition. The unit "degree Celsius" is equal to the unit "kelvin", but "degree Celsius" is a special name in place of "kelvin" for expressing Celsius temperature. A temperature interval or a Celsius temperature difference can be expressed in degrees Celsius as well as in kelvins.

mole: The mole is the amount of substance of a system which contains as many elementary entities as there are atoms in 0,012 kilogram of carbon 12. When the mole is used, the elementary entities must be specified and may be atoms, molecules, ions, electrons, other particles, or specified groups of such particles.

[14th CGPM (1971), Resolution 3]

candela: The candela is the luminous intensity in a given direction of a source which emits monochromatic radiation of frequency 540×10^{12} hertz and of which the radiant intensity in that direction is 1/683 watt per steradian.

[16th CGPM (1979), Resolution 3]

Supplementary units

radian: The radian is the plane angle between two radii of a circle which cut off on the circumference an arc equal in length to the radius.

(ISO 31/1-1978)

steradian: The steradian is the solid angle which, having its vertex in the centre of a sphere, cuts off an area of the surface of the sphere equal to that of a square with sides of length equal to the radius of the sphere.

(ISO 31/1-1978)

INTERNATIONAL STANDARD
NORME INTERNATIONALE

ISO 1219

INTERNATIONAL ORGANIZATION FOR STANDARDIZATION·МЕЖДУНАРОДНАЯ ОРГАНИЗАЦИЯ ПО СТАНДАРТИЗАЦИИ·ORGANISATION INTERNATIONALE DE NORMALISATION

Fluid power systems and components — Graphic symbols

First edition — 1976-08-01

Transmissions hydrauliques et pneumatiques — Symboles graphiques

Première édition — 1976-08-01

UDC/CDU 621.5/.6 : 744.4 : 003.62   Ref. No./Réf. nº : ISO 1219-1976 (E/F)

Descriptors : hydraulic fluid power, pneumatic fluid power, hydraulic equipment, pneumatic equipment, measuring instruments, control equipment, mounting diagrams, graphic symbols/**Descripteurs** : transmission hydraulique, transmission pneumatique, matériel hydraulique, matériel pneumatique, instrument de mesurage, matériel de commande, schéma d'utilisation, symbole graphique.

Price based on 23 pages/Prix basé sur 23 pages

ISO 1219 (E/F)

FOREWORD

ISO (the International Organization for Standardization) is a worldwide federation of national standards institutes (ISO Member Bodies). The work of developing International Standards is carried out through ISO Technical Committees. Every Member Body interested in a subject for which a Technical Committee has been set up has the right to be represented on that Committee. International organizations, governmental and non-governmental, in liaison with ISO, also take part in the work.

Draft International Standards adopted by the Technical Committees are circulated to the Member Bodies for approval before their acceptance as International Standards by the ISO Council.

International Standard ISO 1219 was drawn up by Technical Committees ISO/TC 10, *Technical Drawings* and ISO/TC 131, *Fluid power systems and components*. It was submitted directly to the ISO Council, in accordance with clause 6.12.1 of the Directives for the technical work of ISO.

This International Standard cancels and replaces ISO Recommendation ISO/R 1219-1970, which had been approved by the Member Bodies of the following countries :

Australia	Greece	South Africa, Rep. of
Belgium	Hungary	Spain
Brazil	India	Sweden
Canada	Israel	Switzerland
Chile	Italy	Thailand
Czechoslovakia	Japan	Turkey
Denmark	Norway	United Kingdom
Egypt, Arab Rep. of	Netherlands	U.S.A.
Finland	New Zealand	U.S.S.R.
France	Poland	Yugoslavia
Germany	Romania	

No Member Body had expressed disapproval of the document.

© International Organization for Standardization, 1976 •

Printed in Switzerland

AVANT-PROPOS

L'ISO (Organisation Internationale de Normalisation) est une fédération mondiale d'organismes nationaux de normalisation (Comités Membres ISO). L'élaboration de Normes Internationales est confiée aux Comités Techniques ISO. Chaque Comité Membre intéressé par une étude a le droit de faire partie du Comité Technique correspondant. Les organisations internationales, gouvernementales et non gouvernementales, en liaison avec l'ISO, participent également aux travaux.

Les Projets de Normes Internationales adoptés par les Comités Techniques sont soumis aux Comités Membres pour approbation, avant leur acceptation comme Normes Internationales par le Conseil de l'ISO.

La Norme Internationale ISO 1219 a été établie par les Comités Techniques ISO/TC 10, *Dessins Techniques* et ISO/TC 131, *Transmissions hydrauliques et pneumatiques*. Elle fut soumise directement au Conseil de l'ISO, conformément ua paragraphe 6.12.1 des Directives pour les travaux techniques de l'ISO.

Cette Norme Internationale annule et remplace la Recommandation ISO/R 1219-1970, qui avait été approuvée par les Comités Membres des pays suivants :

Afrique du Sud, Rép. d'	France	Roumanie
Allemagne	Grèce	Royaume-Uni
Australie	Hongrie	Suède
Belgique	Inde	Suisse
Brésil	Israël	Tchécoslovaquie
Canada	Italie	Thaïlande
Chili	Japon	Turquie
Danemark	Norvège	U.R.S.S.
Égypte, Rép. arabe d'	Nouvelle-Zélande	U.S.A.
Espagne	Pays-Bas	Yougoslavie
Finlande	Pologne	

Aucun Comité Membre n'avait désapprouvé le document.

© **Organisation Internationale de Normalisation, 1976** •

Imprimé en Suisse

Fluid power systems and components — Graphic symbols

Transmissions hydrauliques et pneumatiques — Symboles graphiques

0 INTRODUCTION

In fluid power systems, power is transmitted and controlled through a fluid (liquid or gas) under pressure within a circuit.

Graphic symbols are used in diagrams of hydraulic and pneumatic equipment and accessories for fluid power transmission.

1 SCOPE AND FIELD OF APPLICATION

This International Standard establishes principles for the use of symbols and specifies the symbols to be used in diagrams of hydraulic and pneumatic transmission systems and components.

The use of these symbols does not preclude the use of other symbols commonly used for pipe-work in other technical fields.

2 REFERENCE

ISO 5598, *Fluid power — Vocabulary*.[1]

3 DEFINITIONS

For definitions of terms used, see ISO 5598.

4 IDENTIFICATION STATEMENT (Reference to this International Standard)

Use the following statement in test reports, catalogues and sales literature when electing to comply with this International Standard :

"Graphic symbols shown in accordance with ISO 1219, *Fluid power systems and components — Graphic symbols*".

0 INTRODUCTION

Dans les sytèmes de transmissions hydrauliques et pneumatiques, l'énergie est transmise et commandée par un fluide (liquide ou gazeux) sous pression circulant dans un circuit.

Les symboles graphiques sont utilisés dans les schémas des systemes de transmissions hydrauliques et pneumatiques.

1 OBJET ET DOMAINE D'APPLICATION

La présente Norme Internationale établit les principes pour l'utilisation des symboles et spécifie les symboles à utiliser dans les schémas des systèmes de transmissions hydrauliques et pneumatiques.

L'utilisation de ces symboles n'exclut pas celle d'autres symboles communément utilisés pour des schémas de canalisations dans d'autres domaines techniques.

2 RÉFÉRENCE

ISO 5598, *Transmissions hydrauliques et pneumatiques — Vocabulaire*.[1]

3 DÉFINITIONS

Pour les définitions, voir ISO 5598.

4 PHRASE D'IDENTIFICATION (Référence à la présente Norme Internationale)

Utiliser la phrase d'identification suivante dans les rapports d'essai, catalogues et documentation commerciale lorsqu'on veut se référer à la présente Norme Internationale :

«Symboles graphiques conformes à l'ISO 1219, *Transmissions hydrauliques et pneumatiques — Symboles graphiques*».

---

[1] In preparation.

[1] En préparation.

5 GENERAL (BASIC AND FUNCTIONAL SYMBOLS)

The symbols for hydraulic and pneumatic equipment and accessories are *functional* and consist of one or more *basic symbols* and in general of one or more *functional symbols*. The symbols are neither to scale nor in general orientated in any particular direction. The relative sizes of symbols in combination should correspond approximately to those in clauses 11 and 12.

5 GÉNÉRALITÉS (SIGNES DE BASE ET DE FONCTION)

Les symboles pour appareils hydromécaniques et pneumatiques et leurs accessoires sont *fonctionnels* et se composent d'un ou de plusieurs *signes de fonction*. Les symboles n'ont pas d'échelle ni, en général, de sens d'orientation déterminé. Les rapports de grandeur des combinaisons de symboles devraient correspondre environ à ceux des exemples des chapitres 11 et 12.

	Description	Application	Symbol / Signe	Dénomination	Emploi
5.1	**Basic symbols**			**Signes de base**	
5.1.1	*Line* :		1)	*Trait* :	
5.1.1.1	— continuous	} flow lines		— continu	} conduites
5.1.1.2	— long dashes		$L > 10E$	— interrompu long	
5.1.1.3	— short dashes		$L < 5E$	— interrompu court	
5.1.1.4	— double	Mechanical connections (shafts, levers, piston-rods)	$D < 5E$	— double	Liaisons mécaniques (arbres, leviers, tiges de pistons)
5.1.1.5	— long chain thin (optional use)	Enclosure for several components assembled in one unit		— mixte fin (emploi facultatif)	Encadrement de plusieurs appareils réunis dans un seul bloc ou dans une unité de montage
5.1.2	*Circle, semi-circle*			*Cercle, demi-cercle*	
5.1.2.1		As a rule, energy conversion units (pump, compressor, motor)	○		En principe, appareils de transformation de l'énergie (pompe, compresseur, moteur)
5.1.2.2		Measuring instruments	○		Appareils de mesure
5.1.2.3		Non-return valve, rotary connection, etc.	○		Clapet de non-retour, joint tournant, etc.
5.1.2.4		Mechanical link, roller, etc.	○		Articulation, galet, etc.
5.1.2.5		Semi-rotary actuator	D		Appareil oscillant
5.1.3	*Square, rectangle*	As a rule, control valves (valve) except for non-return valves	□	*Carré, rectangle*	En principe, appareils de distribution ou de régulation (soupape, distributeur) à l'exclusion des clapets de non-retour
5.1.4	*Diamond*	Conditioning apparatus (filter, separator, lubricator, heat exchanger)	◇	*Losange*	Appareils de conditionnement (filtre, séparateur, lubrificateur, échangeur de chaleur)

1) L = Length of dash
E = Thickness of line
D = Space between lines

1) L = Longueur du trait
E = Épaisseur du trait
D = Espace des traits

	Description	Application	Symbol Signe	Dénomination	Emploi
5.1.5	*Miscellaneous symbols*		1)	*Signes divers*	
5.1.5.1		Flow line connection	$d \approx 5E$		Connexion de conduites
5.1.5.2		Spring	⋙		Ressort
5.1.5.3		Restriction :			Étranglement :
5.1.5.3.1		— affected by viscosity	⩔		— sensible à la viscosité
5.1.5.3.2		— unaffected by viscosity	⋁⋀		— non sensible à la viscosité
5.2	*Functional symbols*			*Signes de fonction*	
5.2.1	*Triangle :*	The direction of flow and the nature of the fluid		*Triangle :*	Sens du flux et nature du fluide
5.2.1.1	— solid	Hydraulic flow	▼	— plein	Flux hydraulique
5.2.1.2	— in outline only	Pneumatic flow or exhaust to atmosphere	▽	— vide	Flux pneumatique ou son évacuation à l'air libre
5.2.2	*Arrow*	Indication of :	↑ ↑ ↓	*Flèche*	Indication de :
5.2.2.1		— direction			— sens de déplacement
5.2.2.2		— direction of rotation	↶ ↷		— sens de rotation
5.2.2.3		— path and direction of flow through valves	↓⌐ ⌐↓ ⌐ ↓ ↓ ↓ ↓		— voie et sens du flux dans les soupapes ou distributeurs
		For regulating apparatus as in 7.4 both representations with or without a tail to the end of the arrow are used without distinction			Dans les appareils de réglage 7.4, les deux représentations avec ou sans trait latéral à la queue de la flèche sont employées indifféremment
		As a general rule the line perpendicular to the head of the arrow indicates that when the arrow moves the interior path always remains connected to the corresponding exterior path			D'une façon générale le trait perpendiculaire à la pointe de la flèche indique que dans le déplacement de la flèche, la voie intérieure reste toujours reliée à la voie extérieure qui lui correspond
5.2.3	*Sloping arrow*	Indication of the possibility of a regulation or of a progressive variability	↗	*Flèche oblique*	Indication de la possibilité d'un réglage ou de variabilité progressive

1) E — Thickness of line

1) E — Épaisseur du trait

ISO 1219 (E/F)

	Description	**Use of the equipment or explanation of the symbol**	**Symbol Symbole**	**Dénomination**	**Fonction de l'appareil ou explication du symbole**
6	**ENERGY CONVERSION**			**6 TRANSFORMATION DE L'ÉNERGIE**	Dans les dénominations, le terme «cylindrée» est admis dans un sens général pour tout appareil volumétrique.
6.1	**Pumps and compressors**	To convert mechanical energy into hydraulic or pneumatic energy		**Pompes et compresseurs**	Appareils transformant l'énergie mécanique en énergie hydraulique ou pneumatique
6.1.1	*Fixed capacity hydraulic pump* :			*Pompe hydraulique à cylindrée fixe* :	
6.1.1.1	— with one direction of flow			— à un sens de flux	
6.1.1.2	— with two directions of flow			— à deux sens de flux	
6.1.2	*Variable capacity hydraulic pump* :			*Pompe hydraulique à cylindrée variable* :	
6.1.2.1	— with one direction of flow	The symbol is a combination of 6.1.1.1 and 5.2.3 (sloping arrow)		— à un sens de flux	Le symbole est une combinaison de 6.1.1.1 et de 5.2.3 (flèche oblique)
6.1.2.2	— with two directions of flow	The symbol is a combination of 6.1.1.2 and 5.2.3 (sloping arrow)		— à deux sens de flux	Le symbole est une combinaison de 6.1.1.2 et de 5.2.3 (flèche oblique)
6.1.3	*Fixed capacity compressor (always one direction of flow)*			*Compresseur à cylindrée fixe (toujours à un sens de flux)*	
6.2	**Motors**	To convert hydraulic or pneumatic energy into rotary mechanical energy		**Moteurs**	Appareils transformant l'énergie hydraulique ou pneumatique en énergie mécanique rotative
6.2.1	*Fixed capacity hydraulic motor* :			*Moteur hydraulique à cylindrée fixe* :	
6.2.1.1	— with one direction of flow			— à un sens de flux	
6.2.1.2	— with two directions of flow			— à deux sens de flux	
6.2.2	*Variable capacity hydraulic motor* :			*Moteur hydraulique à cylindrée variable* :	
6.2.2.1	— with one direction of flow	The symbol is a combination of 6.2.1.1 and 5.2.3 (sloping arrow)		— à un sens de flux	Le symbole est une combinaison de 6.2.1.1 et de 5.2.3 (flèche oblique)
6.2.2.2	— with two directions of flow	The symbol is a combination of 6.2.1.2 and 5.2.3 (sloping arrow)		— à deux sens de flux	Le symbole est une combinaison de 6.2.1.2 et de 5.2.3 (flèche oblique)

	Description	Use of the equipment or explanation of the symbol	Symbol Symbole	Dénomination	Fonction de l'appareil ou explication du symbole
6.2.3	Fixed capacity pneumatic motor :			Moteur pneumatique à cylindrée fixe :	
6.2.3.1	– with one direction of flow			– à un sens de flux	
6.2.3.2	– with two directions of flow			– à deux sens de flux	
6.2.4	Variable capacity pneumatic motor :			Moteur pneumatique à cylindrée variable :	
6.2.4.1	– with one direction of flow	The symbol is a combination of 6.2.3.1 and 5.2.3 (sloping arrow)		– à un sens de flux	Le symbole est une combinaison de 6.2.3.1 et de 5.2.3 (flèche oblique)
6.2.4.2	– with two directions of flow	The symbol is a combination of 6.2.3.2 and 5.2.3 (sloping arrow)		– à deux sens de flux	Le symbole est une combinaison de 6.2.3.2 et de 5.2.3 (flèche oblique)
6.2.5	Oscillating motor :			Moteur oscillant :	
6.2.5.1	– hydraulic			– hydraulique	
6.2.5.2	– pneumatic			– pneumatique	
6.3	Pump/motor units	Unit with two functions, either as pump or as rotary motor		Pompes-moteurs	Appareils à deux fonctions, soit pompe, soit moteur rotatif
6.3.1	Fixed capacity pump/motor unit :			Pompe-moteur à cylindrée fixe :	
6.3.1.1	– with reversal of the direction of flow	Functioning as pump or motor according to direction of flow		– à inversion du sens de flux	Avec inversion du sens de flux pour fonctionner soit en pompe, soit en moteur
6.3.1.2	– with one single direction of flow	Functioning as pump or motor without change of direction of flow		– à un seul sens de flux	Sans inversion du sens de flux pour fonctionner soit en pompe, soit en moteur
6.3.1.3	– with two directions of flow	Functioning as pump or motor with either direction of flow		– à deux sens de flux	Dans les deux sens du flux pour fonctionner soit en pompe, soit en moteur
6.3.2	Variable capacity pump/motor unit :			Pompe-moteur à cylindrée variable :	
6.3.2.1	– with reversal of the direction of flow	The symbol is a combination of 6.3.1.1 and 5.2.3 (sloping arrow)		– à inversion du sens de flux	Le symbole est une combinaison de 6.3.1.1 et de 5.2.3 (flèche oblique)
6.3.2.2	– with one single direction of flow	The symbol is a combination of 6.3.1.2 and 5.2.3 (sloping arrow)		– à un seul sens de flux	Le symbole est une combinaison de 6.3.1.2 et de 5.2.3 (flèche oblique)
6.3.2.3	– with two directions of flow	The symbol is a combination of 6.3.1.3 and 5.2.3 (sloping arrow)		– à deux sens de flux	Le symbole est une combinaison de 6.3.1.3 et de 5.2.3 (flèche oblique)

	Description	Use of the equipment or explanation of the symbol	Symbol / Symbole		Dénomination	Fonction de l'appareil ou explication du symbole
6.4	Variable speed drive units	Torque converter. Pump and/or motor are of variable capacity. Remote drives, see 12.2			Variateurs	Convertisseur de couple. Pompe et/ou moteur sont à cylindrée réglable. Transmission, voir 12.2
6.5	Cylinders	Equipment to convert hydraulic or pneumatic energy into linear energy			Vérins	Appareils transformant l'énergie hydraulique ou pneumatique en énergie mécanique à mouvement rectiligne
6.5.1	Single acting cylinder :	Cylinder in which the fluid pressure always acts in one and the same direction (on the forward stroke)	Detailed / Détaillé	Simplified / Simplifié	Vérin à simple effet :	Vérin dans lequel la pression du fluide s'exerce dans un seul et même sens (course aller)
6.5.1.1	– returned by an unspecified force	General symbol when the method of return is not specified			– à rappel par force non définie	Symbole général lorsque le mode d'obtention de la course retour n'est pas précisé
6.5.1.2	– returned by spring	Combination of the general symbol 6.5.1.1 and 5.1.5.2 (spring)			– à rappel par ressort	Combinaison du symbole général 6.5.1.1 et de 5.1.5.2 (ressort)
6.5.2	Double acting cylinder :	Cylinder in which the fluid pressure operates alternately in both directions (forward and backward strokes)			Vérin à double effet :	Vérin dans lequel la pression du fluide s'exerce alternativement dans les deux sens (course aller et retour)
6.5.2.1	– with single piston rod				– à simple tige	
6.5.2.2	– with double-ended piston rod				– à double tige	
6.5.3	Differential cylinder	The action is dependent on the difference between the effective areas on each side of the piston			Vérin différentiel	Le fonctionnement du vérin résulte de la différence des aires effectives de chaque côté du piston
6.5.4	Cylinder with cushion :				Vérin avec amortisseur :	
6.5.4.1	– with single fixed cushion	Cylinder incorporating fixed cushion acting in one direction only			– fixe d'un côté	Vérin comportant un amortisseur fixe agissant dans un seul sens
6.5.4.2	– with double fixed cushion	Cylinder with fixed cushion acting in both directions			– fixe des deux côtés	Vérin comportant un amortisseur fixe agissant dans les deux sens
6.5.4.3	– with single adjustable cushion	The symbol is a combination of 6.5.4.1 and 5.2.3 (sloping arrow)			– réglable d'un côté	Le symbole est une combinaison de 6.5.4.1 et de 5.2.3 (flèche oblique)
6.5.4.4	– with double adjustable cushion	The symbol is a combination of 6.5.4.2 and 5.2.3 (sloping arrow)			– réglable des deux côtés	Le symbole est une combinaison de 6.5.4.2 et de 5.2.3 (flèche oblique)

	Description	Use of the equipment or explanation of the symbol	Symbol / Symbole		Dénomination	Fonction de l'appareil ou explication du symbole
6.5.5	*Telescopic cylinder :*				*Vérin télescope :*	
6.5.5.1	– single acting	The fluid pressure always acts in one and the same direction (on the forward stroke)			– à simple effet	La pression du fluide s'exerce dans un seul et même sens (course aller)
6.5.5.2	– double acting	The fluid pressure operates alternately in both directions (forward and backward strokes)			– à double effet	La pression du fluide s'exerce alternativement dans les deux sens (course aller et course retour)
6.6	**Pressure intensifiers :**	Equipment transforming a pressure x into a higher pressure y	Detailed Détaillé	Simplified Simplifié	**Multiplicateurs de pression :**	Appareil transformant une pression x en une pression supérieure y
6.6.1	– *for one type of fluid*	E.g. a pneumatic pressure x is transformed into a higher pneumatic pressure y			– à une seule nature de fluide	Par exemple, une pression x pneumatique est transformée en une pression supérieure y pneumatique
6.6.2	– *for two types of fluid*	E.g. a pneumatic pressure x is transformed into a higher hydraulic pressure y			– à deux natures de fluides	Par exemple, une pression x pneumatique est transformée en une pression supérieure y hydraulique
6.7	Air-oil actuator	Equipment transforming a pneumatic pressure into a substantially equal hydraulic pressure or vice versa			**Échangeur de pression hydraulique-pneumatique**	Appareil transformant une pression pneumatique en une pression hydraulique théoriquement égale, ou inversement

7 CONTROL VALVES

7 DISTRIBUTION ET RÉGULATION DE L'ÉNERGIE

7.1	**Method of representation of valves (except 7.3 and 7.6)**	Made up of one or more squares 5.1.3 and arrows *In circuit diagrams hydraulic and pneumatic units are normally shown in the unoperated condition*		**Principes de représentation des appareils (excepté 7.3 et 7.6)**	Composition d'une ou de plusieurs cases 5.1.3 et de flèches *Dans les schémas d'ensemble, les appareils hydrauliques ou pneumatiques sont normalement représentés en position de repos*
7.1.1	*One single square*	Indicates unit for controlling flow or pressure, having in operation an infinite number of possible positions between its end positions so as to vary the conditions of flow across one or more of its ports, thus ensuring the chosen pressure and/or flow with regard to the operating conditions of the circuit		*Une case*	Indique un appareil de réglage de débit ou de pression susceptible d'avoir, en service, entre ses deux positions extrêmes, une infinité de stades qui, en faisant varier les conditions d'écoulement à travers la ou les voies de l'appareil, permettent d'assurer dans les conditions de fonctionnement du circuit, la valeur voulue de pression et/ou de débit

	Description	Use of the equipment or explanation of the symbol	Symbol Symbole	Dénomination	Fonction de l'appareil ou explication du symbole
7.1.2	*Two or more squares*	Indicate a directional control valve having as many distinct positions as there are squares. The pipe connections are normally represented as connected to the box representing the unoperated condition (see 7.1). The operating positions are deduced by imagining the boxes to be displaced so that the pipe connections correspond with the ports of the box in question		*Plusieurs cases*	Indiquent un appareil de distribution de débit ou de pression susceptible d'avoir autant de positions distinctes qu'il y a de cases. Les conduites sont normalement représentées aboutissant à la case de la position de repos (voir 7.1). On obtient les autres positions par déplacement des cases jusqu'à ce que les orifices aboutissent aux conduites correspondantes
7.1.3	*Simplified symbol for valves in cases of multiple repetition*	The number refers to a note on the diagram in which the symbol for the valve is given in full	3	*Symbole simplifié d'appareil en cas de représentation multiple*	Le numéro se réfère à un repère sur le dessin, sous lequel l'appareil est représenté de façon détaillée
7.2	**Directional control valves**	Units providing for the opening (fully or restricted) or the closing of one or more flow paths (represented by several squares)		**Distributeurs**	Appareils assurant l'ouverture à plein débit ou avec étranglement ou la fermeture d'une ou de plusieurs voies de passage (représentation par plusieurs cases)
7.2.1	*Flow paths :*	Square containing interior lines		*Voies ou canaux :*	Cases comprenant les voies intérieures
7.2.1.1	– one flow path			– 1 voie	
7.2.1.2	– two closed ports			– 2 orifices fermés	
7.2.1.3	– two flow paths			– 2 voies	
7.2.1.4	– two flow paths and one closed port			– 2 voies, 1 orifice fermé	
7.2.1.5	– two flow paths with cross connection			– 2 voies en connexion transversale	
7.2.1.6	– one flow path in a by-pass position, two closed ports			– 1 voie en by-pass, 2 orifices fermés	
7.2.2	*Non-throttling directional control valve*	The unit provides distinct circuit conditions each depicted by a square		*Distributeur sans étranglement*	L'appareil comporte plusieurs positions distinctes caractérisées chacune par une case
7.2.2.1		Basic symbol for 2-position directional control valve			Signe de base d'un distributeur à 2 positions distinctes

	Description	Use of the equipment or explanation of the symbol	Symbol Symbole	Dénomination	Fonction de l'appareil ou explication du symbole
7.2.2.2		Basic symbol for 3-position directional control valve			Signe de base d'un distributeur à 3 positions distinctes
7.2.2.3		A transitory but significant condition between two distinct positions is optionally represented by a square with dashed ends			Représentation facultative de passage à un stade intermédiaire entre deux positions distinctes, par une case délimitée par des traits interrompus
		A basic symbol for a directional control valve with two distinct positions and one transitory intermediate condition			Signe de base d'un distributeur à 2 positions distinctes et un stade intermédiaire de passage
7.2.2.4	Designation : The first figure in the *designation* shows the number of ports (excluding pilot ports) and the second figure the number of distinct positions			Désignation : Le premier chiffre de la *désignation* indique le nombre d'orifices (les orifices de pilotage ne sont pas comptés) le second chiffre marque le nombre de positions distinctes	
7.2.2.5	Directional control valve 2/2 :	Directional control valve with 2 ports and 2 distinct positions		Distributeur 2/2 :	Distributeur à 2 orifices et 2 positions distinctes
7.2.2.5.1	– with manual control			– à commande manuelle	
7.2.2.5.2	– controlled by pressure operating against a return spring (e.g. on air unloading valve)			– à commande par pression avec ressort de rappel (par exemple soupape de décharge)	
7.2.2.6	Directional control valve 3/2 :	Directional control valve with 3 ports and 2 distinct positions		Distributeur 3/2 :	Distributeur à 3 orifices et 2 positions distinctes
7.2.2.6.1	– controlled by pressure in both directions			– à commande par pression (des deux côtés)	
7.2.2.6.2	– controlled by solenoid with return spring	Indicating an intermediate condition (see 7.2.2.3)		– à commande électromagnétique avec rappel par ressort	Avec représentation du passage à un stade intermédiaire (voir 7.2.2.3)

ISO 1219 (E/F)

	Description	Use of the equipment or explanation of the symbol	Symbol / Symbole	Dénomination	Fonction de l'appareil ou explication du symbole
7.2.2.7	Directional control valve 4/2 :	Directional control valve with 4 ports and 2 distinct positions	Detailed Détaillé	Distributeur 4/2 :	Distributeur à 4 orifices et 2 positions distinctes
7.2.2.7.1	– controlled by pressure in both directions by means of a pilot valve (with a single solenoid and spring return)		Simplified Simplifié	– à commande par pression des deux côtés accouplés à un distributeur pilote (à commande électromécanique avec rappel par ressort)	
7.2.2.8	Directional control valve 5/2 :	Directional control valve with 5 ports and 2 distinct positions		Distributeur 5/2 :	Distributeur à 5 orifices et 2 positions distinctes
7.2.2.8.1	– controlled by pressure in both directions			– à commande par pression des deux côtés	
7.2.3	*Throttling directional control*	The unit has 2 extreme positions and an infinite number of intermediate conditions with varying degrees of throttling All the symbols have parallel lines along the length of the boxes. For valves with mechanical feedback see 9.3		*Distributeur à étranglement*	L'appareil comporte 2 positions extrêmes et une infinité de stades intermédiaires correspondant à des degrés variés d'étranglement Tous les symboles ont des lignes parallèles le long des cases. Pour les distributeurs avec rétroaction mécanique, voir 9.3
7.2.3.1		Showing the extreme positions			Montre les positions extrêmes
7.2.3.2		Showing the extreme positions and a central (neutral) position			Montre les positions extrêmes et la position centrale (zéro)
7.2.3.3	– with 2 ports (one throttling orifice)	For example : Tracer valve plunger operated against a return spring		– à 2 orifices (1 étranglement)	Par exemple : Soupape à palpeur, commandée par poussoir avec ressort de rappel
7.2.3.4	– with 3 ports (two throttling orifices)	For example : Directional control valve controlled by pressure against a return spring		– à 3 orifices (2 étranglements)	Par exemple : Distributeur à commande par pression avec ressort de rappel
7.2.3.5	– with 4 ports (four throttling orifices)	For example : Tracer valve, plunger operated against a return spring		– à 4 orifices (4 étranglements)	Par exemple : Soupape à palpeur, commandée par poussoir avec ressort de rappel
7.2.4	*Electro-hydraulic servo valve :* *Electro-pneumatic servo valve :*	A unit which accepts an analogue electrical signal and provides a similar analogue fluid power output		*Servo-distributeur électrohydraulique :* *Servo-distributeur électropneumatique :*	Appareil qui reçoit un signal électrique analogique et fournit un signal de sortie hydraulique ou pneumatique similaire
7.2.4.1	– single stage	– with direct operation		– à 1 étage	– à action directe

- 185 -

	Description	Use of the equipment or explanation of the symbol	Symbol Symbole	Dénomination	Fonction de l'appareil ou explication du symbole
7.2.4.2	– two stage with mechanical feedback	– with indirect pilot operation		– à 2 étages avec asservissement mécanique	– à action indirecte mais avec pilotage
7.2.4.3	– two stage with hydraulic feedback	– with indirect pilot operation		– à 2 étages avec asservissement hydraulique	– à action indirecte mais avec pilotage
7.3	Non-return valves, shuttle valve, rapid exhaust valve	Valves which allow free flow in one direction only		Clapets de non-retour, sélecteur, soupape d'échappement rapide	Appareils permettant le passage libre dans un sens seulement
7.3.1	Non-return valve :			Clapet de non-retour :	
7.3.1.1	– free	Opens if the inlet pressure is higher than the outlet pressure		– sans ressort	Ouverture si la pression d'entrée est supérieure à la pression de sortie
7.3.1.2	– spring loaded	Opens if the inlet pressure is greater than the outlet pressure plus the spring pressure		– avec ressort	Ouverture si la pression d'entrée est supérieure à la pression de sortie plus la pression du ressort
7.3.1.3	– pilot controlled	As 7.3.1.1 but by pilot control it is possible to prevent :		– piloté	Comme 7.3.1.1, mais suppression possible par pilotage :
7.3.1.3.1		– closing of the valve			– de la fermeture
7.3.1.3.2		– opening of the valve			– de l'ouverture
7.3.1.4	– with restriction	Unit allowing free flow in one direction but restricted flow in the other		– avec étranglement	Appareil permettant le passage libre dans un sens et son étranglement dans l'autre sens
7.3.2	Shuttle valve	The inlet port connected to the higher pressure is automatically connected to the outlet port while the other inlet port is closed		Sélecteur de circuit	L'orifice mis sous pression est relié automatiquement avec la sortie pendant que l'autre entrée est fermée
7.3.3	Rapid exhaust valve	When the inlet port is unloaded the outlet port is freely exhausted		Soupape d'échappement rapide	En cas de décharge de la conduite d'entrée, la conduite de sortie est mise à l'air libre
7.4	Pressure control valves	Units ensuring the control of pressure. Represented by one single square as in 7.1.1 with one arrow (the tail to the arrow may be placed at the end of the arrow). For interior controlling conditions see 9.2.4.3		Appareils de réglage de la pression	Appareils assurant le réglage de la pression. Représentation à une seule case comme 7.1.1 avec une flèche (trait latéral éventuel à la queue de la flèche). Pour les voies intérieures de commandes, voir 9.2.4.3

	Description	Use of the equipment or explanation of the symbol	Symbol / Symbole	Dénomination	Fonction de l'appareil ou explication du symbole
7.4.1	Pressure control valve :	General symbols		Appareil de réglage de la pression :	Symboles généraux
7.4.1.1	– 1 throttling orifice normally closed		or / ou	– normalement fermé à 1 étranglement	
7.4.1.2	– 1 throttling orifice normally open		or / ou	– normalement ouvert à 1 étranglement	
7.4.1.3	– 2 throttling orifices, normally closed			– normalement fermé à 2 étranglements	
7.4.2	Pressure relief valve (safety valve) :	Inlet pressure is controlled by opening the exhaust port to the reservoir or to atmosphere against an opposing force (for example a spring)		Limiteur de pression (soupape de sûreté) :	Limitation de la pression à l'orifice d'entrée par ouverture de l'orifice de décharge à l'air libre ou relié au réservoir en utilisant un effort antagoniste (par exemple : ressort)
7.4.2.1	– with remote pilot control	The pressure at the inlet port is limited as in 7.4.2 or to that corresponding to the setting of a pilot control		– piloté par commande à distance	La pression à l'orifice d'entrée est limitée comme en 7.4.2 ou à celle qui correspond au réglage du distributeur-pilote
7.4.3	Proportional pressure relief	Inlet pressure is limited to a value proportional to the pilot pressure (see 9.2.4.1.3)		Limiteur proportionnel de pression	La pression d'entrée est limitée à une valeur qui est proportionnelle à la pression de pilotage (voir 9.2.4.1.3)
7.4.4	Sequence valve	When the inlet pressure overcomes the opposing force of the spring, the valve opens permitting flow from the outlet port	or / ou	Soupape de séquence	À une pression d'entrée plus élevée que l'effort antagoniste du ressort, le passage vers d'autres appareils est libre par ouverture de l'orifice de sortie
7.4.5	Pressure regulator or reducing valve (reducer of pressure) :	A unit which, with a variable inlet pressure, gives substantially constant output pressure provided that the inlet pressure remains higher than the required outlet pressure		Réducteur de pression ou détendeur	Appareil permettant d'obtenir, avec une pression d'entrée variable, une pression de sortie sensiblement constante. La pression d'entrée doit cependant rester toujours supérieure à la pression de sortie désirée
7.4.5.1	– without relief port			– sans orifice de décharge	
7.4.5.2	– without relief port with remote control	As in 7.4.5.1 but the outlet pressure is dependent on the control pressure		– sans orifice de décharge, réglé à distance	Comme 7.4.5.1, mais la valeur de la pression de sortie est fonction de la pression de réglage
7.4.5.3	– with relief port			– avec orifice de décharge	

	Description	Use of the equipment or explanation of the symbol	Symbol Symbole		Dénomination	Fonction de l'appareil ou explication du symbole
7.4.5.4	— with relief port, with remote control	As in 7.4.5.3 but the outlet pressure is dependent on the control pressure			— avec orifice de décharge, réglé à distance	Comme 7.4.5.3, mais la valeur de la pression de sortie est fonction de la pression de réglage
7.4.6	Differential pressure regulator	The outlet pressure is reduced by a fixed amount with respect to the inlet pressure			Régulateur de pression différentiel	La pression de sortie est réduite d'une valeur sensiblement constante par rapport à la pression d'entrée qui lui est supérieure
7.4.7	Proportional pressure regulator	The outlet pressure is reduced by a fixed ratio with respect to the inlet pressure (see 9.2.4.1.3)			Régulateur de pression proportionnel	La pression de sortie est réduite dans un rapport constant avec la pression d'entrée (voir 9.2.4.1.3)
7.5	Flow control valves	Units ensuring control of flow. Excepting 7.5.3 positions and method of representation as 7.4			Appareils de réglage du débit	Appareils assurant le réglage du débit. Positions et principe de représentation comme 7.4, excepté 7.5.3
7.5.1	Throttle valve :	Simplified symbol (Does not indicate the control method or the state of the valve)			Réducteur de débit : — variable	En représentation simplifiée (Mode de commande et position non représentés)
7.5.1.1	— with manual control	Detailed symbol (indicates the control method or the state of the valve)			— à commande manuelle	En représentation détaillée (Indication du mode de commande et de la position)
7.5.1.2	— with mechanical control against a return spring (braking valve)				— à commande mécanique avec ressort de rappel (soupape de freinage)	
7.5.2	Flow control valve :	Variations in inlet pressure do not affect the rate of flow	Detailed Détaillé	Simplified Simplifié	Régulateur de débit :	Le débit est maintenu sensiblement indépendant des variations de pression
7.5.2.1	— with fixed output				— à débit fixe	
7.5.2.2	— with fixed output and relief port to reservoir	As 7.5.2.1 but with relief for excess flow			— à débit fixe avec retour au réservoir	Comme 7.5.2.1, mais avec évacuation de l'excédent de débit
7.5.2.3	— with variable output	As 7.5.2.1 but with arrow 5.2.3 added to the symbol of restriction			— à débit réglable	Comme 7.5.2.1, mais le symbole d'étranglement est complété par une flèche 5.2.3
7.5.2.4	— with variable output and relief port to reservoir	As 7.5.2.3 but with relief for excess flow			— à débit réglable avec retour au réservoir	Comme 7.5.2.3, mais avec évacuation de l'excédent de débit

	Description	Use of the equipment or explanation of the symbol	Symbol Symbole	Dénomination	Fonction de l'appareil ou explication du symbole
7.5.3	*Flow dividing valve*	The flow is divided into two flows in a fixed ratio substantially independent of pressure variations		*Diviseur de débit*	Le débit d'alimentation est divisé en deux débits dans un rapport donné et cela à peu près indépendamment des variations de la pression
7.6	Shut-off valve	Simplified symbol		*Robinet d'isolement*	Symbole simplifié

8 ENERGY TRANSMISSION AND CONDITIONING

8 TRANSMISSION DE L'ÉNERGIE ET CONDITIONNEMENT

8.1	**Sources of energy**			*Sources d'énergie*	
8.1.1	*Pressure source*	Simplified general symbol		*Source de pression*	Symbole général simplifié
8.1.1.1	*Hydraulic pressure source*			*Source de pression hydraulique*	
8.1.1.2	*Pneumatic pressure source*	Symbols to be used when the nature of the source should be indicated		*Source de pression pneumatique*	Symboles à utiliser lorsque la nature de la source doit être indiquée
8.1.2	*Electric motor*	Symbol 113 in IEC Publication 117.2		*Moteur électrique*	Symbole 113 de la Publication CEI 117.2
8.1.3	*Heat engine*			*Moteur thermique*	
8.2	**Flow lines and connections**			*Conduites et connexions*	
8.2.1	*Flow line :*			*Conduite :*	
8.2.1.1	— working line, return line and feed line			— de travail, de retour, d'alimentation	
8.2.1.2	— pilot control line			— de pilotage	
8.2.1.3	— drain or bleed line			— de fuite, de purge, ou de décharge	
8.2.1.4	— flexible pipe	Flexible hose, usually connecting moving parts		— flexible	Tuyau reliant généralement des éléments mobiles
8.2.1.5	— electric line			— électrique	
8.2.2	*Pipeline junction*			*Raccordement de conduites*	
8.2.3	*Crossed pipelines*	not connected		*Croisement de conduites*	sans connexion
8.2.4	*Air bleed*			*Purge d'air*	

	Description	Use of the equipment or explanation of the symbol	Symbol Symbole	Dénomination	Fonction de l'appareil ou explication du symbole
8.2.5	*Exhaust port :*			*Orifice d'évacuation d'air :*	
8.2.5.1	— plain with no provision for connection			— lisse non connectable	
8.2.5.2	— threaded for connection			— taraudé pour connexion	
8.2.6	*Power take-off :*	On equipment or lines, for energy take-off or measurement		*Prise :*	Sur appareils ou conduites pour prise de puissance ou pour le mesurage
8.2.6.1	— plugged			— bouchée	
8.2.6.2	— with take-off line			— avec conduite branchée	
8.2.7	*Quick-acting coupling :*			*Raccordement rapide :*	
8.2.7.1	— connected, without mechanically opened non-return valve			— accouplé, sans clapet de non-retour ouvert mécaniquement	
8.2.7.2	— connected, with mechanically opened non-return valves			— accouplé, avec clapet de non-retour ouvert mécaniquement	
8.2.7.3	— uncoupled, with open end			— désaccouplé, à conduite ouverte	
8.2.7.4	— uncoupled, closed by free non-return valve (see 7.3.1.1)			— désaccouplé, à conduite fermée par clapet de non-retour non taré (voir 7.3.1.1)	
8.2.8	*Rotary connection :*	Line junction allowing angular movement in service		*Raccord rotatif :*	Raccordement de conduite pouvant tourner en service
8.2.8.1	— one way			— à 1 voie	
8.2.8.2	— three way			— à 3 voies	
8.2.9	*Silencer*			*Silencieux*	
8.3	**Reservoirs**			**Réservoirs**	
8.3.1	*Reservoir open to atmosphere :*			*Réservoir à l'air libre :*	
8.3.1.1	— with inlet pipe above fluid level			— à conduite débouchant au-dessus du niveau du fluide	

	Description	Use of the equipment or explanation of the symbol	Symbol Symbole	Dénomination	Fonction de l'appareil ou explication du symbole
8.3.1.2	– with inlet pipe below fluid level			– à conduite débouchant au-dessous du niveau de fluide	
8.3.1.3	– with a header line			– à conduite en charge	
8.3.2	Pressurized reservoir			Réservoir sous pression	
8.4	Accumulators	The fluid is maintained under pressure by a spring, weight or compressed gas (air, nitrogen, etc.)		Accumulateurs	Le fluide est tenu sous pression par un ressort, un poids ou la compressibilité d'un gaz (air, azote, etc.)
8.5	**Filters, water traps, lubricators and miscellaneous apparatus**			**Filtres, purgeurs, lubrificateurs et appareillage divers**	
8.5.1	Filter or strainer			Filtre, crépine	
8.5.2	Water trap :			Purgeur :	
8.5.2.1	– with manual control			– à commande manuelle	
8.5.2.2	– automatically drained			– automatique	
8.5.3	Filter with water trap :			Filtre avec purgeur :	
8.5.3.1	– with manual control	Combination of 8.5.1 and 8.5.2.1		– à commande manuelle	Combinaison de 8.5.1 et de 8.5.2.1
8.5.3.2	– automatically drained	Combination of 8.5.1 and 8.5.2.2		– automatique	Combinaison de 8.5.1 et de 8.5.2.2
8.5.4	Air dryer	A unit drying air (for example by chemical means)		Déshydrateur	Appareil assurant le séchage de l'air (par exemple par des moyens chimiques)
8.5.5	Lubricator	Small quantities of oil are added to the air passing through the unit, in order to lubricate equipment receiving the air		Lubrificateur	Pour la lubrification des appareils, de petites quantités d'huile sont ajoutées à l'air passant au travers du lubrificateur
8.5.6	Conditioning unit	Consisting of filter, pressure regulator, pressure gauge and lubricator		Groupe de conditionnement	Ensemble composé d'un filtre, d'un régulateur de pression avec manomètre et d'un lubrificateur
8.5.6.1		– Detailed symbol			– Symbole détaillé
8.5.6.2		– Simplified symbol			– Symbole simplifié

	Description	Use of the equipment or explanation of the symbol	Symbol Symbole	Dénomination	Fonction de l'appareil ou explication du symbole
8.6	Heat exchangers	Apparatus for heating or cooling the circulating fluid		Échangeurs de chaleur	Appareils assurant le refroidissement ou le réchauffement du fluide en circulation
8.6.1	Temperature controller	The fluid temperature is maintained between two predetermined values. The arrows indicate that heat may be either introduced or dissipated		Régulateur de température	Appareil contrôlant la température du fluide entre deux valeurs préétablies. Les flèches symbolisent l'apport et l'évacuation de la chaleur
8.6.2	Cooler	The arrows in the diamond indicate the extraction of heat		Refroidisseur (Réfrigérant)	Les flèches dans le losange symbolisent l'évacuation de chaleur
8.6.2.1		— without representation of the flow lines of the coolant			— sans représentation des conduites du fluide de refroidissement
8.6.2.2		— indicating the flow lines of the coolant			— avec représentation des conduites du fluide de refroidissement
8.6.3	Heater	The arrows in the diamond indicate the introduction of heat		Réchauffeur	Les flèches dans le losange symbolisent l'apport de chaleur

9 CONTROL MECHANISMS
9 COMMANDES

9.1	Mechanical components			Éléments mécaniques	
9.1.1	Rotating shaft :	The arrow indicates rotation		Arbre tournant :	La flèche symbolise la rotation
9.1.1.1	— in one direction			— dans un sens	
9.1.1.2	— in either direction			— dans deux sens	
9.1.2	Detent	A device for maintaining a given position		Dispositif de maintien en position	Pour maintenir une position systématique d'un appareil
9.1.3	Locking device	* The symbol for unlocking control is inserted in the square		Dispositif de verrouillage	* Le symbole de commande de déverrouillage sera indiqué dans le carré
9.1.4	Over-centre device	Prevents the mechanism stopping in a dead centre position		Basculeur	Empêche l'immobilisation d'un appareil au point mort
9.1.5	Pivoting devices :			Mécanisme d'articulation :	
9.1.5.1	— simple			— simple	

	Description	Use of the equipment or explanation of the symbol	Symbol Symbole	Dénomination	Fonction de l'appareil ou explication du symbole
9.1.5.2	– with traversing lever			– avec levier transversal	
9.1.5.3	– with fixed fulcrum			– avec point fixe	
9.2	Control methods	The symbols representing control methods are incorporated in the symbol of the controlled apparatus to which they should be adjacent. For apparatus with several squares the actuation of the control makes effective the square adjacent to it		Modes de commande	Les symboles représentant les modes de commande font partie du symbole des appareils commandés auxquels ils doivent être accolés. Pour les appareils à plusieurs cases, l'action de la commande met en service la case qui lui est adjacente
9.2.1	Muscular control :	General symbol (without indication of control type)		Commande manuelle :	Symbole de base (sans indication du mode de commande)
9.2.1.1	– by push button			– par bouton poussoir	
9.2.1.2	– by lever			– par levier	
9.2.1.3	– by pedal			– par pédale	
9.2.2	Mechanical control :			Commande mécanique :	
9.2.2.1	– by plunger or tracer			– par poussoir ou palpeur	
9.2.2.2	– by spring			– par ressort	
9.2.2.3	– by roller			– par galet	
9.2.2.4	– by roller, operating in one direction only			– par galet avec effet dans une seule direction	
9.2.3	Electrical control :			Commande électrique :	
9.2.3.1	– by solenoid :			– par électro-aimant :	
9.2.3.1.1		– with 1 winding			– à 1 enroulement
9.2.3.1.2		– with 2 windings operating in opposite directions			– à 2 enroulements agissant en sens contraire

ISO 1219 (E/F)

	Description	Use of the equipment or explanation of the symbol	Symbol Symbole	Dénomination	Fonction de l'appareil ou explication du symbole
9.2.3.1.3		– with 2 windings operating in a variable way progressively, operating in opposite direction			– à 2 enroulements à action variable progressive agissant en sens contraire
9.2.3.2	– by electric motor			– par moteur électrique	
9.2.4	*Control by application or release of pressure*			*Commande par application ou par baisse de la pression*	
9.2.4.1	Direct acting control :			Commande directe :	
9.2.4.1.1	– by application of pressure			– par application de la pression	
9.2.4.1.2	– by release of pressure			– par baisse de la pression	
9.2.4.1.3	– by different control areas	In the symbol the larger rectangle represents the larger control area, i.e. the priority phase		– par aires de commande différentes	Dans le symbole, le grand rectangle représente l'aire de commande la plus grande, c'est-à-dire le côté prioritaire
9.2.4.2	Indirect control, pilot actuated :	General symbol for pilot directional control valve		Commande indirecte par distributeur-pilote actionné :	Symbole général pour distributeur-pilote
9.2.4.2.1	– by application of pressure			– par application de la pression	
9.2.4.2.2	– by release of pressure			– par baisse de la pression	
9.2.4.3	Interior control paths	The control paths are inside the unit		Voies intérieures de commande	Les voies de commande sont situées à l'intérieur de l'appareil
9.2.5	*Combined control :*			*Commande combinée :*	
9.2.5.1	– by solenoid and pilot directional valve	The pilot directional valve is actuated by the solenoid		– par électro-aimant et distributeur-pilote	Le distributeur-pilote est actionné par l'électro-aimant
9.2.5.2	– by solenoid or pilot directional valve	Either may actuate the control independently		– par électro-aimant ou distributeur-pilote	L'une des commandes peut agir seule indépendamment de l'autre

	Description	Use of the equipment or explanation of the symbol	Symbol Symbole	Dénomination	Fonction de l'appareil ou explication du symbole
9.3	Mechanical feedback	The mechanical connection of a control apparatus moving part to a controlled apparatus moving part is represented by the symbol 5.1.1.4 which joins the two parts connected. (For examples see 11.1.2 and 12.1.1)	1) 2) 1) Controlled apparatus Appareil commandé 2) Control apparatus Appareil de commande	Rétroaction mécanique	La liaison mécanique d'une partie mobile de l'appareil de commande avec une partie mobile de l'appareil commandé est représentée par le symbole 5.1.1.4 reliant les deux éléments intéressés (voir exemples d'utilisation 11.1.2 et 12.1.1)

10 SUPPLEMENTARY EQUIPMENT

10 APPAREILS COMPLÉMENTAIRES

10.1	Measuring instruments			Appareils de mesurage	
10.1.1	Pressure measurement :			Mesurage de la pression :	
10.1.1.1	– pressure gauge	The point on the circle at which the connection joins the symbol is immaterial		– Manomètre	Position de la connexion indifférente
10.1.2	Temperature measurement :			Mesurage de la température :	
10.1.2.1	– Thermometer	The point on the circle at which the connection joins the symbol is immaterial		– Thermomètre	Position de la connexion indifférente
10.1.3	Measurement of flow :			Mesurage du débit :	
10.1.3.1	– Flow meter			– Débitmètre	
10.1.3.2	– Integrating flow meter			– Compteur totalisateur	
10.2	Other apparatus			Autres appareils	
10.2.1	Pressure electric switch			Contact électrique à pression	

ISO 1219 (E/F)

11 EXAMPLES OF ASSEMBLIES OF EQUIPMENT	11 EXEMPLES D'APPAREILS GROUPÉS
In circuit diagrams, symbols normally represent equipment in the unoperated condition. However, any other condition can be represented, if clearly stated.	Les schémas de base représentent normalement les appareils dans la position de repos (zéro). Les symboles peuvent cependant être utilisés pour représenter les appareils dans toute autre position, si celle-ci est clairement indiquée.

	Description and interpretation of the examples	Symbol / Symbole	Dénominations et interprétation des exemples
11.1	Driven assemblies (pumps)		Groupes générateurs de pression
11.1.1	A two-stage pump driven by an electric motor with a pressure relief valve in the second stage and a proportioning pressure relief valve which maintains the pressure of the first stage at for example half the pressure of the second stage		Pompe à deux étages, entraînée par moteur électrique, avec limiteur de la pression de sortie et limiteur proportionnel de pression piloté par la pression de sortie qui maintient la pression du premier étage par exemple à la moitié de la pression du deuxième étage
11.1.2	A variable displacement pump driven by an electric motor, control being by a servo-motor with differential cylinder and a tracer valve, with two throttling orifices and mechanical feedback		Pompe à cylindrée réglable, entraînée par moteur électrique et combinée avec des appareils de commande comprenant un servo-moteur à vérin différentiel avec soupape asservie à palpeur à deux étranglements
11.1.3	A single-stage air compressor driven by an electric motor which is automatically switched on and off as the receiver pressure falls and rises		Compresseur d'air à un étage, entraîné par moteur électrique qui est enclenché et interrompu automatiquement, en fonction de la pression dans le réservoir
11.1.4	A two-stage air compressing assembly driven by an internal combustion motor which idles or takes up the load with the switching over of a 3/2 directional control valve, depending on the receiver pressure		Générateur de pression d'air à deux étages entraîné par moteur à combustion interne. Commande de marche à vide ou de charge par commutation d'un distributeur 3/2 dépendant de la pression dans le réservoir
11.2	Driving assemblies (motors)		Groupes moteurs
11.2.1	A motor driven in either direction of rotation, with pressure relief valves and flushing valve		Moteur à deux sens de rotation avec limiteurs de pression et soupape de purge

	Description and interpretation of the examples	Symbol / Symbole	Dénominations et interprétation des exemples
11.3	**Control and regulating assemblies**		**Appareils de distribution et de régulation groupés**
11.3.1	A control unit by which the piston of a cylinder is automatically moved back and forth		Groupe de commande par lequel le piston d'un vérin exécute automatiquement des courses aller et retour
11.3.2	A group of two 6/3 directional control valves which are connected to separate non-return valves and to a common pressure relief valve. When both directional control valves are in the neutral position, the flow is returned to the reservoir		Groupe de deux distributeurs 6/3, munis chacun d'un clapet de non-retour avec limiteur de pression commun. Quand les deux distributeurs se trouvent en position zéro, le fluide retourne au réservoir

12 EXAMPLES OF COMPLETE INSTALLATIONS

In circuit diagrams, symbols normally represent equipment in the unoperated condition. However, any other condition can be represented, if clearly stated.

12 EXEMPLES D'INSTALLATIONS COMPLÈTES

Les schémas de base représentent normalement les appareils dans la position de repos (zéro). Les symboles peuvent cependant être utilisés pour représenter les appareils dans toute autre position, si celle-ci est clairement indiquée.

12.1	Installations	Installations
12.1.1	*Copying control*	*Commande à copier*
	Key 1 = Tool 2 = Template 3 = Machine frame	Légende 1 = Outil 2 = Gabarit 3 = Chassis de la machine
12.1.2	*Clutch operating control*	*Commande d'un embrayage*
12.2	Remote drives	Transmissions
12.2.1	*Reversible drive*	*Transmission réversible*

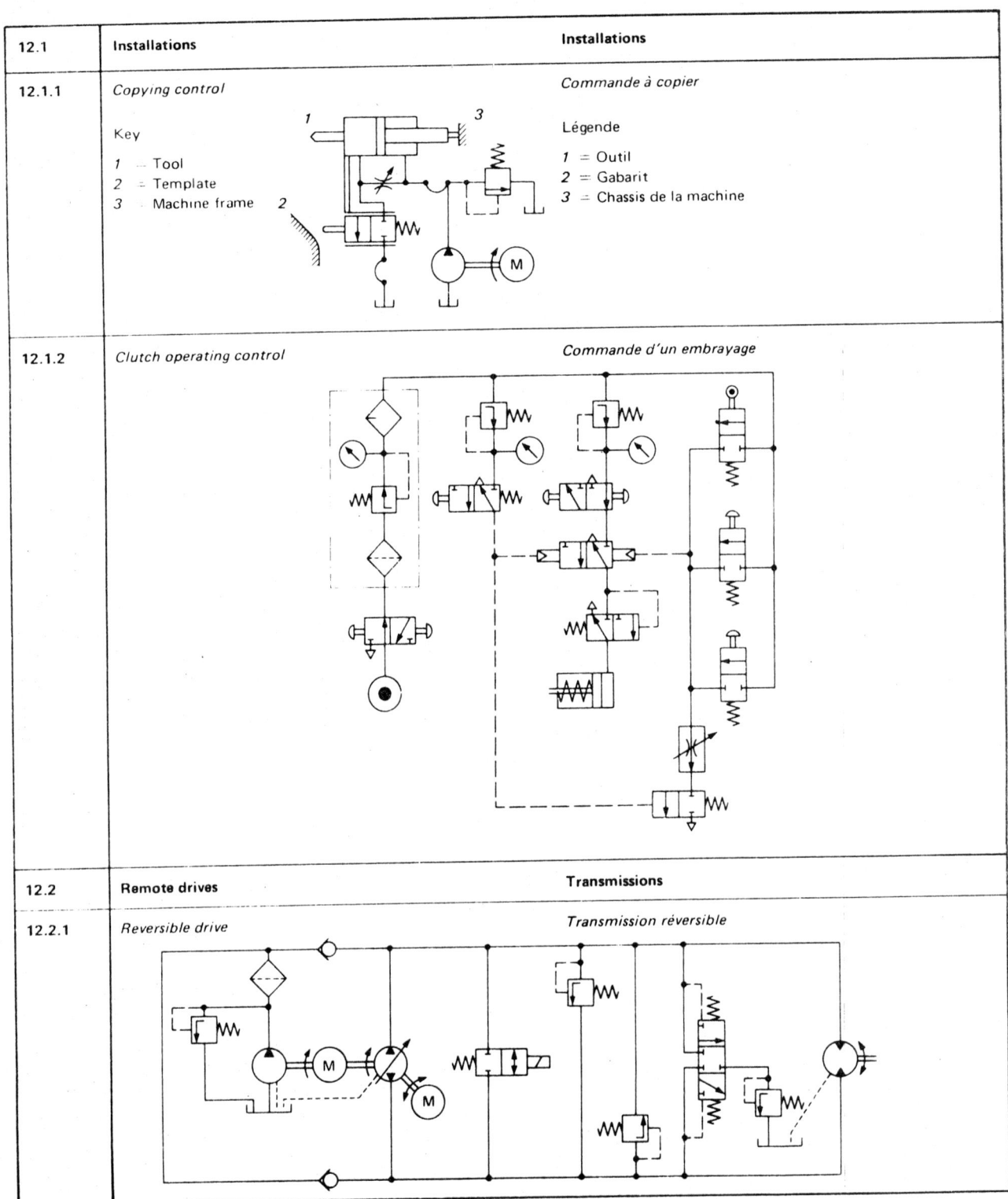

INTERNATIONAL STANDARD  2944

INTERNATIONAL ORGANIZATION FOR STANDARDIZATION ·МЕЖДУНАРОДНАЯ ОРГАНИЗАЦИЯ ПО СТАНДАРТИЗАЦИИ· ORGANISATION INTERNATIONALE DE NORMALISATION

Fluid power systems and components — Nominal pressures

Transmissions hydrauliques et pneumatiques — Gamme de pressions nominales

First edition — 1974-07-01

UDC 621.8.032 : 532.11

Ref. No. ISO 2944-1974 (E)

Descriptors : fluid power, hydraulic fluid power, specifications, components, mechanical properties, pressure.

ISO 2944

FOREWORD

ISO (the International Organization for Standardization) is a worldwide federation of national standards institutes (ISO Member Bodies). The work of developing International Standards is carried out through ISO Technical Committees. Every Member Body interested in a subject for which a Technical Committee has been set up has the right to be represented on that Committee. International organizations, governmental and non-governmental, in liaison with ISO, also take part in the work.

Draft International Standards adopted by the Technical Committees are circulated to the Member Bodies for approval before their acceptance as International Standards by the ISO Council.

International Standard ISO 2944 was drawn up by Technical Committee ISO/TC 131, *Fluid power systems and components,* and circulated to the Member Bodies in November 1972.

It has been approved by the Member Bodies of the following countries :

Australia	Hungary	Romania
Austria	India	South Africa, Rep. of
Belgium	Italy	Sweden
Brazil	Japan	Switzerland
Bulgaria	Mexico	Thailand
Czechoslovakia	Netherlands	Turkey
Finland	New Zealand	United Kingdom
France	Poland	U.S.A.
Germany	Portugal	U.S.S.R.

No Member Body expressed disapproval of the document.

© International Organization for Standardization, 1974 •

Printed in Switzerland

Fluid power systems and components — Nominal pressures

0 INTRODUCTION

In fluid power systems, power is transmitted and controlled by a fluid (liquid or gas) under pressure within an enclosed circuit. Systems and components are generally designed and marketed for a specific fluid pressure.

1 SCOPE AND FIELD OF APPLICATION

This International Standard establishes a series of nominal pressures from which to choose values used in other International Standards related to fluid power.

It provides a standardized series from which to choose values applied to individual fluid power systems and/or components.

The nominal pressures in this International Standard are for positive gauge pressures used with fluid power systems and/or components. Fluid power includes the engineering sciences of hydraulics, pneumatics and fluidics.

NOTE - See 3.1 and 4.3 for explanations of nominal pressure.

2 REFERENCES

ISO ...., *Fluid power — Vocabulary.* [*]

ISO 1000, *SI units and recommendations for the use of their multiples and of certain other units.*

3 DEFINITIONS

3.1 nominal pressure: A pressure value assigned to a component or a system for the purpose of convenient designation.

NOTE — This definition is intended solely to complete this document. A more comprehensive definition for general purposes may be established subsequently.

3.2 For definitions of other terms used, see ISO ....

4 UNITS

4.1 The pressure unit used is the bar.

1 bar = 100 kPa[**] ≈ 14.5 lbf/in^2

4.2 Express nominal pressures as "pressure of ..... bar".

4.3 Assume the nominal pressure to be "gauge" pressure (i.e. the pressure above atmospheric) when no modifier is given.

5 NOMINAL PRESSURES

Select from values in the table.

TABLE — Nominal pressures — Gauge pressures in bar

0,01	0,10	1,0	10	100	1 000
(0,012 5)	(0,125)	(1,25)	(12,5)	(125)	
0,016	0,16	1,6	16	160	
(0,02)	(0,2)	(2,0)	(20)	200	
0,025	0,25	2,5	25	250	
(0,031 5)	(0,315)	(3,15)	(31,5)	315	
0,04	0,4	4,0	40	400	
(0,05)	(0,5)	(5,0)	(50)	500	
0,063	0,63	6,3	63	630	
(0,08)	(0,8)	(8,0)	(80)	800	

NOTE — Non-preferred values are in parentheses.

6 IDENTIFICATION STATEMENT (Reference to this International Standard)

Use the following statement in test reports, catalogues and sales literature when electing to comply with this standard:

"Nominal pressures determined in accordance with ISO 2944, *Fluid power systems and components — Nominal pressures.*"

---

[*] In preparation.

[**] 1 Pa 1 N/m^2.

ISO 2944

For easy application and optimum performance of fluid power systems, it is desirable to have all components in an individual fluid power system, or subsystem thereof, rated to the same preselected nominal pressures. Preselected pressures are desired so that all International Standards relating to fluid power systems can have a common reference. It is desirable that these nominal pressures cover the widest possible range with the smallest reasonable number of steps. This would minimize the number of recommended nominal (system) pressures, improve component interchangeability, and standardize production.

Nominal pressures appearing in this document were selected from the series of preferred numbers in accordance with ISO 3, *Preferred numbers — Series of preferred numbers*, ISO 17, *Guide to the use of preferred numbers and of series of preferred numbers* and ISO 497, *Guide to the choice of series of preferred numbers and of series containing more rounded values of preferred numbers*. These nominal pressures will provide the basis for all pressures used in subsequent documents developed by ISO/TC 131 and its subcommittees.

More specifically, the nominal pressures appearing in this document were selected from the R 5 and R 10 series of preferred numbers as follows :

preferred values to 160 bar inclusive : R 5

preferred values above 200 bar : R 10

non-preferred values below 200 bar : R 10